Advanced Textbooks in Control and Signal Processing

Other titles published in this series:

Genetic Algorithms
K.F. Man, K.S. Tang and S. Kwong

Neural Networks for Modelling and Control of Dynamic Systems
M. Nørgaard, O. Ravn, L.K. Hansen and N.K. Poulsen

Fault Detection and Diagnosis in Industrial Systems
L.H. Chiang, E.L. Russell and R.D. Braatz

Soft Computing
L. Fortuna, G. Rizzotto, M. Lavorgna, G. Nunnari, M.G. Xibilia and R. Caponetto

Statistical Signal Processing
T. Chonavel

Discrete-time Stochastic Processes (2nd Edition)
T. Söderström

Parallel Computing for Real-time Signal Processing and Control
M.O. Tokhi, M.A. Hossain and M.H. Shaheed

Multivariable Control Systems
P. Albertos and A. Sala

Control Systems with Input and Output Constraints
A.H. Glattfelder and W. Schaufelberger

Analysis and Control of Non-linear Process Systems
K. Hangos, J. Bokor and G. Szederkényi

Model Predictive Control (2nd Edition)
E.F. Camacho and C. Bordons

Digital Self-tuning Controllers
V. Bobál, J. Böhm, J. Fessl and J. Macháček

Principles of Adaptive Filters and Self-learning Systems
A. Zaknich

Control of Robot Manipulators in Joint Space
R. Kelly, V. Santibáñez and A. Loría

Robust Control Design with MATLAB®
D.-W. Gu, P.H. Petkov and M.M. Konstantinov

Control of Dead-time Processes
J.E. Normey-Rico and E.F. Camacho

Robotics
B. Siciliano and L. Sciavicco
Publication due October 2007

B. Hrúz and M.C. Zhou

Modeling and Control of Discrete-event Dynamic Systems

with Petri Nets and Other Tool

Springer

Branislav Hrúz, PhD
Department of Automatic Control Systems
Faculty of Electrical Engineering
  and Information Technology
Slovak University of Technology
812 19 Bratislava 1
Slovak Republic

MengChu Zhou, PhD
Department of Electrical
  and Computer Engineering
New Jersey Institute of Technology
Newark, NJ 07102
USA

British Library Cataloguing in Publication Data
Hruz, B. (Branislav)
  Modeling and control of discrete-event dynamical systems :
  with Petri nets and other tools. - (Advanced textbooks in
  control and signal processing)
  1. Discrete-time systems - Mathematical models 2. Automatic
  control - Mathematical models 3. Petri nets
  I. Title II. Zhou, MengChu
  003.8
ISBN-13: 9781846288722

Library of Congress Control Number: 2007929991

Advanced Textbooks in Control and Signal Processing ISSN 1439-2232

ISBN-13: 978-1-84628-872-2   e-ISBN-13: 978-1-84628-877-7

Printed on acid-free paper

9 8 7 6 5 4 3 2 1

Springer Science+Business Media
springer.com

To our wives Mira and Fang Chen

Series Editors' Foreword

The topics of control engineering and signal processing continue to flourish and develop. In common with general scientific investigation, new ideas, concepts and interpretations emerge quite spontaneously and these are then discussed, used, discarded or subsumed into the prevailing subject paradigm. Sometimes these innovative concepts coalesce into a new sub-discipline within the broad subject tapestry of control and signal processing. This preliminary battle between old and new usually takes place at conferences, through the Internet and in the journals of the discipline. After a little more maturity has been acquired by the new concepts then archival publication as a scientific or engineering monograph may occur.

A new concept in control and signal processing is known to have arrived when sufficient material has evolved for the topic to be taught as a specialised tutorial workshop or as a course to undergraduate, graduate or industrial engineers. *Advanced Textbooks in Control and Signal Processing* are designed as a vehicle for the systematic presentation of course material for both popular and innovative topics in the discipline. It is hoped that prospective authors will welcome the opportunity to publish a structured and systematic presentation of some of the newer emerging control and signal processing technologies in the textbook series.

In society today, much of the modern technological infrastructure is event driven where the system outcome is often dependent on a benign sequence of desirable actions taking place. In industry, the action sequence may be highly structured and deterministic but, elsewhere in the community, the action sequence may be fuzzy or even stochastic, making the constraint of a safe system control sequence a much more difficult task. Many of the sequential event-driven systems found today, may be modelled as discrete-event dynamic systems (DEDS). The characterising features of DEDS are discrete states that capture the status to change value at discrete time points. In DEDS, the (logical) conditions that lead to individual or sets of events being activated to generate a sequence of changing system states is an important part of the system and its mathematical model. This viewpoint contrasts with much of the standard control literature which is often dominated by the exhaustive treatment of systems described by linear or nonlinear ordinary differential equations systems or even, occasionally, spatially-dependent partial differential equation systems.

As the references in this book show, the tools to describe DEDS, analyse their performance and generate control algorithms have been under development since the late 1970s. However, it appears that around 1990 there was a flurry of publications as Petri nets and other techniques began to receive serious consideration. Now, in 2007, sufficient development has taken place for this course textbook on the *Modeling and Control of Discrete-event Dynamic Systems* to enter the *Advanced Textbooks in Control and Signal Processing* series. Professors Branislav Hrúz and MengChu Zhou have had many years of experience of teaching courses in the methods of DEDS and we are pleased to welcome their new volume into the series.

This textbook is comprised of three groups of chapters. The first group, Chapters 1–5, is concerned with establishing the basis mathematical tools for the modelling and control of DEDS. This includes chapters on the application of mathematical graph theory, the ideas of formal language concepts and finite automata. The concepts and structure of control for DEDS appears in Chapter 5.

Graphical techniques for DEDS then dominate the second group of chapters, 6–11. These techniques start with flow diagram methods in Chapter 6. This short chapter is followed by four key chapters, 7–10, on Petri nets. These chapters detail the basics (Chapter 7), and the properties (Chapter 8) of Petri nets, and then move on to Grafcet in Chapter 9 and the timed, colored, fuzzy and adaptive varieties of Petri nets in Chapter 10. A brief look at statecharts and their link to Petri nets in Chapter 11 closes this second group of chapters.

The final grouping of three chapters looks at what might be termed DEDS tasks. Chapter 12 has a strong implementation focus with a useful section on ladder logic diagrams and comments on how Petri nets might contribute to this widespread programmable-logic-controller programming paradigm. The problems of supervisory control and job scheduling are considered in the last two chapters of this final group.

The practitioner will be interested to see the applicability of the DEDS techniques illustrated by the wide range of systems used as examples in the textbook. There is a large group of examples based on the problems of flexible manufacturing systems (FMS). These are based on different cell structures using components like 'pick and place' robots, milling units, conveyor belt systems, storage units or bins, and workpiece sorting units. Fortunately, all these components are easily understood by the non-manufacturing specialist and so provide good accessible introductory examples. Manufacturing industry transportation systems based on automatic guided vehicles (AGVs) are also used in examples. Real system complexity is soon experienced by the reader when FMS and AGV systems are interlinked.

Other examples used in the textbook include crane-based loading systems, tank-filling systems, distributed computer systems, motor- and motion-control systems. The discussion of a two-tank-filling system (given in Chapter 8) provides an alternative view of a control problem often treated in classical control engineering textbooks. The problem of modelling the operation of a pedestrian crossing and the human resources problem of when three participants will make it to a meeting give a fascinating illustration of the potential of DEDS techniques to model and analyse problems in fields far removed from manufacturing.

Industrial Control Centre   *M.J. Grimble*
Glasgow, Scotland, U.K.   *M.A. Johnson*
December 2006

Preface

For whatsoever doth make manifest is light
Ephesians 5.13

This book presents results of research achieved in friendly collaboration across borders and moreover between continents and emphasizes a belief in engineering science being for the benefit of mankind the world over. This aspect of the book's ethos is epitomized by the authors' profiles, one being from Central Europe and one from the USA.

Motivation

A number of years ago research work on a woodworking process control raised our interest in discrete event dynamic systems (DEDS). We remembered that the process was an automatic production of laminar parquetry precasts. Work-piece preparation and composition included many discrete events and concurrent processes. Since then, we have started a systematic study of DEDS. Each school year since 1993, we have given lecture courses on DEDS within the Master program at the Department of Automatic Control Systems, Faculty of Electrical Engineering and Information Technology of the Slovak University of Technology in Bratislava, and undergraduate and graduate programs in the Department of Electrical and Computer Engineering, New Jersey Institute of Technology, Newark, NJ 07102, USA, respectively.

The presented textbook contains most of the lecture material gradually elaborated in the courses of the past ten years. Our teaching activities have been accompanied by significant research and student projects in the field of DEDS, mainly on various topics concerning Petri nets used for modeling, analysis, performance evaluation, discrete-event control, supervisory control, and job scheduling of manufacturing processes, automatic guided vehicles in flexible manufacturing, assembly/disassembly processes, computer networking, and workflow management. Other discrete event models and their applications under our study include statecharts, ladder logic diagrams, finite state machines, digraphs, and Grafcet.

While performing the teaching and research activities, we have felt a strong need for a textbook that systematically and comprehensively introduces the mathematical background and various modeling tools for the purpose of DEDS analysis, performance evaluation, control, and scheduling. Thus students and researchers of various background can easily learn and grasp the essence of DEDS that is of growing importance. Their demand and the needs arising from our teaching, research and development activities motivate us to write this present book.

Contents

The mentioned lectures and this book particularly concentrate on Petri nets and their use in the modelling and control design for DEDS. They serve as a basis for extending to other tools and approaches such as Grafcet, statecharts, supervisory control theory and job scheduling. The textbook contains the necessary mathematics and computer science material. It includes discrete mathematics, formal languages, and finite automata. They are essential for non-computer science/engineering students to master the subjects of DEDS. Such material helps one describe and understand the nature of DEDS as well as the methods to describe and govern them. Standard and reactive flow diagrams are then introduced. The substance and properties of Petri nets and other tools useful for the modeling of DEDS have been built up systematically. Advanced Petri net tools include timed, stochastic, colored, fuzzy, and adaptive Petri nets. Petri net-related tools include Grafcet (also terms Sequential Function Charts), and statecharts. Various aspects concerning control design methods are followed consistently throughout the textbook. Theoretical aspects are illustrated and explained using numerous problem-solving examples dealing with various computer-integrated systems. We summarize the contents of all fourteen chapters as follows.

Chapter 1 introduces the concept of systems and states, continuous, discrete-time, and discrete-event systems, the definition and properties of DEDS, and some system examples. Basic transition systems are described in detail as the most fundamental representation of DEDS.

Chapter 2 presents directed graphs, subgraphs, and directed paths and circuits in them. Examples are given to illustrate these concepts.

Chapter 3 introduces the concept of formal languages and their classification. They form the basis for many theoretical developments in both computer science and supervisory control theory of DEDS.

Chapter 4 discusses DEDS control system including specifications and control functions.

Chapter 5 introduces the concept of finite automata or state machines. It discusses through examples how they can be used to describe a real system and how control specification can be described with their help. Non-deterministic finite automata are also presented.

Chapter 6 discusses the standard flow diagrams used in software development and reactive flow diagrams for DEDS.

Chapter 7 introduces the idea of Petri nets, their basic definition, matrix representation, and various classes. It also discusses how they can be interpreted for control purposes.

Chapter 8 presents the important properties of Petri nets and their implications in modelled systems. Analysis methods based on reachability trees are elaborated. Examples are given. The structural properties of Petri nets are also discussed.

Chapter 9 presents Grafcet (also named sequential function chart) – its presence in industry is significant, especially among automatic control and automation equipment companies. Its comparison with Petri nets is given.

Chapter 10 introduces the advanced concepts resulting from the study of Petri nets and industrial needs in exploring their utility. They include deterministic and stochastic timed Petri nets for performance evaluation purposes. Colored Petri nets are used to specify complex systems with many similar subsystems, components or specifications. Fuzzy Petri nets combine fuzzy set theory and Petri nets to describe uncertainty embedded in many practical applications. Adaptive Petri nets further embed learning capability into Fuzzy Petri nets.

Chapter 11 presents the idea of statecharts and their applications to complex system design.

Chapter 12 introduces modeling methodology, conflict resolution, ladder logic diagrams, and control program design for DEDS.

Chapter 13 presents the essential concepts of supervisory control theory based on automata and Petri nets. Several fundamental approaches are presented.

Chapter 14 discusses the job scheduling problems and the use of Petri nets for such purposes. The solution method based on max-plus algebra is also introduced.

Aim and Use of this Textbook

This book aims to introduce to students, engineers and researchers the fundamentals of various discrete event modelling tools, as well as applications. The discrete mathematics and related background material are included. It is suitable for class use and can be easily tailored to meet the different needs from senior undergraduate and graduate students. In an introductory course to DEDS for engineering students, the following contents are suggested:

Chapters 1–8, and 11

For a more advanced course in DEDS and for students with required mathematics and entry-level knowledge of DEDS, the following contents should be offered:

Chapters 1, 2, 4, 7–14.

For computer science and engineering students, such materials as discrete mathematics, formal language, and automata can be skipped or only their brief review is needed.

Acknowledgements

We would like to express our deep gratitude to all who helped us and supported our effort in writing this book, especially our wives, children and parents. We would acknowledge the professionalism demonstrated by the staff members of

Springer. We especially thank Mr. Oliver Jackson and Prof. Michael Johnson for their great help. We are grateful to Ms. Angela Böhl and Mr. Torsten Hartmann, LE-TeX Jelonek, Schmidt & Voeckler GbR for her careful proofreading and his nice cover design of this book, respectively.

The first author thanks, in particular, Prof. Štefan Kozák, Ph.D., head of the Department of Automatic Control Systems, Institute of Robotics and Industrial Informatics at Slovak University of Technology and other colleagues from the Institute, especially Leo Mrafko, his Ph.D. student and assistant, to Alena Kozáková, Ph.D., for her careful reading and correcting the manuscript, Jana Flochová, Ph.D., Milan Struhar, Ph.D., as well as colleagues from other departments of his faculty Prof. Norbert Frištacký, Ph.D. (*in memoriam*), and Prof. Mikuláš Huba, Ph.D. He would like to thank his friends from abroad, Prof. Dr. Antti Niemi from the Control Engineering Laboratory of the Helsinki University of Technology, Prof. Dr. Hanns Peter Jörgl from the Institute of the Process and Machine Automation of the Vienna University of Technology, and Prof. Zdeněk Hanzálek, Ph.D. from the Faculty of Electrical Engineering of the Czech Technical University in Prague for their valuable advice, support and numerous inspiring scientific discussions.

The second author thanks his doctoral thesis advisor Dr. Frank DiCesare, Prof.-Emeritus, and committee members (now friends), Prof. Alan Desrochers, and Prof. Arthur Sanderson, Rensselaer Polytechnic Institute, Troy, NY, USA. It is they who introduced him to this exciting field and helped advance his professional career. He would also like to thank many pioneers in the area of Petri nets including Prof. N. Viswanadham, National Singapore University, Prof. T. Murata, University of Illinois at Chicago, Prof. K. Lautenbach, University of Koblenz, and Prof. Y. Narahari, Indian Institute of Science. He is also grateful for the great professional help and friendship of Prof. P.B. Luh, University of Connecticut, Prof. F. Lewis, University of Texas at Arlington, Prof. J. Tien, Rensselaer Polytechnic Institute, Prof. W. Gruver, Simon Fraser University, and Prof. T.T. Lee, National Chiao-Tung University. During the past many years, he has enjoyed the happy collaboration and friendship with Prof. MuDer Jeng, National Taiwan Ocean University, M.P. Fanti, Polytechnic di Baris, Prof. Jiacun Wang, University of Monmouth, Prof. Naiqi Wu, Guangdong University of Technology, Prof. Zhiwu Li, Xidian University, and Prof. Yushun Fan, Tsinghua University. I owe a great deal to many graduate advisees, especially, Dr. K. Venkatesh of AIG, Dr. H. Xiong of Alcatel-Lucent, Prof. Ying Tang of Rowan University, Prof. Meimei Gao of Seton Hall University, and Dr. Jianqiang Li of NEC. The second author's close friend and colleague, Prof. Fei-Yue Wang, University of Arizona, has been very instrumental and helpful in this project and many others. The work is partially supported by the China National Science Foundation under Grant 60574066, Chinese Academy of Sciences under Grant No. 2F05NO1, New Jersey Commission of Science and Technology, and Chang Jiang Scholar Program, Ministry of Education, PRC.

Bratislava, Slovak Republic
Newark, NJ, USA
June 2007

Branislav Hrúz
MengChu Zhou

Contents

Notation

iff	abbreviation for "if and only if"
$\wedge, \vee, \bar{x}, \Rightarrow$	logical operators: conjunction, disjunction, negation, and implication, respectively
$\Leftrightarrow$	logical implication in both directions
$(o_1, o_2, ..., o_n)$ or (o_{1-n})	ordered n-tuple of n objects $o_1, o_2, ..., o_n$
$A = \{a_1, a_2, ..., a_n\}$	set of n elements $a_1, a_2, ..., a_n$
$\emptyset$	empty set
$|A|$	the number of elements (cardinality) in set A
$\cap, \cup, \backslash$	set intersection, union, and difference
$A \subset B$	set A is a proper subset of B and $A = B$ is excluded
$A \subseteq B$	A is a subset of B
$A \times B$	Cartesian product of sets A and B
$R \subseteq A \times B$	binary relation from set A to set B
$f : A \rightarrow B$	function f, which maps elements of set A into set B

$x \in A$	x belongs to set A
$x \notin A$	x is not an element of set A
N	set of natural numbers $N = \{0,1,2,...\}$
N^+	set of positive integers, *i.e.*, $N^+ = N \setminus \{0\}$
I	set of integers
R	set of real numbers
R^+	set of positive real numbers

$$\mathbf{u} = \begin{pmatrix} u_1 \\ u_2 \\ \vdots \\ u_n \end{pmatrix}$$

vector $\mathbf{u}$ with entries $u_1, u_2, ..., u_n$

$$\mathbf{u} \leq \mathbf{v}, \mathbf{u} = \begin{pmatrix} u_1 \\ u_2 \\ \vdots \\ u_n \end{pmatrix}, \mathbf{v} = \begin{pmatrix} v_1 \\ v_2 \\ \vdots \\ v_n \end{pmatrix}$$

iff $u_k \leq v_k$ for all $k = 1,2,...,n$

$\mathbf{u} < \mathbf{v}$

iff
$\left(u_k \leq v_k \text{ for all } k = 1,..,n \right) \wedge \left(u_k < v_k \text{ at least for one } k \right)$

$$\mathbf{z} = \mathbf{u} + \mathbf{v}, \mathbf{z} = \begin{pmatrix} z_1 \\ z_2 \\ \vdots \\ z_n \end{pmatrix}$$

sum of vectors $z_k = u_k + v_k$, for all $k = 1,2,...,n$

$$\mathbf{u}(t) = \begin{pmatrix} u_1(t) \\ u_2(t) \\ \vdots \\ u_n(t) \end{pmatrix}$$

vector time variable

$N^{|P|}$

set of $|P|$-tuples consisting of $|P|$ natural numbers

$N^{|P|\mathrm{T}}$

set of transposed $|P|$-tuples consisting of $|P|$ integer numbers, *i.e.*, the set of integer number $|P|$-vectors

$$\mathbf{A} = \begin{pmatrix} a_{11} & . & . & . & a_{1m} \\ . & . & . & . & . \\ . & . & . & . & . \\ . & . & . & . & . \\ a_{n1} & . & . & . & a_{nm} \end{pmatrix}$$

$n \times m$ matrix with n rows and m columns

$\tilde{a} = a_{i_1} a_{i_2} ... a_{i_n}$

a sequence (string, formal word) of elements $a_{i_1}, a_{i_2}, ..., a_{i_n}$ in this order

Σ

the set of events

Σ^*

the set of all sequences (strings, formal words) created from elements of set Σ and the empty sequence $\tilde{\varepsilon}$

Σ^+

set Σ^* without the empty sequence $\tilde{\varepsilon}$, *i.e.*, $\Sigma^+ = \Sigma^* \setminus \tilde{\varepsilon}$

$|\tilde{\alpha}|$

length of the sequence (or state path or event path) $\tilde{\alpha}$ given as the number of elements in sequence (or state path or event path) $\tilde{\alpha}$

$\tilde{\varepsilon}$

empty string, *i.e.*, the string for which $|\tilde{\varepsilon}| = 0$

$^{\bullet}t \; (t^{\bullet})$

set of pre-places (post-places) of the Petri net transition t

$^{\bullet}p \; (p^{\bullet})$

set of input (output) transitions of the Petri net place p

$\mathbf{m}[t > \mathbf{m}'$

transition t of a given Petri net is enabled or fireable at marking $\mathbf{m}$ represented in the vector form and its firing results in marking $\mathbf{m}'$

$R_{PN}(\mathbf{m}_0)$ reachability set of Petri net PN given initial marking $\mathbf{m}_0$

$\oplus, \otimes$ operations of the max-plus algebra

1

Basic Description of Discrete-event Dynamic Systems

1.1 Introduction

People observe various phenomena of nature and endeavor to comprehend them. The first step in that is a reflection of the phenomena by imagination and description. The reflexive process is a process of abstraction. In this process, the notion of "system" is of basic importance.

A system is defined to be a group of objects separated from the universe and having mutual relations.

Different physical entities can constitute system objects. If time is included among the system objects, their temporal properties or the system dynamics can be considered. The system dynamics is given by the time behavior of the system objects. The behavior is called the process.

A real physical system is represented by an ideal system created by human thinking and understanding. Mathematical representations of real systems are the most abstract and precise descriptions. Since the very beginning of its existence, mankind strives not only to know and to describe natural systems but also to govern and control them.

Control of a system is based on knowledge about the particular system. This knowledge is developed *via* abstraction based on observation of the system. The observation is realized by measurements and if possible, by experimentation with the system. Two main abstractions are to be distinguished, namely:

1. The notion of a continuous system and
2. The notion of a discrete system.

A natural question arises about the substance of these abstractions. A continuous system is specified by a set of continuous variables, a set of continuous functions over the respective domains of these variables, and by derivatives of the variables and functions. Such a system is called a continuous-variable dynamic system (CVDS). One can find a good systematic survey of the CVDS control theory in the book by Jörgl (1993). A discrete system is specified by a set of discrete variables and relations defined on them. A hybrid system is a combination of both.

Sometimes the relations of system variables are not treated with respect to time. Then they describe the static behavior of the studied systems. However, time is mostly involved in the analysis and synthesis of systems and dynamic system behavior is considered.

Figure 1.1 illustrates the classification of continuous and discrete systems considering dynamic behavior. In order to simplify the illustration, a system with one variable is depicted. Figure 1.1.a shows the case when the continuous variable $x(t)$ is a continuous function over a continuous time interval. The function domain and co-domain are real numbers. A system is continuous if it is defined by continuous variables and continuous functions such as the function depicted in Figure 1.1.a. A discrete system is given by discrete variables and discrete functions or relations as illustrated by Figure 1.1.b. The system in Figure 1.1.b consists of one object in the form of one variable that takes on values from the set of real numbers in discrete time points.

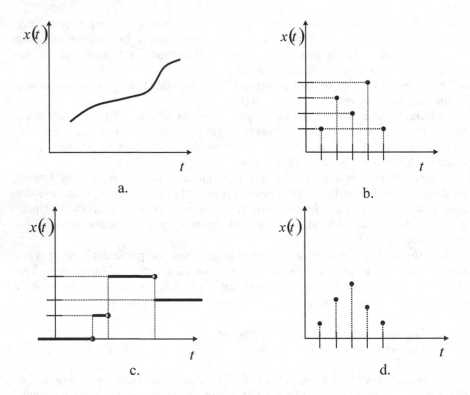

Figure 1.1. Properties of one-variable system

Figures 1.1.c and 1.1.d show the mixed/hybrid cases when the system is semi-continuous. Usually, a system has more than one object or variable. Then a set of variables can be aggregated into one or more vector variables. Note that the case depicted in Figure 1.1.b can be understood either as a discrete representation of a continuous system or as a representation of the system that is discrete by its nature. The role of the semantics or interpretation is obvious. Therefore, the discrete or digital representation of continuous systems has to be distinguished from the representation of systems that are discrete in their nature and substance. The discrete representation of a continuous system is obtained by sampling continuous variables at discrete time points.

Continuity and discreteness of a system is one aspect of the view on system properties. Another aspect is that CVDS are time-driven systems. The reason for the dynamic development of system states is time. On the other hand, discrete systems can be time-driven or event-driven.

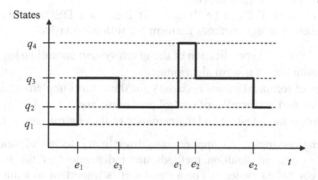

Figure 1.2. An event-driven system

Let us compare Figures 1.1 and 1.2. The discrete variable q describes the state of the system. There are four states q_1, q_2, q_3, and q_4, q_{1-4} for short, and three events e_1, e_2, and e_3, e_{1-3} for short. Figure 1.2 shows that the state change is event-driven. The events occur at discrete time points and the state changes depend on the events only. Such systems are called discrete event dynamic systems or DEDS for short (Ho 1991; Ho and Cassandras 1983). They are also called discrete event systems (Ramadge and Wonham 1987; Zhou and DiCesare 1993; Jafari 1995; Bogdan *et al.* 2006). As mentioned earlier, Figure 1.1.b has a double meaning. It can represent either time-driven CVDS or event-driven DEDS when the events occur at discrete time points and cause the change of system states as depicted in Figure 1.1.b.

The applied system analysis and synthesis methods depend on the system nature. In this textbook we will study systems that are fully discrete in their nature and event-driven, *i.e.*, discrete event dynamic systems. Their name expresses their specific character. DEDS are characterized by a set of states which the system can take, and by the set of events that cause the state changes at discrete time points. The events may take place asynchronously as opposed to the synchronous nature in a discrete time system. The change of states and occurrence of events are the essence of the DEDS dynamic behavior.

A primary task of the DEDS theory is creating a DEDS model. Without such a model it would be impossible to analyze and control DEDS just as it is true in classic CDVS control theory. Obviously we are interested in a model that is sufficiently general and includes the DEDS dynamics. There are two ways to consider the dynamics:

1. To specify values of the system variables and system relations in defined discrete time points; and
2. To specify time order of the states or events.

The latter case means that time is not explicitly expressed and only the precedence relations for DEDS states and events are given. The order of events can be determined by means of their indexing. In other words, it is given which event happens before some other events. Such an approach is more abstract and avoids problems related to the time relativity.

The control of DEDS can be designed if there is a DEDS model available. Control engineering design methods perform the following tasks:

- Formulation and specification of the given system control tasks;
- Determination of control algorithms;
- Design of technical means necessary for the control implementation;
- Creation and verification of control programs; and
- Implementation, testing and maintenance of the control system function.

Control engineering is an applied interdisciplinary technical science. To a considerable extent, the solution methods are independent of the technological substance of controlled systems. For a control it is important to achieve such an influence of various agents on the system that parameters and behavior of the system are as required (Kozák 2002; Jörgl 1993). The system behavior and various influences on it are given by physical, chemical, biological or other quantity values. What is important from the viewpoint of control is the information the quantities carry, but not their physical substance.

Automatic control is based on the information manifestations of the system. In other words, a system is described by means of information about the spatial location of objects, time, system parameters, properties, characteristics, *etc*. Time is substantial for the dynamics of events. As mentioned earlier, the time evolution of system variables is called the process and in the context of DEDS, it is called the discrete process.

1.2 Discrete Variables and Relations

The notion of DEDS has been specified in the previous section. It is based on the discrete character of the individual variables and relations. It is useful to study the property of discreteness in some detail.

Definition 1.1. Let D be a finite set of n elements, *i.e.*,

$$D = \{d_1, d_2, ..., d_n\} \tag{1.1}$$

Let v be a variable taking on values only from set D, *i.e.*,

$$v = d_i \in D \tag{1.2}$$

then v is a discrete variable.

Definition 1.2. Let two non-empty finite sets D and E be given:

$$D = \{d_1, d_2, ..., d_n\} \tag{1.3}$$
$$E = \{e_1, e_2, ..., e_m\} \tag{1.4}$$

A binary relation R from D into E is defined by

$$R \subseteq D \times E \tag{1.5}$$

where symbol $\times$ denotes the Cartesian product.

If a relation is defined on the sets for which $D = E = A$ then $R \subseteq A \times A$ and we say that R is a binary relation on A. The relation R can be empty. If, *e.g.*, $(d_2, e_3) \in R$ we write $d_2 \, R \, e_3$. Functions or mappings are subsets of relations. They are special relation cases as formally given next.

Definition 1.3. Let a binary relation R from $D = \{d_1, d_2, ...d_n\}$ into $E = \{e_1, e_2, ..., e_m\}$ be given. Let for any two elements of $D \times E$

$$(d_i, e_j) \in D \times E, \quad (d_k, e_l) \in D \times E,$$
$$i \in \{1, 2, ..., n\}, j \in \{1, 2, ..., m\}, k \in \{1, 2, ..., n\}, l \in \{1, 2, ..., m\} \tag{1.6}$$

If the following implication holds true

$$(d_i = d_k \ and \ e_j \neq e_l) \Rightarrow ((d_i, e_j) \notin R \ and \ (d_k, e_l) \notin R) \tag{1.7}$$

then the relation R is a discrete function or a discrete mapping notated f defined on the domain

$$DOM = \{d_{i_1}, d_{i_2}, ..., d_{i_s}\} \tag{1.8}$$

where DOM is the set of all first elements of the pairs (d_i, e_j) belonging to the relation R. A co-domain of the function is set $CDOM$ that consists of all the second elements of the pairs (d_i, e_j) belonging to the relation R

$$CDOM = \{e_{i_1}, e_{i_2}, ..., e_{i_u}\} \tag{1.9}$$

We write

$$e_j = f(d_i), d_i \in DOM, e_j \in CDOM \tag{1.10}$$

The right-hand side of Equation (1.7) is an AND conjunction of two propositions. If the premise is true, they both are true. It means that both ordered pairs (d_i, e_j) and (d_k, e_l) cannot belong to the relation R. However, one of them can be in R. Another formulation of this can be as follows. A function is a binary relation from set $D = \{d_1, d_2, ..., d_n\}$ into set $E = \{e_1, e_2, ..., e_m\}$ if there are no two ordered pairs $(d_p, e_r), (d_p, e_v)$ in R such that $e_r \neq e_v$.

1.3 Discrete Processes

Let a finite set Σ be given as

$$\Sigma = \{e_1, e_2, ..., e_n\} \tag{1.11}$$

The set Σ is called the event set. We assume that an event $e_{i_k} \in \Sigma$ occurs at the time point τ_{i_k}. Let a sequence of events be given as

$$\tilde{\sigma} = e_{i_1}, e_{i_2}, ..., e_{i_k}, ..., e_{i_N} \tag{1.12}$$

where $e_{i_1} \in \Sigma$ occurs in the discrete time point τ_{i_1}, $e_{i_2} \in \Sigma$ in time point τ_{i_2}, $etc.$, e_{i_k} in time point τ_{i_k}, $etc.$, and e_{i_N} in time point τ_{i_N}, $\tau_{i_1} \langle \tau_{i_2} \langle ..... \langle \tau_{i_k} \langle ..... \langle \tau_{i_N}$. The sequence $\tilde{\sigma}$ is called a discrete process. In this particular case when elements of a sequence are events we speak about the event string.

Figure 1.3 shows layout of a manufacturing system including a milling machine M, a grinding machine G and three belt conveyors C1–C3. The parts to be processed in the manufacturing system come into the system irregularly with various gaps as a sequence one by one part. Maximum three parts can be fed up on the conveyor C1. Input of a part is detected by a photo-sensor P11. The part is stopped by a stopper at the end of C1. Presence of the part at the end of the conveyor is signalized by a photo-sensor P12. If the milling machine M is free and a part is available at the end of C1, the part is transferred by the transportation means T1 into the milling machine. After milling the part is transferred by T2 onto the conveyor C2. The photo-sensor P21 detects input of the part on the conveyor C2. In the conveyor section between the sensors P21 and P22 there can be maximum two parts. The same mechanism holds for loading of the grinding machine G. Maximum four parts can be loaded on the conveyor section P31–P32.

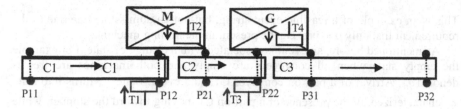

Figure 1.3. Manufacturing system layout

In a manufacturing system, typical events are the input of a part into a conveyor section, arrival of a part in some position on the conveyor, start of an operation, *e.g.*, start of milling, end of an operation, *e.g.*, end of milling.

As an example, consider the following event set:

$$\Sigma = \{s_{C1}, e_{C1}, m_{C1}, m_{C2}, m_{C3}, s_M, e_M, s_{C2}, e_{C2}, g_{C2}, g_{C3}, s_G, e_G, s_{C3}, e_{C3}\} \quad (1.13)$$

where s_{C1} stands for input of a part on conveyor C1, e_{C1} means arrival of a part at the end of the conveyor C1, m_{C1} means the transfer of a part from the conveyor C1 into the milling machine, s_M is the start and e_M the end of milling, m_{C2} is the transfer of a part from M on C2. Similarly, g_{C2} and g_{C3} denote transfers from C2 in G and from G on C3, respectively. The other events are denoted similarly.

Suppose that the manufacturing system is empty in its initial state. Both machines and conveyors are free. A possible sequence of events starting from the initial state is

$$\tilde{\sigma}_1 = s_{C1} e_{C1} m_{C1} s_M s_{C1} e_M m_{C2} s_{C2} e_{C1} m_{C1} s_M e_{C2} g_{C2} s_G \quad (1.14)$$

Event s_{C1} occurs at the time point τ_1, event e_{C1} at τ_2 ... whereas $\tau_1 \langle \tau_2 \langle ..... .$

Let us consider another event sequence example:

$$\tilde{\sigma}_2 = s_{C1} e_{C1} s_{C1} s_{C1} m_{C1} s_M e_{C1} e_M m_{C2} s_{C2} m_{C1} s_M e_{C2} g_{C2} e_M s_{C1} \quad (1.15)$$

Consider the following sequence starting from the initial state when the system is without parts (empty system):

$$\tilde{\sigma}_3 = s_{C1} e_{C1} e_{C1} \quad (1.16)$$

It represents an example of a technologically unfeasible event sequence in the given system. Consider a sequence from the beginning:

$$\tilde{\sigma}_4 = s_{C1} e_{C1} m_{C1} s_M s_{C1} e_{C1} m_{C1} \quad (1.17)$$

This is an example of a feasible event string, but not an admissible one due to the requirement that only one part can be present in the milling machine.

As mentioned before, an event is associated with a change of state. For example, the empty state when all conveyors are empty and both machines are free is denoted q_0. Arrival of a part on conveyor C1 is an event. State q_0 turns into state q_1 characterized by the presence of a part on C1 moving toward the stopper, while other conveyors and machines are still free.

The manufacturing system in Figure 1.3 is a serially arranged production line. A serial-parallel production cell example is shown in Figure 1.4. Suppose that four kinds of semi-products are produced from one kind of parts coming in *via* conveyor C1 and transported through the cell via conveyors C2–C4. Table 1.1 describes the options how to produce them.

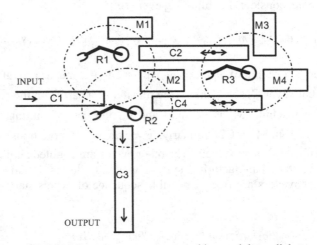

Figure 1.4. Manufacturing system arranged in a serial-parallel structure

Table 1.1. Job options in the manufacturing system

Operation	A	B	C	D
1	M1	M3	M2	M1
2	M2 or M3	M2 or M4	M4	M3
3	M4	M3 or M4	M3 or M4	M3 or M4
4		M4		M2

The system is flexible in that there are several ways to finish the production tasks having the job alternatives given in Table 1.1. The optimal route of the processed parts is to be found. A related problem to this is the job scheduling. Both problems can be solved with respect to the given optimality criterion, *e.g.*, to minimize the overall production work-span (work-time), also called makespan and completion time. The operation times have to be available for that task. A joblist breakdown with respect to operation time specifications is given in Table 1.2. Time durations for the semi-products A, B, C, and D are denoted by a, b, c, and d, respectively. The scheduling problem will be treated in detail later.

Individual events of the system depicted in Figure 1.4 can be specified as before:

$$\Sigma = \left\{ \begin{array}{l} a_{C1}, b_{C1}, a_{C2}, b_{C2}, a_{C3}, b_{C3}, a_{C4}, b_{C4}, \\ s_{M1}, e_{M1}, \ldots, s_{M4}, e_{M4}, \\ s_{R1C1M1}, e_{R1C1M1}, s_{R1C1C2}, e_{R1C1C2}, s_{R1C1M2}, e_{R1C1M2}, s_{R1C1C4}, e_{R1C1C4}, s_{R1M1C2}, e_{R1M1C2}, \\ s_{R1M1M2}, e_{R1M1M2}, s_{R1M2C2}, e_{R1M2C2}, \\ s_{R2C1M2}, e_{R2C1M2}, s_{R2C1C4}, e_{R2C1C4}, s_{R2M2C4}, e_{R2M2C4}, s_{R2C4M2}, e_{R2C4M2}, s_{R2C4C3}, \\ e_{R2C4C3}, s_{R2M2C3}, e_{R2M2C3}, \\ s_{R3C2M3}, e_{R3C2M3}, s_{R3M3C2}, e_{R3M3C2}, s_{R3C2M4}, e_{R3C2M4}, s_{R3M4C2}, e_{R3M4C2}, \\ s_{R3C2C4}, e_{R3C2C4}, s_{R3C4C2}, e_{R3C4C2}, s_{R3M3M4}, e_{R3M3M4}, s_{R3M3C4}, e_{R3M3C4}, s_{R3C4M3}, \\ e_{R3C4M3}, s_{R3M4C4}, e_{R3M4C4}, s_{R3C4M4}, e_{R3C4M4} \end{array} \right\}$$

(1.18)

Table 1.2. Job duration times specifications

Oper ation	Products															
	A				B			C			D				
	Machines				Machines			Machines			Machines				
	M1	M2	M3	M4	M2	M3	M4	M2	M3	M4	M1	M2	M3	M4		
1	a_{11}					b_{13}		c_{12}			d_{11}					
2		a_{22}	a_{23}		b_{22}		b_{24}			c_{24}			d_{23}			
3				a_{34}		b_{33}	b_{34}		c_{33}	c_{34}			d_{33}	d_{34}		
4							b_{44}					d_{42}				

Occurrence of a part at one side of the conveyor is denoted as event a, on the other side as b. s_{R1C1M1} is the start of the part transfer via Robot R1 from conveyor C1 to the machine M1, e_{R1C1M1} is the end of the transfer. The system control depends on the conveyor capacities.

It is assumed that the transfer times consumed by the robots and conveyors are negligible because they are much smaller than operation times. If such an assumption is not acceptable times of transfer operations can be considered separately. In the latter case Table 1.2 would be extended by further time specifications. Sometimes the transportation times can be included in the operation times of the processing machines.

1.4 Basic Properties of DEDS and their Specification

Characteristic properties of DEDS can be best illustrated on examples. DEDS include flexible manufacturing systems, digital computers, local or global computer networks, operation centers, and transportation systems on surface or in air. Various properties of DEDS events are to be studied:

- Event synchronization
- Concurrency
- Parallelism
- Conflict
- Mutual exclusion
- Deadlock
- System liveness
- Reversibility
- State reachability
- Event scheduling

Mankind strives not only to observe the natural phenomena but also to govern, to control and to benefit from them. Many various tools for the specification and analysis of DEDS are used nowadays. In addition, there is a need for tools that are suitable for the design of DEDS control. They should be able to specify the required properties of DEDS and to ensure the real-time reactivity of the controlled DEDS.

Basically the tools can be divided in three groups:

- Graphical tools
- Algebraic tools
- Formal language-based tools

The graphical tools are frequently used due to their transparency and ability to provide rich visual information. The main graphical tools include:

- State-transition diagrams or finite-automata
- Reactive (real-time) flow diagrams
- Statecharts
- Petri nets
- Grafcet
- Ladder logic diagrams

The algebraic tools are the following:

- Boolean algebra
- Algebraic expressions based on the respective state space
- Temporal logic
- Max-plus algebra

The tools based on formal languages are as follows:

- Formal language models

- Standard programming languages combined with real-time operating systems
- Real-time programming languages

The different tools listed above are not equivalent with respect to the application field. For example, max-plus algebra is effective especially for the analysis and control of job scheduling, while Grafcet is useful for the specification of the sequence control (Frištacký *et al.* 1981, 1990; Zhou and Venkatesh 1998; Zhou 1995).

The first two basic groups can serve as intermediate means between the requirements imposed on the system and the control that ensures them. A final specification and implementation of control requires the third group, namely a programming language. The specified control will then be implemented on appropriate hardware components, *e.g.,* personal computer, process computer, programmable logic controller, *etc.* From a graphical or algebraic specification a control program can be generated automatically. Also the transformations among the different specification tools are useful.

On the other hand, sometimes and for someone there is no need to use any intermediate means and it is possible to write a control program directly. However, for most people the opposite is true. A program formulated in any procedural language is a string of instructions to be performed separately and to force the system to behave as required. Intermediate means help to avoid the programming incorrectness.

Each specification tool listed above is based on the concept of the system state and state transitions described earlier as events. This fact can be illustrated by the following generally valid system behavior scheme:

$$\dots \ \to STATE \to TRANSITION \to STATE \to TRANSITION \to \ \dots$$

Because of the generality of this scheme, the "state and transition" concept is dealt with in more detail in the following section.

1.5 Basic Transition System

Various DEDS can be described uniquely by means of the so-called basic transition model proposed by Manna and Pnueli (1991), which serves us as a general description framework. It is defined by the quadruple

$$SYST = (\Pi, Q, \Sigma, \Theta) \tag{1.19}$$

where

$\Pi = \{u_1, u_2, ..., u_n\}$ is the finite set of state variables;

Q is the set of states where each state is given by the particular values of the variables from set Π. This value assignment is called the interpretation of variables belonging to the set Π ;

Σ is the set of transitions whereby a transition $e \in \Sigma$ is a partial function $e: Q \xrightarrow{C_e} 2^Q$. Note that 2^Q is the power set defined as a set of all subsets created from set Q including the empty set $\varnothing$, C_e is a condition imposed on a transition e so that e can occur only if C_e is fulfilled, and C_e can be empty (meaning no condition); and

Θ is the set of initial conditions of the system. It includes states in which the execution of potential events can start.

The modeling power of the Manna and Pnueli model is that any correct specification by means of any tool described earlier can be transformed into a basic transition system. In other words, suitable transformations can be established between different system specifications. We can see that transitions in this model correspond to events introduced before. The time is not explicitly expressed in a basic transition. Rather, a possible sequence of events or an event precedence relation is used.

The function $Q \xrightarrow{C_e} 2^Q$ defining an event e is quite abstract. A standard particular case by excluding system control (if only controlled system is represented) is when one state is mapped into one another state due to the condition C_e or because there is only one-to-one mapping. The case when a state is mapped into state subsets presents indeterminism and its significance is purely theoretical. An example of that is the indeterministic finite automaton (see Section 5.4). In practical system control the indeterminism should be removed. If control is included the function $Q \xrightarrow{C_e} 2^Q$ can map a state to more states. See Chapter 4 for more details.

A general form of the specification of an event e is a transition relation given as an assertion for each transition e:

$$\rho_e\left(\Pi, \Pi'\right) \tag{1.20}$$

which relates the interpretation of state variables given as a state s with the interpretation of state variables given as a succeeding state s'. Under assertions we understand Boolean expressions extended by quantificators $\exists, \forall$, etc. In other words, in each state the relation $\rho_e\left(\Pi, \Pi'\right)$ determines the next state or states after transition e takes place. The following form describing the transition relation e can be used:

$$\rho_e\left(\Pi, \Pi'\right) = C_e\left(\Pi\right) \wedge \left(u'_1 = ex_1\right) \wedge \left(u'_2 = ex_2\right) \wedge \ldots \wedge \left(u'_n = ex_n\right) \tag{1.21}$$

where $C_e\left(\Pi\right)$ is an assertion stating a condition for state s under which transition e is enabled and the system comes over into state s'; ex_{1-n} are logic expressions. Assertion $C_e\left(\Pi\right)$ is constructed over state variables such that if the variable values lead to a true Boolean value from $C_e\left(\Pi\right)$, e is enabled and state variables u'_{1-n}

are given the values according to expressions ex_{1-n}. These expressions are built up of state variables u_{1-n}. Notation ex_i is shortened in Equation (1.21). It has the following meaning:

$$u'_i = lex_{i1} \text{ if and only if } lex_{i2} \text{ is true} \tag{1.22}$$

State variables in DEDS are discrete ones. If they are logical variables or expressions built up of logical variables, then Equation (1.21) can be put together directly based on them. Other than logic variables can be represented by means of a set of the auxiliary logic variables further used in Equation (1.21).

Figure 1.5 shows an example of a simple discrete event dynamic system. The system is an input portion of a flexible manufacturing system. The parts to be processed are transported into the system by belt conveyor C1. They arrive as an irregular stream. There are different gaps between individual parts. A video system VS scans each part when the latter enters the VS range (detected by sensor P_0). It evaluates the parameters of shape and quality of an incoming part and sorts it in two groups. These two groups are routed *via* turntable TT_1. Intervals between individual parts are so that a new part comes in the range of sensor P_0 when the preceding part is already on conveyor C2 or C3. The parts of the first group are placed on conveyor C2 while those of the second group on conveyor C3.

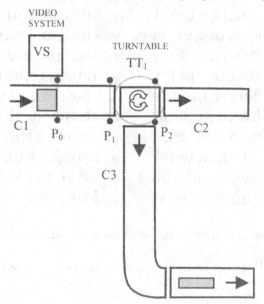

Figure 1.5. A manufacturing system

Now, let us model the system described in the example in a form given by Equation (1.19). We have

$$SYST = (\Pi, Q, \Sigma, \Theta) \tag{1.23}$$

$$\Pi = \left\{ \begin{aligned} &P_0, P_1, P_2; \gamma_{01}, \gamma_{02}; T1; ETT1H, ETT1V; \\ &TT1H, TT1V \end{aligned} \right\} \tag{1.24}$$

where P_0-P_2 are logic variables corresponding to sensors P$_0$-P$_2$, respectively. If a part is under sensor P$_0$ then $P_0 = 1$, etc. Note that the variables are written in italic in order to distinguish them from the corresponding sources of the variables. γ_{01} and γ_{02} serve for the group distinction: γ_{01}=1 and γ_{02}=0 for the first group and γ_{01}=0 and γ_{02}=1 for the second one. $T1$ is a state variable signaling the presence of a part in turntable TT1 when $T1 = 1$; otherwise $T1 = 0$. $ETT1H$ and $ETT1V$ indicate the turntable horizontal and vertical positions, respectively. $TT1H$, and $TT1V$ are commands to set the turntable horizontally or vertically. Before the start of the system operation, conveyors C1–C3 are switched on and remain in this state during the operation. Let there be the logic C1–C3 corresponding to the conveyor state; $C1 = 1$ if conveyor C1 is switched on and similarly for C2 and C3.

The set of states is as follows:

$$\begin{aligned}
Q = \{ q_0 &= (0,0,0; 0,0; 0; 1,0; 1,0), & q_{12} &= (0,0,1; 0,1; 1; 0,1; 0,1), \\
q_1 &= (1,0,0; 0,0; 0; 1,0; 1,0), & q_{13} &= (0,0,0; 0,0; 0; 0,1; 0,1), \\
q_2 &= (1,0,0; 1,0; 0; 1,0; 1,0), & q_{14} &= (1,0,0; 0,0; 0; 0,1; 0,1), \\
q_3 &= (0,0,0; 1,0; 0; 1,0; 1,0), & q_{15} &= (1,0,0; 0,1; 0; 0,1; 0,1), \\
q_4 &= (0,1,0; 1,0; 0; 1,0; 1,0), & q_{16} &= (0,0,0; 0,1; 0; 0,1; 0,1), \\
q_5 &= (0,0,0; 1,0; 1; 1,0; 1,0), & q_{17} &= (0,1,0; 0,1; 0; 0,1; 0,1), \\
q_6 &= (0,0,1; 1,0; 1; 1,0; 1,0), & q_{18} &= (0,0,1; 0,1; 1; 0,1; 0,1), \\
q_7 &= (1,0,0; 0,1; 0; 1,0; 1,0), & q_{19} &= (1,0,0; 1,0; 0; 0,1; 0,1), \\
q_8 &= (0,0,0; 0,1; 0; 1,0; 1,0), & q_{20} &= (0,0,0; 1,0; 0; 0,1; 0,1), \\
q_9 &= (0,1,0; 0,1; 0; 1,0; 1,0), & q_{21} &= (0,1,0; 1,0; 0; 0,1; 0,1), \\
q_{10} &= (0,0,0; 0,1; 1; 1,0; 1,0), & q_{22} &= (0,0,0; 1,0; 1; 0,1; 0,1), \\
q_{11} &= (0,0,1; 0,1; 1; 1,0; 0,1), & q_{23} &= (0,0,1; 0,1; 1; 1,0; 1,0) \}
\end{aligned} \tag{1.25}$$

The set of transitions Σ is given by the set of the following partial functions:

$$e_1 : e_1(q_0) = q_1, \quad e_2 : e_2(q_1) = q_2, \quad e_3 : e_3(q_2) = q_3, \quad e_4 : e_4(q_3) = q_4, \dots etc. \tag{1.26}$$

All functions can be given *via* Tables 1.3 and 1.4. Function values are in the table cells. Let the initial conditions Θ for the occurrence of the events from set Σ are these: the system is in state q$_0$ and $C1 = C2 = C3 = 1$. Establishing the system in q$_0$ is indicated by the logic variable *INIT*.

Table 1.3. The first part of the transition set functions

	Function argument											
	q_0	q_1	q_2	q_3	q_4	q_5	q_6	q_7	q_8	q_9	q_{10}	q_{11}
e_1	q_1	-	-	-	-	-	-	-	-	-	-	-
e_2	-	q_2	-	-	-	-	-	-	-	-	-	-
e_3	-	-	q_3	-	-	-	-	-	-	-	-	-
e_4	-	-	-	q_4	-	-	-	-	-	-	-	-
e_5	-	-	-	-	q_5	-	-	-	-	-	-	-
e_6	-	-	-	-	-	q_6	-	-	-	-	-	-
e_7	-	-	-	-	-	-	q_0	-	-	-	-	-
e_8	-	q_7	-	-	-	-	-	-	-	-	-	-
e_9	-	-	-	-	-	-	-	q_8	-	-	-	-
e_{10}	-	-	-	-	-	-	-	-	q_9	-	-	-
e_{11}	-	-	-	-	-	-	-	-	-	q_{10}	-	-
e_{12}	-	-	-	-	-	-	-	-	-	-	q_{11}	-
e_{13}	-	-	-	-	-	-	-	-	-	-	-	q_{12}
e_{14}	-	-	-	-	-	-	-	-	-	-	-	-
e_{15}	-	-	-	-	-	-	-	-	-	-	-	-
e_{16}	-	-	-	-	-	-	-	-	-	-	-	-
e_{17}	-	-	-	-	-	-	-	-	-	-	-	-
e_{18}	-	-	-	-	-	-	-	-	-	-	-	-
e_{19}	-	-	-	-	-	-	-	-	-	-	-	-
e_{20}	-	-	-	-	-	-	-	-	-	-	-	-
e_{21}	-	-	-	-	-	-	-	-	-	-	-	-
e_{22}	-	-	-	-	-	-	-	-	-	-	-	-
e_{23}	-	-	-	-	-	-	-	-	-	-	-	-
e_{24}	-	-	-	-	-	-	-	-	-	-	-	-
e_{25}	-	-	-	-	-	-	-	-	-	-	-	-
e_{26}	-	-	-	-	-	-	-	-	-	-	-	-

Table 1.4. The second part of the transition set functions

| | Function argument | | | | | | | | | | | |
	q_{12}	q_{13}	q_{14}	q_{15}	q_{16}	q_{17}	q_{18}	q_{19}	q_{20}	q_{21}	q_{22}	q_{23}
	-	-	-	-	-	-	-	-	-	-	-	-
	-	-	-	-	-	-	-	-	-	-	-	-
	-	-	-	-	-	-	-	-	-	-	-	-
	-	-	-	-	-	-	-	-	-	-	-	-
	-	-	-	-	-	-	-	-	-	-	-	-
	-	-	-	-	-	-	-	-	-	-	-	-
	-	-	-	-	-	-	-	-	-	-	-	-
	-	-	-	-	-	-	-	-	-	-	-	-
	-	-	-	-	-	-	-	-	-	-	-	-
	-	-	-	-	-	-	-	-	-	-	-	-
	-	-	-	-	-	-	-	-	-	-	-	-
	-	-	-	-	-	-	-	-	-	-	-	-
	q_{13}	-	-	-	-	-	-	-	-	-	-	-
	-	q_{14}	-	-	-	-	-	-	-	-	-	-
	-	-	q_{15}	-	-	-	-	-	-	-	-	-
	-	-	-	q_{16}	-	-	-	-	-	-	-	-
	-	-	-	-	q_{17}	-	-	-	-	-	-	-
	-	-	-	-	-	q_{18}	-	-	-	-	-	-
	-	-	-	-	-	-	q_{13}	-	-	-	-	-
	-	-	q_{19}	-	-	-	-	-	-	-	-	-
	-	-	-	-	-	-	-	q_{20}	-	-	-	-
	-	-	-	-	-	-	-	-	q_{21}	-	-	-
	-	-	-	-	-	-	-	-	-	q_{22}	-	-
	-	-	-	-	-	-	-	-	-	-	q_{23}	-
	-	-	-	-	-	-	-	-	-	-	-	q_6

All possible event sequences in the analyzed manufacturing system are built up from the sequences

$$\tilde{\sigma}_1 = e_1 \qquad \tilde{\sigma}_2 = e_2\, e_3\, e_4\, e_5\, e_6\, e_7 \qquad \tilde{\sigma}_3 = e_8\, e_9\, e_{10}\, e_{11}\, e_{12}\, e_{13}\, e_{14}$$

$$\tilde{\sigma}_4 = e_{15} \qquad \tilde{\sigma}_5 = e_{16}\, e_{17}\, e_{18}\, e_{19}\, e_{20}\, e_{14} \qquad \tilde{\sigma}_6 = e_{21}\, e_{22}\, e_{23}\, e_{24}\, e_{25}\, e_{26}\, e_7$$

$$(1.27)$$

The buildup or concatenations of the event sequences at Equation (1.27) follows the scheme as shown in Figure 1.6. Activities of the system can start when the condition Θ is fulfilled and it means that the first event sequence can be only $\tilde{\sigma}_1$. Equation (1.21) for the investigated system are

$$\rho_{e_1}(\Pi,\Pi') = (C1 \wedge C2 \wedge C3 \wedge INIT) \wedge \left(P_0' = lex_1 \text{ if and only if } lex_2 = 1 \right) \wedge$$

$$\left(P_1' = \overline{lex_1} \text{ if and only if } lex_2 = 1 \right) \wedge \left(P_2' = \overline{lex_1} \text{ if and only if } lex_2 = 1 \right) \wedge$$

$$\left(\gamma_{01}' = \overline{lex_1} \text{ if and only if } lex_2 = 1 \right) \wedge \left(\gamma_{02}' = \overline{lex_1} \text{ if and only if } lex_2 = 1 \right) \wedge$$

$$\left(T_1' = \overline{lex_1} \text{ if and only if } lex_2 = 1 \right) \wedge \left(ETT1H' = lex_1 \text{ if and only if } lex_2 = 1 \right) \wedge$$

$$\left(ETT1V' = \overline{lex_1} \text{ if and only if } lex_2 = 1 \right) \wedge \left(TT1H' = lex_1 \text{ if and only if } lex_2 = 1 \right)$$

$$\wedge \left(TT1V' = \overline{lex_1} \text{ if and only if } lex_2 = 1 \right)$$

$$(1.28)$$

where

$$lex_1 = lex_2,$$

$$lex_2 = \overline{P_0} \wedge \overline{P_1} \wedge \overline{P_2} \wedge \overline{\gamma_{01}} \wedge \overline{\gamma_{02}} \wedge \overline{T1} \wedge ETT1H \wedge \overline{ETT1V} \wedge TT1H \wedge \overline{TT1V} \qquad (1.29)$$

True logic value is 1 and false 0 in Equation (1.29). For event e_2 we have

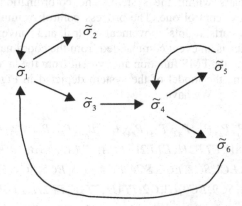

Figure 1.6. Event sequence patterns

$$\rho_{e_2}(\Pi, \Pi') = (C1 \wedge C2 \wedge C3 \wedge INIT) \wedge \left(P_0' = lex_1 \text{ if and only if } lex_2 = 1\right) \wedge$$

$$\left(P_1' = \overline{lex_1} \text{ if and only if } lex_2 = 1\right) \wedge \left(P_2' = \overline{lex_1} \text{ if and only if } lex_2 = 1\right) \wedge$$

$$\left(\gamma_{01}' = lex_1 \text{ if and only if } lex_2 = 1\right) \wedge \left(\gamma_{02}' = \overline{lex_1} \text{ if and only if } lex_2 = 1\right) \wedge$$

$$\left(T_1' = \overline{lex_1} \text{ if and only if } lex_2 = 1\right) \wedge \left(ETT1H' = lex_1 \text{ if and only if } lex_2 = 1\right) \wedge$$

$$\left(ETT1V' = \overline{lex_1} \text{ if and only if } lex_2 = 1\right) \wedge \left(TT1H' = lex_1 \text{ if and only if } lex_2 = 1\right)$$

$$\wedge \left(TT1V' = \overline{lex_1} \text{ if and only if } lex_2 = 1\right)$$

$$(1.30)$$

where

$$lex_1 = lex_2,$$
$$lex_2 = P_0 \wedge \overline{P_1} \wedge \overline{P_2} \wedge \overline{\gamma_{01}} \wedge \overline{\gamma_{02}} \wedge \overline{T1} \wedge ETT1H \wedge \overline{ETT1V} \wedge TT1H \wedge \overline{TT1V} \quad (1.31)$$

Other events would be expressed in a similar way.

Now consider an extension to the above system so that the parts of the first group are processed in a batch of three by R_{A1} or R_{A2}. Both robots perform the similar operations. The parts of the second group to a cell are routed *via* turntable TT_2. They are processed in two by R_B robotic cell. The number of parts is checked by means of photo-sensors P_{A1}–P_{A3} and P_{B1}–P_{B2}, respectively. Gate G1 (G2) goes up when three (two) parts are prepared for the next processing. Transport conveyor capacities are three transported parts for C2–C5, C8–C9 and two for C6–C7, respectively. Only one part can be allowed between sensors P_0 and P_1.

When a triple is prepared under sensor P_{A5}, robot R_{A1} performs the required processing operations and then transfers the triple onto conveyor O1. R_{A2} and R_B operate similarly. The aim of the control is to coordinate and control operations and movement of parts within the system. The co-ordination control level is superior to the process control one. The process control examples are the vision system's detection of parts, robots' movement control, and conveyor speed control. The vision system start is an event commanded from the coordination control level. Other facts concerning the FMS function are evident from the layout in Figure 1.7.

Now, let us outline the model of the system depicted in Figure 1.7 in a form given by Equation (1.19). We have

$$\Pi = \begin{cases} P_0, P_1, P_2, P_{A1}, P_{A2}, P_{A3}, P_{A4}, P_{A5}, P_{A6}, P_{B1}, P_{B2}, P_{B3}, ETT1H, \\ ETT1V, ETT2H, ETT2V, ERA1, ERA2, ERB; \\ SC1, EC1, SC2, EC2, SC4, EC4, SC5, EC5, SC7, EC7, SC8, \\ EC8, SC9, EC9, G1, G2, TT1H, TT1V, TT2H, TT2V, RA1, RA2, RB; \\ \gamma_{01}, \gamma_{02}, \gamma_{11}, \gamma_{12}, \gamma_{21}, \gamma_{22}, LC_{21}, LC_{22}, LC_{31}, LC_{32}, LC_{41}, LC_{42}, LC_{31}, \\ LC_{51}, LC_{52}, LC_{61}, LC_{62}, LC_{71}, LC_{72}, LC_{81}, LC_{82}, LC_{91}, LC_{92} \end{cases}$$

$$(1.32)$$

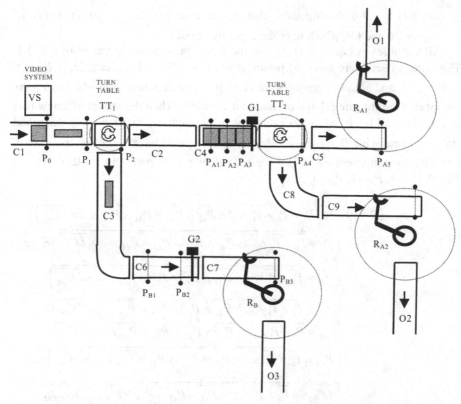

Figure 1.7. Example of a basic transition system: FMS with three robots

where the first part (the first two rows of Equation (1.28)) is inputs fed from the controlled system to the control system, the second part (the third and fourth rows) is outputs from the latter fed into the former and the third part is internal state variables. The variables Π are determined at the higher coordination control level. The model is built for the coordination control purpose. As in the preceding example, variable $ETT1H$ indicates the straight position of turntable TT1, $ETT1V$ its transversal position, *etc.* We assume that initially conveyors C3 and C6 and conveyors in both turntables are switched on and moving during the FMS operation. Conveyor C1 is started by command variable $SC1$ and stopped by $EC1$. Other conveyor variables listed in Equation (1.32) have analogous meanings. The gates are operated by variables $G1$ and $G2$. Variables $TT1H$, $TT1V$, *etc.*, are used to control the turntables, while $RA1$, $RA2$, and RB start robot operations. $ERA1$, $ERA2$, and ERB signals the end of the part processing by robots RA1, RA2, and RB, respectively. Information about routing a part is transferred from γ_{01} and γ_{02} to γ_{11} and γ_{22} when the part moves from sensors P_0 to P_1 (γ_{01} and γ_{02} should be free for the next part), and analogously for γ_{11} and γ_{22} and sensor P_2. LC_{21} and LC_{22} stand for storing the number of parts loaded on C2 so that no part gives $LC_{21} = 0, LC_{22} = 0$; one part gives $LC_{21} = 0, LC_{22} = 1$; two parts $LC_{21} = 1, LC_{22} = 1$; and three parts

$LC_{21} = 1, LC_{22} = 1$. Analogously, this holds true for LC_{ij} , but in terms of conveyors C6 and C7, which have the capacity equal 2.

All variables in Equation (1.32) are the Boolean ones taking values of 0 and 1. The states in set Q are given by pertinent variable values. For example, if $P_{A2} = 1$ and $P_{A3} = 1$ and all other variables are zero, the system is in a state when two parts are located before gate $G1$ and otherwise it is empty. Then the arrival of a new part in $C1$, signaled by $P_0 = 1$, is an event given by mapping the previously described state into one with $P_0 = 1, P_{A2} = 1, P_{A3} = 1$ and all remaining variables being zero. For example, an event $e = WP3$ - the arrival of a next part – in P_0=0, P_1=0, ..., P_{A1}=0, P_{A2}=1, P_{A3}=1, P_{A4}=0, ..., is

$$\rho_{WP3}(\Pi,\Pi')=1 \wedge \left(P_0' = \overline{P_0} \wedge \overline{P_1} \wedge ... \wedge \overline{P_{A1}} \wedge P_{A2} \wedge P_{A3} \wedge \overline{P_{A4}} \wedge ... \wedge \overline{LC_{92}} \right) \wedge$$

$$\left(P_1' = \overline{P_0} \wedge \overline{P_1} \wedge ... \wedge \overline{P_{A1}} \wedge P_{A2} \wedge P_{A3} \wedge \overline{P_{A4}} \wedge ... \wedge \overline{LC_{92}} \right) \wedge ... \wedge$$

$$\left(P_{A1}' = \overline{P_0} \wedge \overline{P_1} \wedge ... \wedge \overline{P_{A1}} \wedge P_{A2} \wedge P_{A3} \wedge \overline{P_{A4}} \wedge ... \wedge \overline{LC_{92}} \right) \wedge$$

$$\left(P_{A2}' = \overline{P_0} \wedge \overline{P_1} \wedge ... \wedge \overline{P_{A1}} \wedge P_{A2} \wedge P_{A3} \wedge \overline{P_{A4}} \wedge ... \wedge \overline{LC_{92}} \right) \wedge$$

$$\left(P_{A3}' = \overline{P_0} \wedge \overline{P_1} \wedge ... \wedge \overline{P_{A1}} \wedge P_{A2} \wedge P_{A3} \wedge \overline{P_{A4}} \wedge ... \wedge \overline{LC_{92}} \right) \wedge$$

$$\left(P_{A4}' = \overline{P_0} \wedge \overline{P_1} \wedge ... \wedge \overline{P_{A1}} \wedge P_{A2} \wedge P_{A3} \wedge \overline{P_{A4}} \wedge ... \wedge \overline{LC_{92}} \right) \wedge ... \wedge$$

$$\left(LC_{92}' = \overline{P_0} \wedge \overline{P_1} \wedge ... \wedge \overline{P_{A1}} \wedge P_{A2} \wedge P_{A3} \wedge \overline{P_{A4}} \wedge ... \wedge \overline{LC_{92}} \right)$$

(1.33)

where Θ represents the condition that the system has to be initialized and empty before the first event can be accepted.

The example illustrates that in a little more complex case the modeling using a basic transition system is complicated, not transparent and very difficult for analysis and control design. In the following chapters of this book, we try to develop systematically theory and a way for practical use of other tools aiming to model, analyze, evaluate, simulate and control DEDS.

1.6 Problems and Exercises

1.1. Cite some CDVS and DEDS examples from your daily life.

1.2. Derive the basic transition system models for Figures 1.3 and 1.4.

1.3. In the system in Figure 1.5, change the assumption that only one part is processed in it so that a part can come in when another part is between sensors P_0 and P_1. Consider capacities of the conveyors.

1.4. Write the expression for the event next to that given by Equation (1.33) when a part moves from sensors P_0 to P_1.

1.5. A robotic cell is depicted in Figure 1.8. A-Parts are loaded into it *via* input conveyor I1. The input has capacity of 1 part. The same holds for input I2 and output O. Robot R2 picks up an A-workpiece from I2 and transfers it onto table T. R1 picks up a B-work-piece from I1 and puts it into the free milling machine M1 or M2. If both are free, M1 is preferred. After the machining, R1 transfers it onto palette P. If there is an A-workpiece on T, R2 transfers it from the palette to T and an assembly starts. After it, R2 transfers the product onto O. a) Analyze the system as DEDS; and b) Create an event set and event strings corresponding to the required behavior of the system, a realizable but not admissible event string and a non-realizable event string.

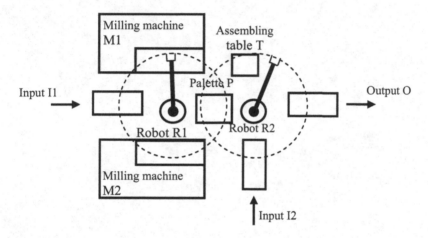

Figure 1.8. A robotic cell with two milling machines

1.6. For the system depicted in Figure 1.4 write a realizable event string corresponding to the technological process A in Table 1.1, a realizable but not admissible event string and a non-realizable event string.

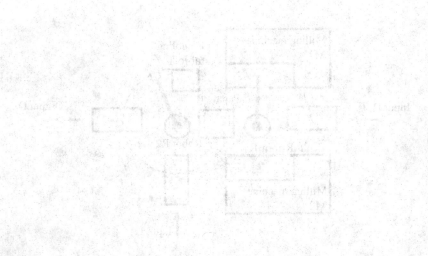

2

Graphs in Modeling DEDS

2.1 Simple Non-labeled Directed Mathematical Graphs

The basic transition model and its derivatives can be very easily and transparently represented graphically. A platform for this representation is a mathematical graph.

Definition 2.1. A simple non-labeled directed mathematical graph is given by an ordered pair

$$G = (A, R) \qquad (2.1)$$

where

      A is a finite non-empty set of elements called nodes or vertices of the graph

      R is a binary relation on A, which can be empty.

As explained in Section 1.2, binary relation R determines a set of ordered pairs chosen from nodes in set A. Definition 2.1 allows for isolated nodes in a graph.

The representation of a mathematical graph based on a set-theoretic way according to Definition 2.1 can be equivalently substituted or transformed into a true graphical form. Let us call it the drawn graphical form. In this form to each node corresponds a circle drawn in a plane and to each element of the relation corresponds an arrow or directed arc. For example, a graph given by

$$G = (\{A_1, A_2, A_3, A_4\}, \ \{(A_1, A_3), (A_1, A_4), (A_3, A_2), (A_4, A_2), (A_4, A_4)\}) \quad (2.2)$$

is equivalently represented in a drawn graphical form as shown in Figure 2.1.

Each circle has its individuality and corresponds to one node. Even if the circles are not denoted with symbols, the drawn graphical form is fully isomorphic with that of Equation (2.2). Of course, it is cumbersome to refer to the left-upper or right-lower circle, *etc.* Therefore, it is convenient and usual to denote the node-circle correspondence as in Figure 2.2. For short we call a simple non-labeled directed graph a non-labeled digraph. Obviously a non-labeled digraph G has a close connection with relation R. It can be said that a non-labeled digraph represents the corresponding binary relation. The bipartite simple non-labeled directed mathematical graphs constitute a special subset of the non-labeled digraphs. They are characterized by a set of nodes consisting of two disjunctive node subsets and directed arcs connecting only nodes from the different node sets. To distinguish the two node subsets graphically, circles and bars or boxes are usually used to represent them, respectively (see Petri nets in Chapter 7).

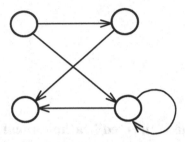

Figure 2.1. Graphical form of a mathematical graph

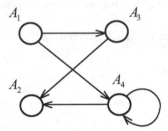

Figure 2.2. Specification of the correspondence node-circle

Many graph properties of digraphs can be analyzed by analogy with non-directed graphs. Non-directed graphs are defined as $G = (A, M)$ where M is a set of non-ordered pairs of graph nodes. In the drawn-graphical form the non-directed edges are used instead of directed arcs.

2.2 Labeled Mathematical Graphs

Definition 2.1 can be further developed as follows.

Definition 2.2. A simple labeled directed mathematical graph is a 6-tuple

$$G = (A, R, f_1, f_2, S_1, S_2) \tag{2.3}$$

where
> A is a finite set of the nodes;
> R is a relation on A, which can be empty;

f_1 is a function $A \rightarrow S_1$ defined if S_1 is defined;

f_2 is a function $R \rightarrow S_2$ defined if S_2 is defined; and

S_1 and S_2 are sets that can be empty.

If both sets S_1 and S_2 are empty, the graph in Definition 2.2 becomes a simple non-labeled digraph.

The labels can denote, *e.g.*, the arc weights given as integers. In Figure 2.3 there is an example when $S_1 = \varnothing$ (the empty set) and $S_2 = N^+$ (the set of positive integers).

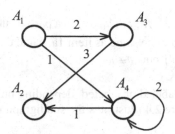

Figure 2.3. A labeled digraph

Commonly, the non-labeled and labeled digraphs are called digraphs. In this context, bipartite simple labeled directed mathematical graphs are also defined. We will see later that Petri nets belong to that kind of mathematical graphs.

There are several ways to represent mathematical graphs equivalently. One frequently used is the incidence matrix. For instance the graph in Figure 2.2 can be equivalently represented by the following incidence matrix:

$$\mathbf{G} = \begin{pmatrix} 0 & 0 & 1 & 1 \\ 0 & 0 & 0 & 0 \\ 0 & 1 & 0 & 0 \\ 0 & 1 & 0 & 1 \end{pmatrix} \tag{2.4}$$

In matrix $\mathbf{G}$, its rows and columns correspond to the nodes such that the first row and first column correspond to node A_1, the second row and second column to A_2, *etc*. In the matrix form, directions of arcs are considered from rows to columns. If a directed arc is present in the graph, the corresponding matrix element equals one, otherwise zero.

2.3 Subgraphs and Components

Subgraphs and graph components defined below play an important role in our next considerations.

Definition 2.3. Consider a digraph $G = (A, R, f_1, f_2, S_1, S_2)$. Then:

1. A digraph $G' = (A', R', f_1', f_2', S_1, S_2)$ is a subdigraph of G if
    a. $A' \subseteq A$
    b. $R' = R \cap (A' \times A')$
    c. $f_1': A' \to S_1$, such that $f_1'(a) = f_1(a)$ for $a \in A'$
    d. $f_2': R' \to S_2$, such that $f_2'(r) = f_2(r)$ for $r \in R'$
    The graph G is also a subdigraph of itself, *i.e.*, $G' = G$.
2. If G' is a subdigraph of G and $G' \ne G$, then G' is a proper subdigraph of G.
3. A digraph $G' = (A', R', f_1', f_2', S_1, S_2)$ is a partial subdigraph of G if
    a. Item c. and d. are the same as item 1a, c and d.
    b. $R' \subset R \cap (A' \times A')$ but $R' \ne R \cap (A' \times A')$

A subdigraph can be non-labeled or labeled, depending on the digraph for which it has been constructed. Figure 2.4 illustrates the subdigraph idea.

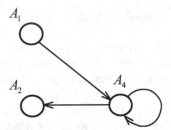

Figure 2.4. A proper subdigraph of the digraph in Figure 2.2

The digraph in Figure 2.4 is a proper subdigraph of the digraph in Figure 2.2. The set of nodes of the proper subdigraph is a proper subset

$$\{A_1, A_2, A_4\} \subset \{A_1, A_2, A_3, A_4\}$$

and all arcs not connecting the nodes of A' are canceled.

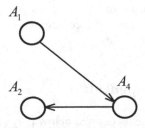

Figure 2.5. A partial subdigraph of the graph in Figure 2.2

On the other hand the digraph in Figure 2.5 is a partial subdigraph of that in Figure 2.2 because arc (A_4, A_4) has been omitted.

2.4 Directed Paths

The notion of a directed path is useful later in this book. We deal with it in the sequel.

Definition 2.4. Consider a digraph $G = (A, R, f_1, f_2, S_1, S_2)$. A sequence of nodes

$$\tilde{a} = a_{i_1} a_{i_2} \ldots a_{i_n} \tag{2.5}$$

is called a path from a_{i_1} to a_{i_n} in G if

1. $a_{i_k} \in A$ for all $k = 1, 2, \ldots, n$
2. $\left(a_{i_k} R a_{i_{k+1}}\right)$ or $\left(a_{i_{k+1}} R a_{i_k}\right)$ holds for all $k = 1, 2, \ldots, n-1$, i.e., a_{i_k} is in relation R with $a_{i_{k+1}}$. In other words, there is an arc either from a_{i_k} to $a_{i_{k+1}}$ or from $a_{i_{k+1}}$ to a_{i_k}. We say that path $\tilde{a}$ goes from a_{i_1} to a_{i_n}.

A path can be non-labeled or labeled, depending on the respective digraph property.

Definition 2.5. If there is a path of the form of Equation (2.5) in a given digraph $G = (A, R, f_1, f_2, S_1, S_2)$ and

1. If for all $k = 1, 2, \ldots, n-1$ either $\left(a_{i_k} R a_{i_{k+1}}\right)$ or $\left(a_{i_{k+1}} R a_{i_k}\right)$ holds, $\tilde{a}$ is called a directed path
2. If $a_{i_1} = a_{i_n}$, then $\tilde{a}$ is called a cycle
3. If $\tilde{a}$ is a directed path and $a_{i_1} = a_{i_n}$ the path is called a directed cycle (cycle for short when no confusion arises)
4. If all nodes in $\tilde{a}$ are distinct except for a_{i_1} and a_{i_n}, then $\tilde{a}$ is called a simple path. In other words, no node repeats in the path. The idea of simplicity can be applied to directed paths and directed cycles to obtain directed simple paths and directed simple cycles, respectively.

Definition 2.6. A digraph $G = (A, R)$ is called connected if for every two nodes $a_i, a_j \in A$ there is a path from a_i to a_j. G is strongly connected if there are directed paths from a_i to a_j and from a_j to a_i.

Connected digraphs have neither isolated nodes nor isolated groups of nodes. The notion of a digraph component is based on the graph connectivity dealt with in the following definition.

Definition 2.7. A strong component of a digraph G is a strongly connected subdigraph of G, which is not a proper subdigraph of any strongly connected subdigraph of G.

The meaning of the last definition is illustrated using the example in Figure 2.6.a. There are two strong components of the digraph depicted in Figures 2.6.b and 2.6.c. A strong component is a maximum strongly connected subdigraph, *i.e.*, not contained in any other strongly connected digraph. Hence, the subdigraph depicted in Figure 2.6.d is not a strong component. In particular, it is a proper subdigraph of the connected subdigraph in Figure 2.6.b and this fact is contradictory to the assumption of Definition 2.7.

Sometimes it is useful to express the multiplicity of arcs in a digraph. This can be done by introducing a weight function over relation R as shown below. Another method is to use the multiset concept. A multiset allows multiple same members. For example $X=\{(A_1, A_2), (A_1, A_2), (A_1, A_2), (A_2, A_3), (A_3, A_3), (A_3, A_1), (A_3, A_1)\}$ is a multiset example in which (A_1, A_2) appears three times and (A_3, A_1) twice in the multiset X.

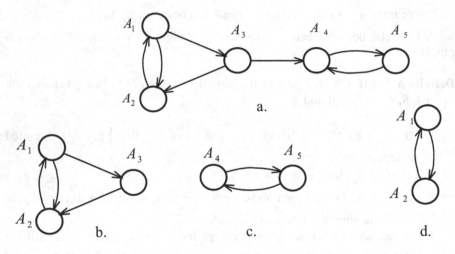

Figure 2.6a-d. A digraph (**a**) and its strong components: (**b, c**) while its subdigraph (**d**) is not its strong component

Definition 2.8. A directed multigraph is a triple:

$$G = (A, R, f) \qquad (2.6)$$

where A and R are the same as in Definition 2.1 and f is a function $f : R \rightarrow N^+$.

A multigraph in the drawn-graphical form can be equivalently represented as those in Figures 2.7 and 2.8. Note that label 1 on an arc can be omitted. Function f is given as follows:

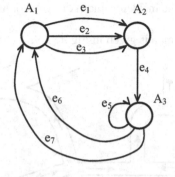

$$f(A_1, A_2) = 3,\ f(A_2, A_3) = 1,\ f(A_3, A_3) = 1,\quad f(A_3, A_1) = 2$$

Figure 2.7. A multigraph

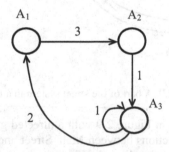

Figure 2.8. Labeling of arcs in a digraph

An important group of connected digraphs are trees. A tree is a non-labeled digraph with the following properties: it has exactly one node (root) with no in-going arcs and all other nodes have exactly one in-going arc. Obviously, a tree does not contain cycles.

2.5 Problems and Exercises

2.1. Let x_{1-4} be your last four digits in your identification number, respectively. Let $g_i=$**sign**(x_i), $i=1$, 2, 3 and 4. For example, **sign**$(0)=0$ and **sign**$(7)=1$. Let $f_i=1-g_i$. Present the graphical representation of the following digraph given the below matrix; and identify its strong component(s) if any:

$$G = \begin{pmatrix} g_1 & g_2 & g_3 & g_4 \\ f_1 & f_2 & f_3 & f_4 \\ g_1 & g_2 & g_3 & g_4 \\ f_1 & f_2 & f_3 & f_4 \end{pmatrix}$$

2.2. Prove that the rank of the incidence matrix of any connected tree cannot be full.

2.3. A part of a town street map is given in Figure 2.9.

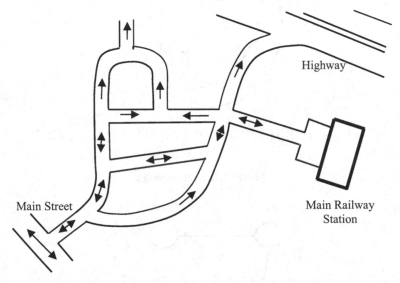

Figure 2.9. A part of the street system in a town

Represent the street system in Figure 2.9 with a directed graph. Find a subdigraph representing possible connections between Main Street and Main Railway Station. Find a strong component of the digraph. Find all simple paths connecting Main Street and Highway. Give some labels to the digraph corresponding to distances and find a shortest path from Main Railway Station to Main Street.

2.4. Given five instructors A1–A5, and five courses C1–C5 to be taught, use a circle to represent an instructor and a box a course. Connect a solid line from an instructor to a course if the instructor is assigned to teach it and a dotted line meaning that the instructor can teach it if needed. Suppose that each instructor is familiar with two and only two course materials (hence can teach at most two courses). Given Figure 2.10a, b, please derive which one can better cover instruction if an emergency happens such that one instructor cannot come. If an instructor is assigned to teach a course and can teach at most two courses, how many courses must s/he be able to teach if any one instructor has an emergency? How about if any two instructors have an emergency? Assume that these five courses are offered Monday through Friday, respectively.

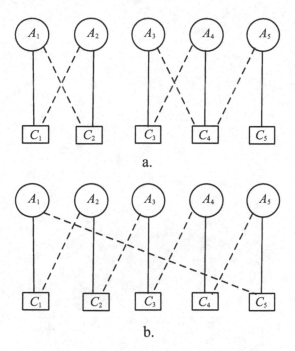

a.

b.

Figure 2.10a, b. Graph showing who are teaching courses and capable of teaching additional courses

3

Formal Languages

3.1 Notion of the Formal Language

We will follow the way of reasoning developed in the preceding chapters. Consider an event set $\Sigma = \{e_1, e_2, ..., e_n\}$ given for a DEDS. Further consider it to be in an initial state q_0. The system behavior can be defined by all possible sequences (strings or words) of events that can occur in it starting from q_0. It is assumed that an event occurs in a discrete point of time. Further it is assumed that just one event occurs in one discrete time point. The set of all finite and infinite sequences, which can be created from the elements of Σ including the empty sequence $\tilde{\varepsilon}$, is denoted as Σ^*. The set that does not include $\tilde{\varepsilon}$ is denoted as Σ^+, *i.e.*, $\Sigma^+ = \Sigma^* \setminus \tilde{\varepsilon}$ where symbol " \ " stands for the set subtraction.

Usually only a part L of all possible sequences Σ^* can occur in a given DEDS. Such a particular behavior of the DEDS is due to a subset L of sequences from Σ^*, *i.e.*, $L \subseteq \Sigma^*$. L is supposed always to include the empty sequence (string, word) and is called a formal language. The formal language L defines the behavior of a DEDS. Our attention is aimed at formal languages with respect to the above-introduced interpretation related to DEDS. A formal definition is useful in order to exactly communicate the idea of a formal language.

Definition 3.1. Let a finite non-empty set of events $\Sigma = \{e_1, e_2, ..., e_n\}$ be given. The formal language L over Σ is a set of sequences formed from the events including the empty sequence $\tilde{\varepsilon}$. Σ is called the alphabet of the formal language. A sequence of L is also called a string or word.

From Definition 3.1 it follows that the empty string $\tilde{\varepsilon}$ always belongs to a formal language.

Definition 3.2. The length of a word (string) α denoted $|\alpha|$ is the number of events in the word. The length of the empty string is 0.

Example 3.1. Consider an alphabet $\Sigma = \{\alpha, \beta\}$. A formal language L_1 is given by

$$L_1 = \{\tilde{\varepsilon}, \alpha, \beta, \alpha\alpha, \alpha\beta, \beta\beta, \beta\alpha\} \tag{3.1}$$

Language L_1 can verbally be determined as a set of all strings over Σ whose lengths do not exceed 2, including the empty string. It is finite.

Example 3.2. Consider the same alphabet as in Example 3.1. Let a language L_2 consist of all strings beginning with some event β. It is expressed as

$$L_2 = \{\tilde{\varepsilon}, \beta, \beta\alpha, \beta\alpha\alpha, \beta\alpha\alpha...\alpha, \beta\beta\alpha, \beta\beta\alpha\alpha,...\} \tag{3.2}$$

This set obviously has an infinite number of members.

We will focus on how a formal language can be utilized for the description, analysis and synthesis of DEDS. An important assumption is that only one event can occur at one discrete time point. Time is not explicitly given for a formal language string. However, the order of events is specified. The connection of a formal language string with a discrete process as described in Section 1.3 can easily be recognized.

Example 3.2 shows that even in very simple cases it is impossible to put down all strings of a formal language. The example also illustrates that the language specification is some rule or rules for the creation of strings. Such kinds of rules are called formal grammars.

3.2 Formal Grammars and Classification of Formal Languages

Any formal language can be defined by a generation rule of its words. The rules for generating them, called formal grammar, are described below.

Definition 3.3. A formal grammar is formally defined by the quadruple

$$G = (V_N, V_T, P, s) \tag{3.3}$$

where

V_N is a finite set of non-terminal elements

V_T is a finite set of terminal elements whereby $V_N \cap V_T = \varnothing$; the union $V = V_N \cup V_T$ is called the alphabet and its elements are called symbols

P is a finite set of rules $P = \{P_1, P_2, ..., P_k\}$ for generating words. The generating rule has the form $\tilde{\alpha} \underset{G}{\mapsto} \tilde{\beta}$, $\tilde{\alpha} \in V^+$, $\tilde{\beta} \in V^*$, $V^+ = V^* \setminus \tilde{\varepsilon}$, where the symbol $\underset{G}{\mapsto}$ indicates that a word transformation is accomplished using the generating rule belonging to grammar G

s is a special non-terminal one in V_N called initial element or symbol.

The fact that the left hand side word of a generating rule is $\tilde{\alpha} \in V^+$ means that $\tilde{\alpha}$ cannot be an empty word (string). On the other hand, a word $\tilde{\beta} \in V^*$ can be an empty one. In other words, an empty string $\tilde{\varepsilon}$ cannot be mapped into a string but there can exist a production rule mapping a non-empty word into an empty one.

The word generating rules are used for formal language generation. A formal language L generated by a formal grammar G is a set of all words, which consists of the words containing only terminal elements of the given grammar and are generated by the repeated application of one or more rules, always beginning at the initial element s. Next words are generated from already created words. Rules can be applied also on a word part. Their application is formally described by the scheme

$$\tilde{\mu} \tilde{\alpha} \tilde{v} \underset{G}{\mapsto} \tilde{\mu} \tilde{\beta} \tilde{v} \tag{3.4}$$

where $\tilde{\mu} \in V^*, \tilde{v} \in V^*$, and as stated before $\tilde{\alpha} \in V^+, \tilde{\beta} \in V^*$ and $\tilde{\alpha} \underset{G}{\to} \tilde{\beta}$. The process starts with s, i.e., $\tilde{\mu} = \tilde{\varepsilon}$ and $\tilde{v} = \tilde{\varepsilon}$. It continues with the new generated words according to Equation (3.4). As already mentioned, a part of word can be transformed into another. If $\tilde{\mu} = \tilde{v} = \tilde{\varepsilon}$, the whole word is being transformed into another one. The word generation can be considered as a replacement rule. The generation sequence formally proceeds as

$$\tilde{\mu}_{11} s \tilde{v}_{11} \underset{G}{\mapsto} \tilde{\mu}_{11} \tilde{\phi}_{11b} \tilde{v}_{11}, \tilde{\mu}_{12} \tilde{\phi}_{12a} \tilde{v}_{12} \underset{G}{\mapsto} \tilde{\mu}_{12} \tilde{\phi}_{12b} \tilde{v}_{12}, ..., \tilde{\mu}_{1m} \tilde{\phi}_{1ma} \tilde{v}_{1m} \underset{G}{\mapsto} \tilde{\mu}_{1m} \tilde{\phi}_{1mb} \tilde{v}_{1m}$$

where $\tilde{\mu}_{11} \tilde{\phi}_{11b} \tilde{v}_{11} = \tilde{\mu}_{12} \tilde{\phi}_{12a} \tilde{v}_{12}, ..., \tilde{\mu}_{1,m-1} \tilde{\phi}_{1,m-1,b} \tilde{v}_{12} = \tilde{\mu}_{1m} \tilde{\phi}_{1ma} \tilde{v}_{1m}$

$$\tilde{\mu}_{21} s \tilde{v}_{21} \underset{G}{\mapsto} \tilde{\mu}_{21} \tilde{\phi}_{21b} \tilde{v}_{21}, \tilde{\mu}_{22} \tilde{\phi}_{22a} \tilde{v}_{22} \underset{G}{\mapsto} \tilde{\mu}_{22} \tilde{\phi}_{22b} \tilde{v}_{22}, ..., \tilde{\mu}_{2n} \tilde{\phi}_{2na} \tilde{v}_{2n} \underset{G}{\mapsto} \tilde{\mu}_{2n} \tilde{\phi}_{2nb} \tilde{v}_{2n}$$

where $\tilde{\mu}_{21} \tilde{\phi}_{21b} \tilde{v}_{21} = \tilde{\mu}_{22} \tilde{\phi}_{22a} \tilde{v}_{22}, ..., \tilde{\mu}_{2,n-1} \tilde{\phi}_{2,n-1,b} \tilde{v}_{2,n-1} = \tilde{\mu}_{2n} \tilde{\phi}_{2na} \tilde{v}_{2n}$

$$\tag{3.5}$$

In Equation (3.5) $\tilde{\mu}_{11} = \tilde{v}_{11} = \tilde{\mu}_{21} = \tilde{v}_{21} = ... = \tilde{\varepsilon}$ because each generation process starts with symbol s. Consider the first row in Equation (3.5). The final word in the row, which closes the particular generation process, is a word for which there is no continuation of the generation process. The second generation process starts again in s. If the sequences starting in s are exhausted, the next possible generation processes start with the words generated in the previous generation ones. If $\tilde{\alpha}_2$ is generated from $\tilde{\alpha}_1$ by the repetitive application of grammar G, a generation sequence can be briefly written as

$$\tilde{\alpha}_1 \underset{G}{\overset{*}{\mapsto}} \tilde{\alpha}_2 \tag{3.6}$$

where symbol * denotes the repeated word generation, and G denotes the used formal grammar. For the first sequence in Equation (3.5) we have

$$s \underset{G}{\overset{*}{\mapsto}} \tilde{\mu}_{1m} \tilde{\phi}_{1mb} \tilde{v}_{1m} \tag{3.7}$$

A language L consists of words that are generated by a given formal grammar in a described way and contains only terminal elements of the given alphabet Σ. If, e.g., $\tilde{\mu}_{1m} \tilde{\phi}_{1mb} \tilde{v}_{1m}$ consists only of terminal elements (of V_T), then $\tilde{\mu}_{1m} \tilde{\phi}_{1mb} \tilde{v}_{1m} \in L$.

Example 3.3. Let a language L be generated by the following formal grammar: $G = (V_N, V_T, P, s)$, $V_N = \{s, A, B\}$, $V_T = \{0,1\}$,

$$P = \{s \mapsto 0A, A \mapsto 1B, B \mapsto 0A, A \mapsto 1\} \tag{3.8}$$

s is the initial symbol.

The only rule containing s is $s \underset{G}{\mapsto} 0A$ and thus the first generation step (the first word generation) should be $s \underset{G}{\mapsto} 0A$. The word $0A$ does not belong to language L generated by grammar G because in the word there is a non-terminal symbol A. Thus two rules can be applied to word $0A$:

$$0A \underset{G}{\mapsto} 01 \tag{3.9}$$

where rule $A \underset{G}{\mapsto} 1$ has been used or

$$0A \underset{G}{\mapsto} 01B \tag{3.10}$$

where rule $A \underset{G}{\mapsto} 1B$ has been used.

Because the word from Equation (3.9) consists only of terminal symbols, the generation Equation (3.9) yields a word belonging to language L, while the generation Equation (3.10) does not. In the next step we have

$$01B \underset{G}{\mapsto} 010A \tag{3.11}$$

which is a similar situation to before, *i.e.,* either

$$010A \underset{G}{\mapsto} 0101 \qquad (3.12)$$

or

$$010A \underset{G}{\mapsto} 0101B \qquad (3.13)$$

We can write $s \underset{G}{\overset{*}{\mapsto}} 0101$. Continuing in this way we obtain the infinite set of words belonging to language L:

$$L = \{01, 0101, 010101, 01010101, ...\} \qquad (3.14)$$

The formal grammar as defined in Equation (3.4) is the most general one. It is called a type 0 grammar and the corresponding generated language is called a type 0 language. If the generation rules in the type 0 grammar are restricted, various grammars are obtained. They can be classified as type 0 or unrestricted grammar, type 1 or context grammar, type 2 or context-free grammar, and type 3 or regular grammar. Correspondingly, there are type 0 or unrestricted, type 1 or context, type 2 or context-free, and type 3 or regular languages. The regular grammar is of the first-rate interest in DEDS.

The regular grammar has the generation rules in the form

$$A \underset{G}{\mapsto} aB \quad \text{or} \quad A \underset{G}{\mapsto} a \qquad (3.15)$$

where $A, B \in V_N$ and $a \in V_T$.

The grammar given in Example 3.3 is regular and thus generates a regular language. According to the type definition, the set of languages with a lower type number contains all languages of a higher one. Hence, their set inclusion property is given:

Type 3 language $\subset$ Type 2 language $\subset$ Type 1 language $\subset$ Type 0 language
i.e.,

Regular language $\subset$ Context-free language $\subset$ Context language $\subset$ Unrestricted language

A formal grammar determines the corresponding language in a generative way. A different approach to specifying a formal language is the recognition way. The idea underlying this approach consists in finding an abstract model that can recognize whether a string over an alphabet belongs to a given language or not. For each language type a model can be constructed capable of recognizing the language. A finite automaton is such a model for the set of regular languages (type 3). The next larger set of formal languages, the context-free languages, can be recognized by push-down automata, *etc.* In this sense, a recognizing model is a specification tool for a formal language being recognized. Recognizing models are built up as the finite-state machines. In a recognizing model, the symbols of a processed string are fed into the model being in some state. The model passes a sequence of states and finishes in a state that determines whether the string belongs to the formal language specified by the model.

Moreover, a recognizing model can generate the represented language in a different way to that of a formal grammar. The generation in this case is based upon the basic transition model described in Section 1.5. The states and allowed transitions among the states are used for the generation of the language words.

In this book we focus on the topics related to regular languages, as they are the most important for modeling and control of DEDS. For this type of the languages it is necessary to know the finite automata, which serve as recognizing models for regular languages.

3.3 Regular Expressions

Regular languages can be well specified using regular expressions. The adjective "regular" stresses a provable fact that for each regular language there is a regular expression that specifies it and *vice versa*.

Definition 3.4. Regular expressions over a finite set Σ are recursively defined as:

a.   The symbol for the empty set $\varnothing$ is a regular expression;
b.   The symbol for the empty element of the set Σ (if included in Σ) is a regular expression;
c.   The symbol of any element of Σ is a regular expression, *i.e.*, if $a \in \Sigma$, then a is a regular expression;
d.   If r and s are regular expressions, so are $r \cup s$, rs, and r^* where $\cup$ means the set union, rs is the concatenation of strings or the sets of strings, and r^* is the iteration of a string or a set.

Three operations in Definition 3.4(d), *i.e.*, union, concatenation and iteration can be applied on strings or sets. Consider the strings

$$\tilde{s} = s_1 s_2 ... s_n , \quad \tilde{r} = r_1 r_2 ... r_m \tag{3.16}$$

The concatenation of strings $\tilde{r}$ and $\tilde{s}$ is string $\tilde{c} = \tilde{r}\tilde{s} = r_1 r_2 ... r_m s_1 s_2 ... s_n$. If S and R are the sets of strings (denote them for better clarity as R and S) the concatenation $RS = \{\tilde{r}\tilde{s} \mid \tilde{r} \in R, \tilde{s} \in S\}$ is the set of concatenated strings from the sets R and S. The iteration or Kleene closure is

$$\tilde{r}^* = \bigcup_{i=0}^{\infty} \tilde{r}^i \tag{3.17}$$

where $\tilde{r}$ is a string and

$\tilde{r}^0 = \tilde{\varepsilon}$ is the empty string

$\tilde{r}^1 = \tilde{r}$

$\tilde{r}^2 = \tilde{r}\tilde{r}$

...

$\tilde{r}^k = \tilde{r}\tilde{r}\ldots\ldots\ldots\tilde{r}$ (k times) $\hspace{4cm}$ (3.18)

In Definition 3.4.d, r is supposed to be a regular expression. From the definition it follows that it can be a string or a set of strings created in the iterative expansion of regular expressions. Consider the case when r is a set of strings. Denote the set by R. The Kleene closure is analogously defined as

$$R^* = \bigcup_{i=0}^{\infty} R^i \hspace{5cm} (3.19)$$

where

$R^0 = \{\varepsilon\}$

$R^1 = R$

$R^2 = RR,$ it is the concatanation of two sets $\hspace{2cm}$ (3.20)

$\hspace{2cm}$ of strings : $RR = \{\tilde{u}\tilde{v} \mid \tilde{u} \in R, \tilde{v} \in R\}$

...

Example 3.4. Let a formal language over $\Sigma = \{\alpha, \beta\}$ be given by the regular expression

$L = \alpha\beta^* \cup \beta$

The language consists of the strings

$L = \{\tilde{\varepsilon}, \beta, \alpha, \alpha\beta, \alpha\beta\beta, \alpha\beta\beta\beta, \alpha\beta\beta\beta\beta, \ldots\}$

i.e., it contains $\tilde{\varepsilon}, \beta, \alpha$, and an infinite number of strings beginning with α and including k times symbol β , where $k = 1,2,3,\ldots$ grows to infinity.

3.4 Problems and Exercises

3.1. Consider for the manufacturing system depicted in Figure 1.3 the capacities of all conveyors to be 1. Specify a formal language representing the behavior of the system.

3.2. Let two languages be given by regular expressions $L_1 = \alpha\beta^*$ and $L_2 = (\alpha \cup \beta)^*$. Determine a language given by the concatenation $L_1 L_2$.

3.3. Let transits between street crossings in Exercise 2.3 define events (the drive from the station to the closest crossing and contrariwise are events, too). Specify a language, which is given by possible transits of one car starting from Main Street.

3.4. A robotic cell contains two machines and two robots as shown in Figure 3.2. Parts are loaded irregularly and sequentially in the cell by two inputs: the first kind by I1 and the second one by I2. Transfer of parts is done by the robots. Machine and conveyor capacities are one. The robot R2 transfers the processed parts on the conveyor C3 whenever is part ready, order is not important.

Explain what is a state and an event in the described system. Explain what it is a formal language and how it can be represented with regular expressions. Define the event set for the system and write several strings of the formal language describing the system.

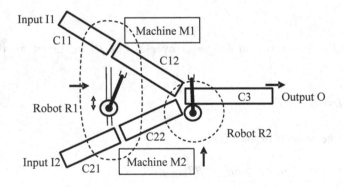

Figure 3.2. Robotic cell with two machines

4

Control of DEDS

4.1 State and Control Variables

Consider the basic transition system Equation (1.19) by Manna and Pnueli described in Section 1.5 as a general model of a DEDS. As pointed out earlier, time is not explicitly considered in the model. The DEDS dynamics depends on events that appear in discrete time points. The events provoke changes of the system states. The relations and mutual influence of events and states with respect to control are studied in this chapter. According to that, let a DEDS be given by the event set

$$\Sigma = \{e_1, e_2, ..., e_n\} \tag{4.1}$$

A sample event path of the system is

$$\tilde{e} = e_{i_1} \ e_{i_2} \ ... \ e_{i_v} \tag{4.2}$$

where e_{i_1} occurs at time point τ_{i_1}, e_{i_2} at time point τ_{i_2}, ..., e_{i_v} at τ_{i_v}, $\tau_{i_1} \langle \tau_{i_2} \langle ... \langle \tau_{i_v}$.

We emphasize that just one event can occur in a discrete time point. The path at Equation (4.2) starts in the system initial state. Let another sequence of time points be determined as

$$\tau_{a_1} = \tau_{i_1} + \frac{\tau_{i_2} - \tau_{i_1}}{2}, \ \tau_{a_2} = \tau_{i_2} + \frac{\tau_{i_3} - \tau_{i_2}}{2}, \ ..., \tau_{a_{v-1}} = \tau_{i_{v-1}} + \frac{\tau_{i_v} - \tau_{i_{v-1}}}{2} \tag{4.3}$$

i.e., time points at Equation (4.3) are in the middle between the time intervals of two consecutive events.

Let the set of state variables be represented by a time-dependent vector variable

$$\mathbf{u}(t) = \begin{pmatrix} u_1(t) \\ u_2(t) \\ \vdots \\ u_b(t) \end{pmatrix} \tag{4.4}$$

Assume the vector component $u_i(t)$, $i = 1, 2, ..., b$, having a value from a finite set U_i of real numbers. The system state dynamics can be expressed by a sequence $\tilde{\mathbf{u}}$, which is a discrete process

$$\tilde{\mathbf{u}} = \mathbf{u}\left(\tau_{a_1}\right) \mathbf{u}\left(\tau_{a_2}\right) \, ... \, \mathbf{u}\left(\tau_{a_{v-1}}\right) = \mathbf{u}_{a_1} \mathbf{u}_{a_2} ... \mathbf{u}_{a_{v-1}} \tag{4.5}$$

where the discrete time points are $\tau_{a_1} \langle \tau_{a_2} \langle ... \langle \tau_{a_{v-1}}$. Changes of variable $\mathbf{u}(t)$ at discrete time points are changes of the system states, which correspond to events. The state changes occur as responses to events. In real systems the responses are delayed some time after points $\tau_{i_1}, \tau_{i_2}, ..., \tau_{i_v}$. Just to come to a conceivable model, assume that all responses finish before points $\tau_{a_1}, \tau_{a_2}, ..., \tau_{a_{v-1}}$ so that new states are fully observable in these points.

We have not yet dealt with the question of how a certain required behavior of the system can be achieved. Various system descriptions and models represent the given system from the observer point of view. The required and for some purpose useful properties and behavior of the system are achieved by the system control performed fully automatically or with human participation. In this book we are interested in both modeling and control. It is necessary to distinguish between the approach aiming at the description of a system as a whole and the approach aiming at the control specification. This difference will always be taken into account in the next chapters. The control function can be extracted from the required behavior of the whole system.

The system control is enabled through purposeful intervening into the system. The interventions are represented by control variables. Assume that they can be represented by a vector variable $\mathbf{w}(t)$. The control variables are time-dependent and react to the actual situation in the system with respect to a required system behavior. In other words, this is the control. The interventions through the control variables correspond to requirements imposed on the system behavior. The degree of agreement of the required system behavior with the actual one is judged by a relevant criterion specifying the control performance.

4.2 Control System and Control Function

From the observer point of view, a system *SYST* can be represented as one including the controlled and control parts as shown in Figure 4.1. The system described by the basic transition system at Equation (1.19) can be decomposed into two subsystems with the feedback structure typical for control: the subsystem S to be controlled and the control system C. Figure 4.1 shows the decomposition where $\tilde{\mathbf{w}}$ is a sequence of control variable values given at discrete time points as follows:

$$\tilde{\mathbf{w}} = \mathbf{w}\left(\tau_{a_1} + \Delta\tau_{a_1}\right) \mathbf{w}\left(\tau_{a_2} + \Delta\tau_{a_2}\right) \dots \mathbf{w}\left(\tau_{a_{r-1}} + \Delta\tau_{a_{r-1}}\right) \tag{4.6}$$

and

$$\mathbf{w}(t) = \begin{pmatrix} w_1(t) \\ w_2(t) \\ \cdot \\ \cdot \\ \cdot \\ w_c(t) \end{pmatrix} \quad \text{and} \quad \mathbf{w}(\tau_{a_k}) = \begin{pmatrix} w_1(\tau_{a_k}) \\ w_2(\tau_{a_k}) \\ \cdot \\ \cdot \\ \cdot \\ w_c(\tau_{a_k}) \end{pmatrix} \tag{4.7}$$

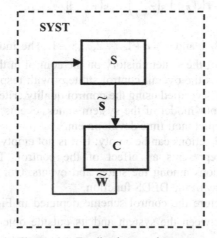

Figure 4.1. Feedback control structure

Let the value $w_i(t)$, $i = 1,2,\dots,c$, of the vector component be from a finite set W_i of real numbers.

State variables of SYST are decomposed in Figure 4.1 into two subsets given by $\mathbf{s}$ and $\mathbf{w}$. The decomposition model of SYST now distinguishes two subsystems S and C. $\mathbf{w}$ is input and $\mathbf{s}$ output for S, $\mathbf{s}$ input and $\mathbf{w}$ output for C.

Very often Equation (4.6) can be simplified considering the assumption that

$$\Delta\tau_{a_1} = \Delta\tau_{a_2} = \dots = \Delta\tau_{a_{r-1}} = \Delta\tau \tag{4.8}$$

i.e., the variable values in Equation (4.6) are given in time points delayed with respect to $\tau_{a_1} - \tau_{a_{v-1}}$ by a fixed time $\Delta \tau$ suitable chosen with respect to the system dynamics so that

$$\tilde{\mathbf{w}} = \mathbf{w}\!\left(\tau_{a_1} + \Delta\tau\right)\ \mathbf{w}\!\left(\tau_{a_2} + \Delta\tau\right)\ \dots\ \mathbf{w}\!\left(\tau_{a_{v-1}} + \Delta\tau\right) \qquad (4.9)$$

In order to avoid model complications assume that

$$\tau_{a_1} + \Delta\tau < \tau_{i_2}$$
$$\tau_{a_2} + \Delta\tau < \tau_{i_3}$$
$$\vdots$$
$$\tau_{a_{v-1}} + \Delta\tau < \tau_{i_v}$$

Control of the system S (Figure 4.1) can be described using a vector function $\mathbf{f}$ in the following way:

$$\mathbf{w}\!\left(\tau_{a_k} + \Delta\tau\right) = \mathbf{f}\!\left(\mathbf{s}\!\left(\tau_{a_k}\right)\!, \mathbf{s}\!\left(\tau_{a_{k-1}}\right)\!, ..., \mathbf{s}\!\left(\tau_{a_{k-r}}\right)\right) \qquad (4.10)$$

where $r=0, 1, 2, \dots, k-1$ and $k = r+1, r+2, ..., v-1$. The index r determines the depth of the influence of the system history on the control in the actual time point. The function $\mathbf{f}$ expresses the overall control strategy with respect to control goals. Fulfilment of the goals is verified using the control quality criteria.

Now, to illustrate our model of the system states, events, and control actions Figure 4.2 is used to depict their time development.

Some of the control actions can be empty. If it is not empty an immediate event after control action represents an effect of the control. There can be more complicated time relations among the states and events. Our assumptions enable one to gain the first view of the DEDS function.

It is useful to structure the control scheme depicted in Figure 4.1 in order to express the relation between the system and its outside objects. Variable $\mathbf{s}\!\left(\tau_{a_k}\right)$ from Figure 4.1 is decomposed into the input variable $\mathbf{x}\!\left(\tau_{a_k}\right)$, output variable $\mathbf{y}\!\left(\tau_{a_k}\right)$ and internal state variable $\mathbf{z}\!\left(\tau_{a_k}\right)$ as shown in Figure 4.3.

Classification of the actual system variables in the introduced variable groups according to the structure in Figure 4.3 is not strictly and uniquely given for a particular system. It depends largely on a control designer.

Presently, the control systems are predominantly realized by control computers. There are various designs or versions. Their main and common feature is the possibility to program their operation by a sequence of instructions – a program. There are two basic kinds of programs:

- Transformation programs
- Reactive programs

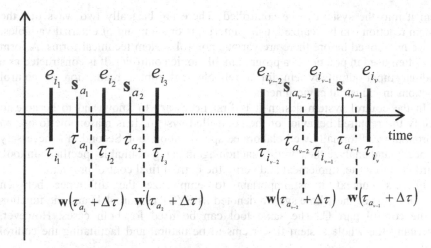

Figure 4.2. Time diagram for states, events and control

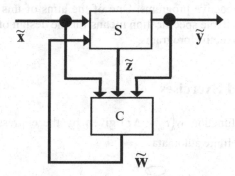

Figure 4.3. Input-output control structure

A transformation program represents a traditional way of computer usage. It produces a final data result starting from some initial data through a sequence of instruction steps that depend only on the situation and states within the program itself. A reactive program has to react to external situation of the system during its execution, *i.e.,* to states and changes occurring outside the program. In other words, during its execution, a reactive program has to be able to accept and detect external variables and external data sources and to produce stimuli for the environment.

The character of reactive programs exactly matches the function of control systems as described above. A reactive program realizes a control function according to Equation (4.10) by processing the system variable $\mathbf{s}(\tau_{a_k})$ in course of time and producing the control actions generated by the computer realizing the control program.

Using the above level of abstraction we have not treated technical details on how reactive programs are implemented in a control system. There are many ways that external variables can be detected and data conveyed into the control system or from it into the system to be controlled. There are basically two ways that the system reaction can be realized: using interrupts or sampling of external variables. As we mentioned before there are various control system technical forms. A form very often used in practice is a programmable logic controller. It is constructed as a modular input-output system for a reliable and robust realization of control functions in industrial environments.

In the control system design it is first necessary to know and to be able to specify the required behavior of the controlled system. It is reasonable to use an appropriate and formally well-elaborated specification tool. Second, it is necessary to specify and analyze the control function again using a suitable specification tool. Third, it is to write, implement and verify the correct final control program.

In this context it is important to emphasize the difference between specifications of the whole system denoted as SYST (Figure 4.1) and the function of the control part C. The same tool can be used for both cases. However, describing the whole system first seems to be natural and facilitating the control design.

A flow diagram is a conventional type of program specification used prior to final program writing. However, flow diagrams are insufficient for the specification of the reactive programs. One of the aims of this book is to present more effective and suitable specification means for the design of the DEDS control realized through the reactive programs.

4.3 Problems and Exercises

4.1. Determine the function $\mathbf{w}\left(\tau_{a_k} + \Delta\tau\right)$ given by the expression (4.10) for the case of deterministic finite automata.

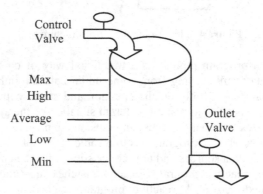

Figure 4.4. Figure for Problem 4.2

4.2. Given a tank as shown in Figure 4.4, determine controlled part, control, and a control strategy such that the tank maintains at the average level as much as possible. The sensory system can detect whether the current level is at Min, Low, Average, High, and Max levels, and whether the outlet valve is off and on (either high or low flow rate). The control on the inlet valve can be controlled so that it is off and on (high, average, and low).

4.3. Analyze the relation of the basic transition system in Section 5.1 with the model of the system control described using discrete processes according the input-output control structure in Figure 4.3.

5

Finite Automata

5.1 Basic Definitions

Finite automata are a classical tool used for many years for DEDS modeling. A finite automaton incorporates both principal system features - system states and system transitions in an abstract form. The basic definition of a finite automaton is given in the sequel.

Definition 5.1. The deterministic finite automaton (DFA) is a quintuple

$$A = (\Sigma, Q, q_0, \delta, F) \qquad (5.1)$$

where

Σ is a nonempty finite set of events,

Q is a nonempty finite set of elements called states,

$q_0 \in Q$ is an initial (or start) state,

δ is a state transition partial function given by $\delta : Q \times \Sigma \to Q$, where the Cartesian product $\times$ means that an ordered pair of elements from Q and Σ is mapped into an element of set Q, and the term "partial function" means that the function δ may not be defined for all ordered pairs that can be created of the sets Q and Σ, and

F is a set of final states given as a subset of Q: $F \subseteq Q$, where F can be the empty set.

There are several modifications of the deterministic finite automaton definition. The definition given above is useful in the context of formal languages because a deterministic finite automaton can serve as a formal language generator or acceptor.

Let a deterministic finite automaton $A = (\Sigma, Q, q_0, \delta, F)$ be given where $\Sigma = \{e_1, e_2, ..., e_n\}$ and $Q = \{q_0, q_1, ..., q_m\}$. Consider sequences of state transitions, which always start in the initial state q_0 and the sequential states are obtained by repeatedly applying the function δ to stepwise generated states. The sequence of state transitions is given by

$$\delta\left(q_0, e_{i_1}\right) = q_{j_1}, \quad \delta\left(q_{j_1}, e_{i_2}\right) = q_{j_2}, \ldots\ldots, \delta\left(q_{j_{k-1}}, e_{i_k}\right) = q_{j_k} \tag{5.2}$$

where $q_0, q_{j_1}, \ldots, q_{j_k}$ are states, i.e., $q_0 \in Q, q_{j_1} \in Q, \ldots, q_{j_k} \in Q$, and the function δ is defined for each pair (state, event) in Equation (5.2). A sequence

$$\tilde{\omega} = q_0 \, q_{j_1} \, q_{j_2} \cdots q_{j_k} \tag{5.3}$$

for which each part of Equation (5.2) is fulfilled is called the state path in DFA associated with the event string $\tilde{\eta} = e_{i_1} \, e_{i_2} \ldots e_{i_k}$. The state path in DFA is oriented. The length (symbol "$|\;|$") of the state path is $|\tilde{\omega}| = k$. There can be several different sequences in a given DFA starting with the same element. A singular case would occur if $\delta\left(q_0, e_i\right)$ is not defined for any element of Σ.

The sequence

$$\tilde{\eta} = e_{i_1} \, e_{i_2} \ldots e_{i_k} \tag{5.4}$$

associated with the state path at Equation (5.3), for which each partial expression in Equation (5.2) holds is called an event path in DFA. Hence, each event path starts with an event applied in the initial state q_0. Now, it is possible to proceed to the definition of the formal language generator.

Definition 5.2. A deterministic finite automaton given by a quintuple $A = \left(\Sigma, Q, q_0, \delta, F\right)$ is called a generator of the formal language L over the alphabet Σ, whereby the language is given by the set of all possible event paths $\tilde{\eta} = e_{i_1} \, e_{i_2} \ldots e_{i_k}$ in the DFA, and a generator of the marked language L_m, where L_m is a set of all possible event paths whose last element of the state path $\tilde{\omega} = q_0 \, q_{j_1} \, q_{j_2} \cdots q_{j_k}$, associated with $\tilde{\eta}$, is from the set F. Formally, L is a set of strings $\tilde{\eta}$:

$$L = \left\{ \tilde{\eta} \,\middle|\, \hat{\delta}\left(q_0, \eta\right) \text{ is defined} \right\} \tag{5.5}$$

where $\hat{\delta}$ is an extended partial transition function $\hat{\delta} : \Sigma^* \times Q \to Q$ obtained from the transition function δ defined by

$$\hat{\delta}\left(q, \tilde{\varepsilon}\right) = q, q \in Q, \tilde{\varepsilon} \text{ is the empty string} \tag{5.6}$$

$$\hat{\delta}(q,\tilde{\eta}) = q_{j_k}, \tilde{\eta} = e_{i_1}e_{i_2}...e_{i_k} \tag{5.7}$$

and

$$\delta(q,e_{i_1}) \quad \text{is defined and } \delta(q,e_{i_1}) = q_{j_1} \tag{5.8}$$

$$\delta(q_{j_1},e_{i_2}) \quad \text{is defined and } \delta(q_{j_1},e_{i_2}) = q_{j_2} \tag{5.9}$$

$$...$$

$$\delta(q_{j_{k-1}},e_{i_k}) \quad \text{is defined and } \delta(q_{j_{k-1}},e_{i_k}) = q_{j_k} \tag{5.10}$$

For the marked language L_m, the ending state $q_{j_k} \in F$.

From Definition 5.1 it follows that a finite automaton is a simple labeled directed graph denoted for short in Chapter 2 as a digraph. Recalling Definition 2.2 we can see that the set of states corresponds to the set of graph nodes: $A = Q$. Relation R is given recursively by the function δ starting from the initial state q_0. Function δ determines the next node q_b by applying the input e_i to the actual node q_a, i.e., $q_b = \delta(q_a, e_i)$. Thus, the function δ yields ordered pairs of states constituting the graph relation. It is used to be given by the state transition table (see the following example). An oriented arc corresponds to (q_a, q_b) with label e_i, i.e., $f_1 = \varnothing$ and $f_2 : R \rightarrow \Sigma$. Any finite automaton can be represented in a drawn graphical form.

Example 5.1. A deterministic finite automaton $A = (\Sigma, Q, q_0, \delta, F)$ is given by

$$\Sigma = \{0,1\} \qquad \delta(q_0,0) = q_1 \qquad F = \{q_2\}$$
$$Q = \{q_0, q_1, q_2\} \qquad \delta(q_1,1) = q_2$$
$$\delta(q_2,0) = q_1$$

This automaton generates a marked language equal to the one from Example 3.3, namely $L = \{\tilde{\varepsilon}, 01, 0101, 010101,.....\}$. It is better understood from the drawn-graphical automaton form in Figure 5.1.

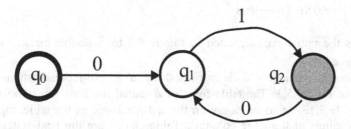

Figure 5.1. Finite automaton generating a language

5.2 Description of the System Behavior Using Finite Automata

Application of finite automata for describing the system behavior will be illustrated through the following example.

Example 5.2. Consider a flexible manufacturing system depicted in Figure 5.2. It consists of two input and one output conveyors, a servicing robot and a processing-assembling center. Workpieces to be processed come irregularly in a one-after-another sequence into the system. The workpieces of type A are delivered *via* conveyor C1 and workpieces of type B *via* conveyor C2. Only one workpiece can be on the input conveyor. A robot R transfers workpieces one by one into the processing center M. The next workpiece can be put on the input conveyor when it has been emptied by the robot. The production technology requires that first one A-workpiece is inserted into M and processed, then one B-workpiece is added into the center M, and last both workpieces are assembled. Afterwards, the assembled product is picked up by the robot and put on the output conveyor C3. The assembled product can be transferred onto C3 only when the output conveyor is empty and ready to receive the next product. The finite automaton describing behavior of the flexible manufacturing system is in Figure 5.3. Following Definition 5.1 we have

$$\Sigma = \left\{ IA,\ IB, T_{AM}, T_{BM}, T_{ABO}, T_{ABN} \right\} \tag{5.11}$$

where

IA denotes input of a workpiece A to conveyor C1,

IB denotes input of a workpiece B to conveyor C2,

T_{AM} represents transfer of workpiece A into machine M by using robot R,

T_{BM} represents analogously for a workpiece B,

T_{ABO} represents transfer of the product AB assembled of A and B from machine M onto output conveyor C3 by robot R, and

T_{ABN} represents transfer of the product AB from conveyor C3 out of the manufacturing cell, *i.e.,* emptying the output conveyor.

The automaton states are

$$Q = \left\{ q_0, q_1, q_2, ..., q_{23} \right\}$$

where q_0 is the initial state depicted in Figure 5.3 by a double circle. For final states we have $F = Q$.

The transition function δ is specified by a graph representation of the automaton in Figure 5.3. The table form of δ called the state transition table is given in Table 5.1. Actual states are on the left-hand side of the table, inputs are given as headings of the table columns. Table entries are the next states, and a hyphen indicates that δ is undefined.

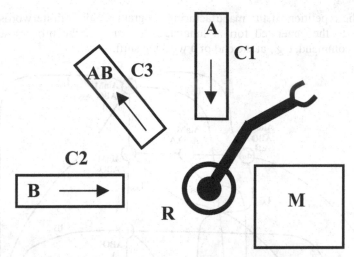

Figure 5.2. Flexible manufacturing system with one robot

The arcs of the graph in Figure 5.3 are labeled with events of Σ. If the automaton is in some state, *e.g.*, q_0, and a workpiece (of type A) arrives in the system (event *IA*), then the automaton passes from q_0 into state q_1. Being in q_1 two events can occur; either a workpiece B arrives (event *IB*) or the workpiece A is transferred into machine M (event T_{AM}). Behavior of the system with respect to other events is represented analogously.

An additional description of the automaton states is given in Figure 5.3. Symbol A indicates that a workpiece A is available at the input after it arrives in the system. Notation using symbol B has an analogous meaning. AinM, BinM represent states when workpieces A and B are inside machine M. ABO denotes that in the given state, the product AB is on the output conveyor.

In the adopted finite automaton representation of the system, it is assumed that just one event occurs at a discrete time point. Technologically simultaneous actions are performed in the consecutive steps at separate discrete time points. The time interval between individual steps can be very short relative to the system dynamics. In practice, the interval is given by the processing time of the considered system control unit. Thus the required parallelism of the system actions can be ensured by relatively quick consecutive (serial) system actions, which can be quite sufficient with respect to time requirements of a particular system.

If we consider the finite automaton in this example as a generator of a formal language, the generated words can be constructed in the following way (recall event paths in DFA):

$$\tilde{\alpha}_1 = IA \quad T_{AM} \quad IB$$
$$\tilde{\alpha}_2 = IA \quad T_{AM} \quad IB \quad T_{BM} \quad T_{ABO} \quad T_{ABN}$$
$$\tilde{\alpha}_3 = IA \quad T_{AM} \quad IA \quad IB \quad T_{BM} \quad T_{ABO} \quad T_{ABN}$$
$$\tilde{\alpha}_4 = IA \quad T_{AM} \quad IA \quad IB \quad T_{BM} \quad T_{ABO} \quad T_{ABN} \quad IB \quad IA \quad etc.$$

A cyclic repetition of the manufacturing program yields infinite words which all belong to the generated formal language. In practice, the process stops on operator's command, *e.g.,* at the end of a working shift.

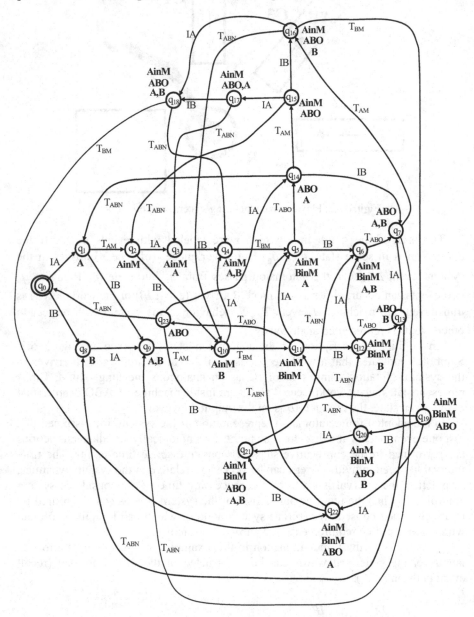

Figure 5.3. Finite automaton for a flexible manufacturing system

Table 5.1. State transition table for the automaton in Figure 5.3

$q_i \setminus e_j$	IA	IB	T_{AM}	T_{BM}	T_{ABO}	T_{ABN}
q_0	q_1	q_8	-	-	-	-
q_1	-	q_9	q_2	-	-	-
q_2	q_3	q_{10}	-	-	-	-
q_3	-	q_4	-	-	-	-
q_4	-	-	-	q_5	-	-
q_5	-	q_6	-	-	q_{14}	(-)
q_6	-	-	-	-	q_7	-
q_7	-	-	q_{16}	-	-	q_9
q_8	q_9	-	-	-	-	-
q_9	-	-	q_{10}	-	-	-
q_{10}	q_4	-	-	q_{11}	-	-
q_{11}	q_5	q_{12}	-	-	q_{23}	-
q_{12}	q_6	-	-	-	q_{13}	-
q_{13}	-	-	-	-	-	q_8
q_{14}	-	q_7	q_{15}	-	-	q_1
q_{15}	q_{17}	q_{16}	-	-	-	q_2
q_{16}	q_{18}	-	-	q_{19}	-	q_{10}
q_{17}	-	q_{18}	-	-	-	q_3
q_{18}	-	-	-	q_{22}	-	q_4
q_{19}	q_{22}	q_{20}	-	-	-	q_{11}
q_{20}	q_{21}	-	-	-	-	q_{12}
q_{21}	-	-	-	-	-	q_6
q_{22}	-	q_{21}	-	-	-	q_5
q_{23}	q_{14}	q_{13}	-	-	-	q_0

5.3 Control Specification Using Finite Automata

The goal of the DEDS control is to generate control variables $\mathbf{w}(t)$ as shown in Chapter 4. In the previous section a finite automaton is used for describing the DEDS behavior. By means of a finite automaton it is also possible to specify the system control, *i.e.*, the function of a control subsystem C (Figure 4.1). We extend Definition 5.1 for that purpose. Let the modified model be called a deterministic finite automaton with outputs – DFAO. Its structure is very close to the one of the finite automaton defined before.

It is natural to proceed as follows. The set of events Σ is constructed as inputs to the control system from system S (Figure 4.1). Definition 5.1 is extended by a new set – the outputs, and the output function.

Definition 5.3. A deterministic finite automaton with outputs (DFAO) is the triple

$$AC = (A, Y, \varphi) \tag{5.12}$$

where A is a deterministic finite automaton, Y is a finite set of outputs, and φ is a function $\varphi : Q \to Y$.

Example 5.3. Consider a set of events $\Sigma = \{0,1,R\}$ and a set of outputs $Y = \{0,1,R\}$. 0 and 1 are binary digits. R is the symbol denoting the event "no digit". The symbol serves as a separator. Let a deterministic finite automaton with outputs be constructed to model a serial conversion of binary numbers into arithmetic binary complements used in the computers for the representation of negative numbers. For instance, the binary number 0 0 1 1 0 1 0 0 is given. The lowest order bit is at the right-hand side. The highest order bit is the seventh one on left and the eighth one is the sign bit. The digits and/or separators appear as a sequence where the first digit element in the sequence is the lowest bit *etc.* $\tilde{\omega} = R00101100RRR$. Now the arithmetic complement is

$$0 \ \ 0\,1\,1\,0\,1\,0\,0 \ \ \to \ \ 1 \ \ 1\,0\,0\,1\,0\,1\,1$$
$$+ \qquad\qquad\qquad 1$$
$$\overline{\qquad\qquad\qquad\qquad}$$
$$1 \ \ 1\,0\,0\,1\,1\,0\,0$$

The sequence $R\,0\,0\,1\,0\,1\,1\,0\,0\,R\,R\,R$ whose elements are ordered as usual in sequences, *i.e.*, the lowest order being assigned to the first left-hand digit element in the sequence should be converted into the sequence $R\,0\,0\,1\,1\,0\,0\,1\,1\,R\,R\,R$. The sequences are inversely written compared with the writing of the binary numbers. Time is not expressed explicitly. Event order is given by event sequences.

The task is to construct DFAO such that the serially arriving binary numbers are converted into serially generated arithmetic complements. The conversion function is described as a DFAO as follows:

$$\Sigma = \{0,1,R\} \qquad Q = \{q_0, q_1, q_2, q_3\} \quad \text{where } q_0 \text{ is the initial state}$$
$$Y = \{0,1,R\} \qquad F = Q$$

The functions δ and φ are specified in the graphical form in Figure 5.4. The control function of the DFAO being in an actual state q_k and receiving an event e_i is to generate an output $y_k \in Y$ and to go into the next state $q_{k+1} = \delta(q_k, e_i)$. The outputs influence the controlled system.

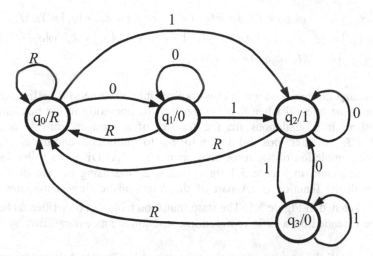

Figure 5.4. Finite automaton of the binary number conversion

Let us return to Example 5.2. For the sake of a better understanding we notate by *XIA* an event related to *IA*, *XIB* to *IB etc*. The DFAO is given as follows:

$$AC = (A, Y, \varphi)$$
$$A = (\Sigma, Q, q_0, \delta, F)$$

where

$$\Sigma = \{XIA, XIB, XT_{AM}, XT_{BM}, XT_{ABO}, XT_{ABN}\}$$
$$Q = \{q_0, q_1, q_2, ..., q_{23}, q_{1x}, q_{2x}, ..., q_{kx}\}$$
$$F = Q$$

and φ is given below.

The meaning of elements in set Σ is slightly different from Equation (5.11). *XIA* denotes a signal from a sensor detecting the arrival of a workpiece *A* on the conveyor *C1*. An analogous meaning has the element *XIB*. Similarly, $XT_{AM}, XT_{BM}, XT_{ABO},$ and XT_{ABN} correspond to signals indicating that a workpiece *A*, a workpiece *B,* and the assembled product are in the machine, on the output conveyor, and leaving the output conveyor, respectively. The set of outputs is

$$Y = \{\lambda, YAM, YBM, YABO, YABN\} \tag{5.13}$$

where λ stands for the empty symbol, *YAM* is a command for the robot to pick up a workpiece from conveyor *C1* and to insert it into machine *M*.

Function φ is defined as follows:

$$\varphi(q_0) = \lambda, \quad \varphi(q_1) = YAM, \quad \varphi(q_2) = \lambda, \quad \varphi(q_3) = \lambda, \quad \varphi(q_4) = YBM,$$
$$\varphi(q_5) = \lambda, \quad \varphi(q_6) = YABO, \quad \varphi(q_7) = YABN, \quad \varphi(q_8) = \lambda, \quad \varphi(q_9) = YAM,$$
$$\varphi(q_{10}) = YBM, \quad \varphi(q_{1x}) = \lambda, \ etc.$$

A command for the transfer of a B-workpiece into machine M is realized after a fixed given time interval when it is certain that the operation in the machine has been finished. It is analogous for the transfer of a product from M or from conveyor $C3$. Another possibility could be to introduce additional signals announcing completion of the respective operations.   A DFAO is to be derived from the automaton in Figure 5.3 for different Σ and using the set of outputs together with the function φ. A part of the deterministic finite automaton with outputs is sketched in Figure 5.5. The state transition table can be written as before. It must be extended by a table representing function φ, as exemplified by Table 5.2.

Finally, recall the regular expressions described in Chapter 3. It can be proved that the languages generated by finite automata are of type 3, *i.e.,* regular languages. Inversely, regular languages are generated by finite automata or by their marked languages. The regular languages can be specified by regular expressions. Therefore, it is possible to apply triangular transformations according to the scheme in Figure 5.6.

The regular expression $L = \alpha\beta^* \cup \beta$ in Example 3.4 is generated according to Definition 5.2 by a deterministic finite automaton depicted in Figure 5.7.

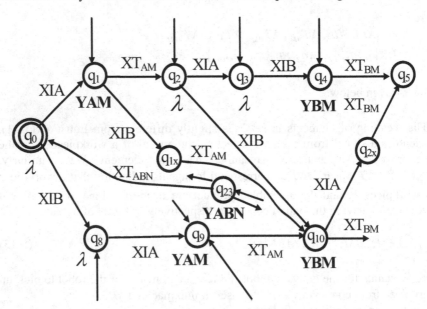

Figure 5.5. A control automaton

Table 5.2. Table for the output function φ

q_0	-
q_1	**AM**
q_2	-
q_3	-
q_4	**BM**
.	.
.	.
.	.
q_8	-
q_9	**AM**
q_{10}	**BM**
.	.
.	.
q_{23}	**ABN**
q_{1x}	-
.	.

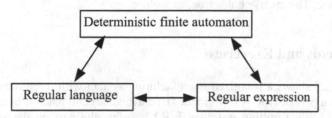

Figure 5.6. Relation between regular language, regular expression and automaton

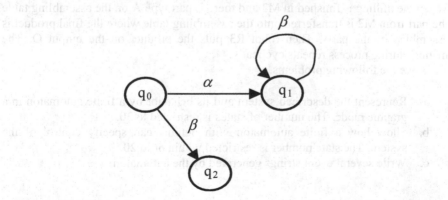

Figure 5.7. Finite automaton generating the language given by a regular expression

5.4 Non-deterministic Finite Automata

Sometimes uncertainties may occur in the system. There is a whole theoretical branch, namely the fuzzy set theory and its applications dealing with uncertainties. In the finite automata, the uncertainty can be represented by a modification of the transition function δ. Instead of the next state a subset of next states is defined.

Definition 5.4. A non-deterministic finite automaton is a quintuple

$$NA = (\Sigma, Q, q_0, \delta, F) \tag{5.14}$$

where all symbols but the transition function δ have the same meaning as in Definition 5.1. In this case

$$\delta: \quad Q \times \Sigma \to 2^Q \tag{5.15}$$

An ordered pair state-event is mapped into a subset of Q. The power set 2^Q is the set of all subsets of Q and always including the empty set.

Non-deterministic finite automata are generators of the formal languages in the sense of Definition 5.2, as well. Both kinds of automata are equivalent with respect to language generation. If a non-deterministic finite automaton generates a language then a deterministic finite automaton can be constructed generating the same language. The inverse holds true, as well.

5.5 Problems and Exercises

5.1. Figure 5.8 shows a robotized manufacturing system with three robots. Robot R1 picks up a part of type A from input I1 (if available) and loads it in the milling machine M1. When milling is finished, R2 transfers the part on the assembling table. Similarly robot R2 picks up from input I2, a part of type B and loads it in M2. When the milling is finished in M2 and there is part type A on the assembling table, the part from M2 is transferred onto the assembling table where the final product is assembled of the parts. Then robot R3 puts the product on the output O. The manufacturing process repeats cyclically.

Solve the following problems.

    a.   Represent the described system and its behavior by a finite automaton in a graphic mode. The number of states is restricted to 30.

    b.   Show how a finite automaton with outputs can specify control of the system. The state number is restricted within or to 20.

    c.   Write several event strings generated by the automaton.

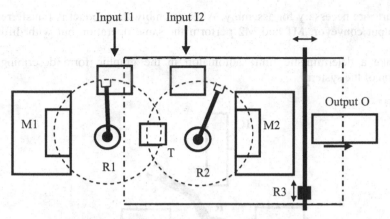

Figure 5.8. Manufacturing system with three robots

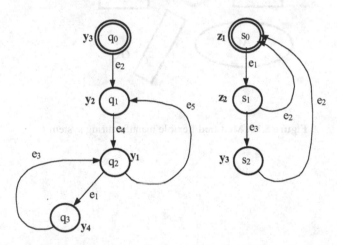

Figure 5.9. Two finite automata with outputs for Exercise 5.2

5.2. Two finite automata with outputs are depicted in Figure 5.9. Specify their event sets, state sets and outputs. Determine the formal languages they are generating. Choose some states as the set for one automaton and specify the marked language, which it generates.

Form the so-called product of automata with the set of states given by the Cartesian product of their state sets using a modeling assumption that only one event occurs at a discrete time point.

5.3. Modify slightly the flexible manufacturing system in Figure 5.2 as depicted in Figure 5.10. The system has one input belt conveyor, two working machines and one assembly center. Parts of one kind are coming into the system one by one *via* the input conveyor. The robot transfers parts in free M1 or M2. Both machines have capacity one part. Then the parts from M1 or M2 are transferred to AC. There two parts are necessary for assembly. After assembly the product is transferred on the output conveyor. M1 and M2 perform the same operation but with different times.

Make a deterministic finite automaton in the graphic form describing the function of the system.

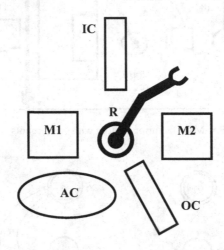

Figure 5.10. Modified flexible manufacturing system

6

Reactive Flow Diagrams

6.1 Standard Flow Diagrams

Flow diagrams, sometimes called flow charts, are popular in programming. They are graphical tools for drawn-graphical visualization of algorithms to be programmed and executed by computers. Flow diagrams make final programming easier and help one minimize programming errors.

Flow diagrams have been developed and used for decades for transformation programs dealt with in Section 4.2. A flow diagram prescribes a sequence of computer operations forced by computer instructions. There are four basic elementary building blocks used in flow diagrams: operational block (Fig. 6.1a), decision block (Fig. 6.1b), start and end block (Fig. 6.1c), and subprogram block (6.1d). The blocks are connected with arrows determining the next operation block. The decision block is equipped with one or more conditions. A continuation of a program depends on the conditions. A cyclic repetition of a same group of operations can be specified by means of a decision block, too.

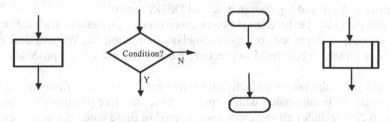

a. Operational block   b. Decision block.   c. Start and end block.   d. Subprogram block.

Figure 6.1. Elementary building blocks of flow diagrams

Application of the flow diagram technique to transformation programs is well known. A different situation arises if flow diagrams are used for reactive programs. In his book Zöbel (1987) analyzes in detail and shows troubles with standard flow diagrams used for control system programming.

6.2. Reactive Flow Diagrams

Reactive programs are able to respond to external stimuli during their execution. Flow diagrams used for reactive program specification must be adapted for that purpose. An important problem regarding how to manage and process concurrent processes is connected with the reactivity property. In a controlled DEDS several technological operations usually run in parallel. In such a system there are signals mediating data about the system events. Such events are called concurrent.

Let us return to Example 5.2 and consider the state q_0 in the finite automaton in Figure 5.5. Both asynchronous and spontaneous concurrent input variables XIA and XIB are to be detected. Similarly XIB and XT_{AM} in the state q_1: arrival of a workpiece B on the input conveyor B (variable XIB) and transfer of a workpiece A into machine M (variable XT_{AM}) are concurrent events. Both events are spontaneous and do not influence each other. XIB may occur earlier than XT_{AM} or *vice versa*. The control system passes into the next state according to the occurred event. As mentioned earlier, in a finite automaton model it is assumed that no two events occur exactly at the same time. It is clear that this assumption is very well substantiated in case of one-processor control system.

The adaptation of a traditional flow diagram technique uses a cyclic repetition of a flow diagram or of its part, and locking or unlocking groups of operations in order to enable real-time processing of concurrent system events (Hrúz 1994). A cyclic repetition of program operations has to be sufficiently quick with respect to the controlled system dynamic requirements.

There are only two ways to ensure fulfilment of the above-mentioned requirements if one control computer, or generally speaking one control processor is used in the system:

a.   Repetitive or cyclic sampling of concurrent variables
b.   Use of the computer interrupt (alarm) system

Both ways can be mixed together in practical DEDS control.

The solution may be different if two or more control processors are used in the system. Of course, the processors must somehow communicate. We will deal with the solution mostly encountered in practice, namely the use of one processor and item a.

We will show the use of a reactive flow diagram to a control finite automaton (deterministic finite automaton with outputs) whose structure paradigm is depicted in Figure 6.2. It includes all situations encountered in finite automata: sequences of states, branching from a state and a feedback. Two related finite automata are used in order to show better the use of reactive flow diagrams for concurrent processes. $\{e_1, e_2, ..., e_5\}$ are inputs, $\{y_1, y_2, y_3, y_4\}$ are outputs, $\{q_0, q_1, q_2, q_3\}$ are the automaton states, and q_0 is the initial state for the first automaton. $\{e_1, e_2, e_3\}$ are inputs, $\{z_1, z_2, y_3\}$ outputs, $\{s_0, s_1, s_2\}$ states, and s_0 is the initial state for the second automaton. States of the whole shuffle automaton are given by the parallel composition of the described automata (treated as sub-automata) states. The set of outputs is union of the sets of the sub-automata outputs.

The finite automaton has been converted into the reactive flow diagram as shown in Figure 6.3, where the sampling or polling technique with the cyclic structure has been utilized. In the reactive flow diagram, the states of a finite automaton are represented by the operation blocks that are either locked or unlocked by means of the auxiliary variables from the set K:

$$K = \{k_1, k_2, k_3, k_4, k_5, k_6, k_7\}.$$

When a transition from one state to another occurs, the block associated with the active state is locked using variables from set K and the block associated with the next state is released. Figure 6.3 explains how branching and looping is solved using the same idea. As mentioned above, another possibility would be to use the interrupt mechanism or to combine it with the cyclic polling technique just described.

Control finite automata outputs are control variables for the controlled system. They are denoted as y or z. Inputs to the automaton are outputs of the controlled system, that represent events e. Branching in the first automaton in state q_2 is solved by a block which is guarded by the variable $k_3 = 1$ while $k_1 = 0$, $k_2 = 0$, $k_4 = 0$ so that the other three remaining blocks corresponding to the first automaton are locked. It also means that the state q_2 is active. Depending on the arrival of the events e_1 or e_5, the next active state will be q_3 or q_1, respectively. Guarding variables associated with the second automaton are k_5, k_6, and k_7. Processes are parallel when, for example, $k_3 = 1$ and $k_5 = 1$. Two blocks are accessible in that case in a reactive flow diagram.

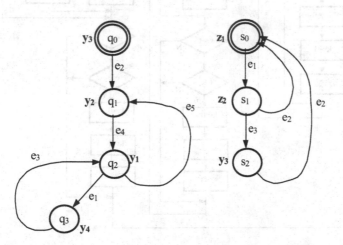

Figure 6.2. A structure paradigm for control finite automata

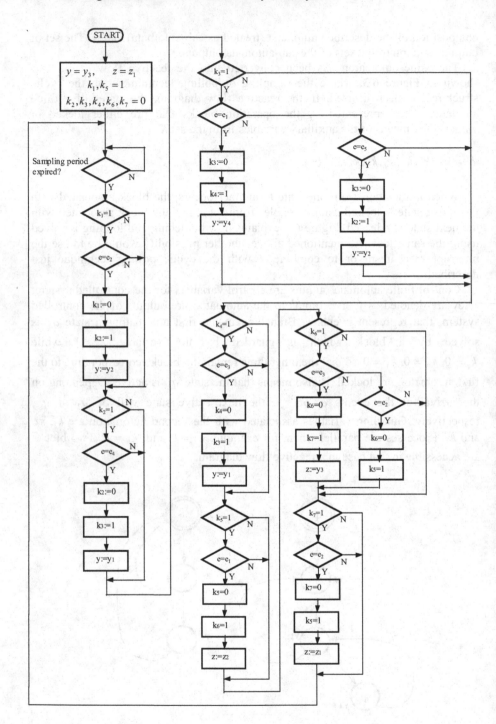

Figure 6.3. Reactive flow diagram for the finite automata in Figure 6.2

As mentioned earlier, the sampling frequency has to be such that the response to the events is sufficiently quick. Expression $y := y_4$ means that the variable y is set value to y_4, $z := z_1$ means that z is set to z_1, *etc.*, and similarly for the variables k_i. The rest is clear from Figures 6.2 and 6.3.

6.3 Problems and Exercises

6.1. Minimum time duration of signals that are external ones for the control systems is decisive for a good control function. Consider a reactivity problem of the control function with respect to temporal properties of external signals using reactive flow diagram with its cyclic repetition as described in this chapter.

6.2. Figure 6.4 shows a robotic manufacturing cell. Workpieces A are processed sequentially with robot R1 and with robot R2 respectively. The sequence results in the product A.
A similar operation sequence is applied for workpieces B resulting in products of the kind B. Workpieces come in the cell irregularly. Specify control of the robotic cell with the use of a reactive flow diagram.

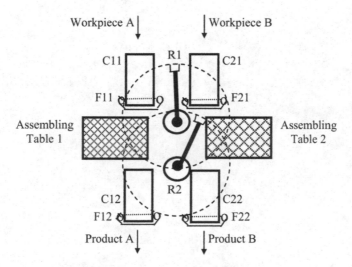

Figure 6.4. A two-robot manufacturing cell

7

Petri Net Models of DEDS

7.1 Notion of Petri Nets

As described in Chapter 5, a finite automaton specifies a system by means of a set of states and a transition function. The arguments of the transition function are the state and event. We can speak about an actual state. The transition function assigns a state to an actual state. The assigned state is a next state while the actual state can be called the active present state. By repeating the assignments, a sequence of actual states is obtained. In the finite automaton there is always only one state active.

A system can often be broken down into subsystems. If it is required to describe activities of subsystems and their mutual relations, a finite automaton model can be cumbrous because each combination of subsystem states needs a separate state of the finite automaton. Another model known as a Petri net removes that inadequacy. Petri nets are named after a German mathematician C. A. Petri who first proposed a model of that kind (C. A. Petri, 1962). With Petri nets the main idea is to represent states of subsystems separately. Then, the distributed activities of a system can be represented very effectively. Many properties of the DEDS, *e.g.*, synchronization, concurrency, and choices can be well presented and analyzed using Petri nets. They can be used not only for the specification of the DEDS behavior but also the control design. However, Petri nets have various other uses. To illustrate them we introduce, *e.g.*, fuzzy reasoning with Petri nets (Gao *et al*. 2003, 2004) or creation of algorithms (Hanzálek 1998a, b). Several supporting programs exist for design and analysis of Petri nets, *e.g.*, PESIM (Češka 1994), MATLAB® Toolbox (Svádová and Hanzálek 2001), CPN analysis tools (Jensen 1997), SPNP (Hirel *et al*. 2000), and many others.

Let a manufacturing cell be configured as shown in Figure 7.1. A workpiece arriving at the cell on the input conveyor is transferred to the milling machine. Both workpiece and robot must be available to perform first the transfer operation of the workpiece into the machine input. Then the milling is taking place in the milling machine. Obviously, the milling machine must be free for that. After milling, the workpiece is moved onto the machine output. Then, if the robot is free, the processed workpiece is transferred by the robot onto the cell output conveyor.

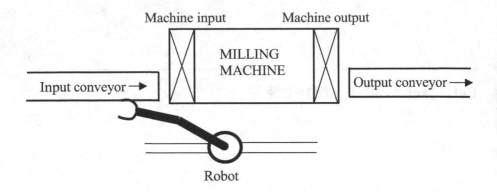

Figure 7.1. A robotic manufacturing cell

Only one workpiece can be located in any of the cell subsystems. So far, technological conditions and the procedure have been described verbally.

Figure 7.2 shows how the function of the manufacturing cell with its subsystems can be presented using a Petri net. Let the circles in Figure 7.2 denoted by p_1, p_2, p_5 and p_7 correspond to four subsystems as follows: input conveyor - p_1, robot - p_2, milling machine - p_5 and output conveyor - p_7. Let the other circles correspond to the following operations: transfer of a workpiece into the milling machine by means of the robot - p_3, milling operation - p_4, transfer of the milled workpiece on the output conveyor - p_6. The circles are called places of Petri nets. The presence or availability of a workpiece at the cell input is modeled by a dot in place p_1. We say that a token is in p_1. Analogously, a token in p_2 (Figure 7.2b) means that the robot is free or available to transfer a workpiece somewhere. Figure 7.2a shows a situation when both conditions for the transfer of a workpiece into the machine are not satisfied. A vertical bar denoted as t_1 is called a transition. It symbolizes an event. In this case, it is the start of the transfer operation. Transition t_2 represents the end of the transfer and start of the milling operation. Clearly, realization of this event requires that the transfer has been performed and the milling machine is available. t_3 denotes the end of the milling and start of the workpiece transfer on the output conveyor; t_4 is the end of the output transfer and arrival of a workpiece on the output conveyor.

The token distribution describes an actual state of the system. It changes through a so-called transition firing. A transition firing is possible if all places before this transition have enough tokens – the transition is said to be enabled. Firing has the following effect: one token is taken from all places before the transition and one token is placed into each place located after the transition. The effect complies with the so-called firing rules just described. According to Figure 7.2b both conditions are met for a workpiece transfer. Figure 7.2c shows the next system state: the robot moves the workpiece from the input conveyor into the milling machine. Figure 7.2d gives the next state when the milling operation is in progress and the robot is again free.

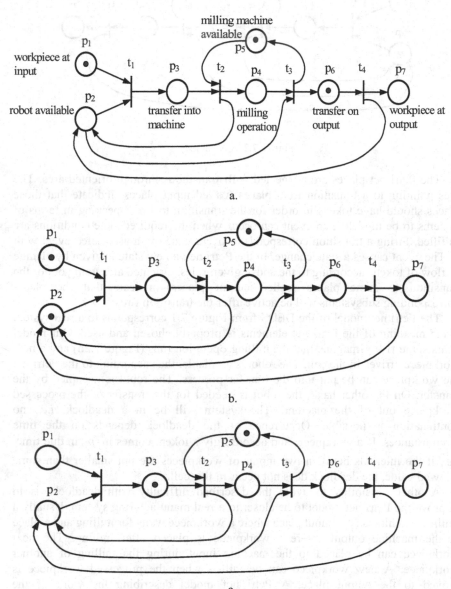

Figure 7.2. Petri net of a manufacturing cell: **a**. input conditions not met; **b**. input conditions met; **c**. workpiece transfer into the milling machine; and **d**. milling operation in progress

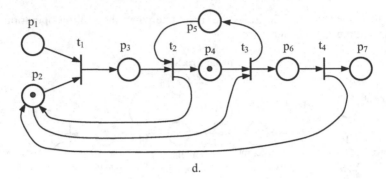

d.

Figure 7.2. (continued)

The Petri net places are connected with transitions through oriented arcs. The arcs pointing to a transition from places (called input places) indicate that those places should have tokens in order for the transition to fire. Speaking in terms of systems to be modeled, an event can occur when the required state conditions are fulfilled. Firing a transition corresponds to occurrence of an associated event with it. The event causes a state change. In the Petri net, a new state is given by change or flow of tokens according to the above given rules. Oriented arcs going out of the transition to some places (called output places) indicate that the place-corresponding subsystems will be active after the transition fires.

The Petri net model of the DEDS from Figure 7.1 corresponds to a real system if the meaning of the Petri net elements is properly chosen and used. Our model bears some risk. Imagine, that the milling operation runs (Figure 7.2d) and a new workpiece arrives in the cell. The robot is available, thus according to the Petri net the workpiece can be put into the transfer process. The robot is occupied by the transfer. On the other hand, the robot is needed for the transfer of the processed workpiece out of the machine. The system will be in a deadlock, *i.e.*, no continuation is possible. Occurrence of the deadlock depends on the time circumstances. If a workpiece, and accordingly a token, comes in p_1 in due time, *i.e.*, if the intervals between two inputs of workpieces are not smaller than some allowed value, the deadlock does not occur in this cell.

Another possibility to avoid the described difficulty with deadlocks is to improve the Petri net model to be closer to a real manufacturing system. Usually a milling machine has an input place where a workpiece waits for milling and a place at the machine output where a workpiece is placed after milling. The next workpiece can be placed to the machine input during the milling of another workpiece. A new workpiece can be milled when the processed workpiece is moved to the output place. A Petri net model describing the work of the manufacturing cell more realistically is shown in Figure 7.3. There are some new places and transitions in Figure 7.3 specifying a correct function of the cell without system deadlocks. If a workpiece appears at the input, a token is placed in p_1. Place p_{32} ensures that the transfer of a next workpiece into the machine starts only when the machine input is free. Similarly, a token in p_{42} means that the machine output is empty. Place p_{61} guards the transfer onto the output conveyor. We can see that Petri nets can well express the distributed activities in the modeled system.

The situation just described is very often encountered in the design practice. If the Petri net in Figure 7.2d is correct with respect to the system behavior, the real system should be re-arranged and improved. Petri nets can thus help discover an ill-shaped technological layout. On the other hand, they may not correctly specify the system. A solution of that case is shown in Figure 7.3 using a corrected net.

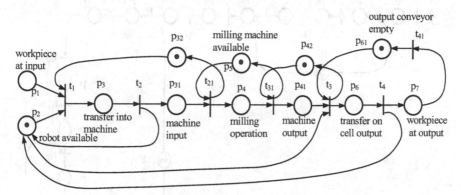

Figure 7.3. An improved Petri net model of the manufacturing cell

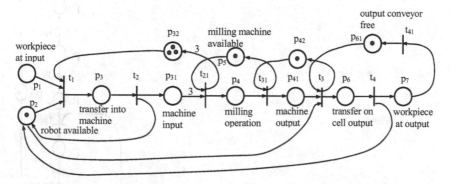

Figure 7.4. Petri net model of the manufacturing cell with the arc weights

Consider now a situation when three parts are to be fed into the milling machine. In the machine the parts are simultaneously processed and assembled in one product. The situation can be modeled using weights of the Petri net arcs. According to modified rules for transition firing, each place before a transition must have at least as many tokens as the weight of the arc connecting the place and the transition. Tokens whose quantity is equal to the arc weight are removed from each input place of the considered transition. On the other hand, after a transition fires, as many tokens are put into each output place as the weight of the outgoing arc from the transition to the place. The weights are positive integers. The use of weights is illustrated in Figure 7.4. A weight is assumed to be 1 if no number is associated with an arc. It was the case in Figures 7.2a–d and 7.3. Other weights than one are given as arc labels. Transition t_{21} can be fired only when the machine is free (a token is in p_5) and three tokens in p_{31}. Then, three tokens are taken away from p_{31}, one from p_5 and one is placed into p_4.

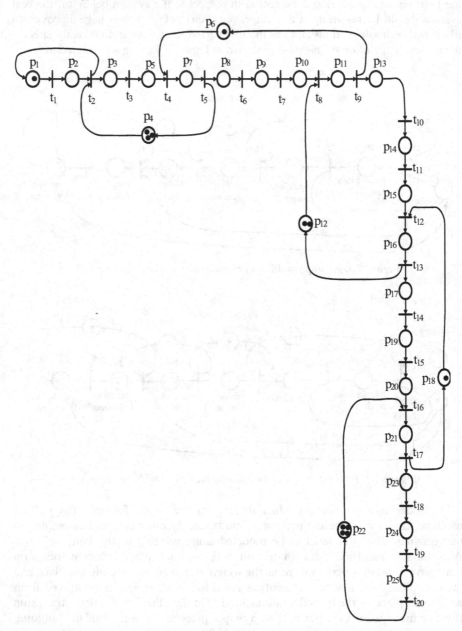

Figure 7.5. Petri net for the manufacturing system from Figure 1.3

We are now able to construct a Petri net for the manufacturing system in Figure 1.3. The Petri net is depicted in Figure 7.5. The place and transition meaning is explained in Tables 7.1 and 7.2.

Table 7.1. Meaning of places of the Petri net in Figure 7.5 by a token presence

p_1	Waiting on workpiece arrival
p_2	Workpiece in the detection area of sensor P11
p_3	Workpiece between sensors P11 and P12
p_4	Capacity of conveyor C1
p_5	Workpiece ready at the end of C1 for the transfer into machine M
p_6	Machine M free
p_7	Transfer of the workpiece into machine M is in progress
p_8	Workpiece waitng for processing in M
p_9	Processing of the workpiece in machine M is running
p_{10}	Workpiece waiting on transfer onto conveyor C2
p_{11}	Workpiece transfer on conveyor C2
p_{12}	Capacity of conveyor C2
p_{13}	Workpiece waiting on processing
p_{14}	Workpiece between sensors P21 a P22
p_{15}	Workpiece at the end of conveyor C2
p_{16}	Transfer of the workpiece into machine G is in progress
p_{17}	Workpiece waiting on processing
p_{18}	Machine G is free
p_{19}	Processing of the workpiece in machine G is running
p_{20}	Workpiece waiting for transfer onto conveyor C3
p_{21}	Transfer of the workpiece onto conveyor C3 in progress
p_{22}	Capacity of conveyor C3
p_{23}	Workpiece at the begin of conveyor C3
p_{24}	Work-piece between sensors P31 and P32
p_{25}	Work-piece at the end of conveyor C3

Table 7.2. Meaning of transitions of the Petri net in Figure 7.5

t_1	Work-piece arrives into the detection area of sensor P11
t_2	Work-piece arrives into the area between sensors P11 and P12
t_3	Work-piece arrives at the end of conveyor C1
t_4	Start of the work-piece transfer from conveyor C1 into machine M
t_5	End of the work-piece transfer into machine M
t_6	Start of the work-piece processing in machine M
t_7	End of the work-piece processing in machine M
t_8	Start of the work-piece transfer from machine M onto conveyor C2
t_9	End of the work-piece transfer from machine M onto conveyor C2
t_{10}	Work-piece arrived in the area between P21 and P22
t_{11}	Work-piece arrived at the end of conveyor C2
t_{12}	Start of the work-piece transfer from conveyor C2 into machine G
t_{13}	End of the work-piece transfer into machine G
t_{14}	Start of the work-piece processing in machine G
t_{15}	End of the work-piece processing in machine M
t_{16}	Start of the work-piece transfer from machine G onto conveyor C3
t_{17}	End of the work-piece transfer from machine G onto conveyor C3
t_{18}	Work-piece arrived in the area between P31 and P32
t_{19}	Work-piece arrived at the end of conveyor C3
t_{20}	Work-piece left conveyor C3 and the manufacturing system

7.2 Basic Definitions

The notion introduced intuitively in the preceding section is systematically analyzed and defined in this section. First, the basic definition of a net is given.

Definition 7.1. The net *NET* is defined by a triple

$$NET = (P, T, F) \tag{7.1}$$

where $P=\{p_1, p_2, ..., p_n\}$ is a finite non-empty set of elements called places, $T=\{t_1, t_2, ..., t_m\}$ is a finite non-empty set of elements called transitions, and F is the union of two binary relations F_1 and F_2: $F = F_1 \cup F_2$. P and T are the disjunctive sets, *i.e.*, $P \cap T = \varnothing$ (empty set). F_1 is a binary relation from P to T : $F_1 \subseteq P \times T$. Analogously $F_2 \subseteq T \times P$ is the binary relation from T to P. F is the set of ordered pairs consisting of a place (transition) at the first position and a transition (place) at the second one. F is called a flow relation. P, T and F are such that the following holds for them:

$\forall p_i \in P: \exists t_j \in T$ such that $(p_i,t_j)\in F$ or $(t_j,p_i)\in F$ and

$\forall t_s \in T: \exists p_r \in P$ such that $(t_s,p_r)\in F$ or $(p_r,t_s)\in F$.

According to the definition, each place is included at least in one ordered pair of F. Similarly, each transition is included in at least one ordered pair of F. In other words there are neither isolated places nor isolated transitions in net NET.

Definition 7.2. A Petri net PN is defined by the triple

$$PN = (NET,W,M_0) \tag{7.2}$$

where NET is a net by Definition 7.1 such that

$$PN = (P,T,F,W,M_0) \tag{7.3}$$

W is the weight function given as $W:F\to N^+$ where N^+ is a set of positive integers. $M_0:P\to N$ is a function called the initial marking whose element $M_0(p)$ is the number of tokens initially in place p where N is a set of non-negative integers.

The numbers to which are mapped the pairs of F are called weights. Obviously, the weights are positive integers. The initial marking is non-negative integers.

Definition 7.3. The function $M:P\to N$ is called the marking of a Petri net. $M(p)$ represents the number of tokens in place p at marking M.

The initial marking is specifically given in the definition of a Petri net. Similarly as in finite automata, it is reasonable to include the initial state in Petri net model definition because any real system begins its activity at an initial state. The different functions $M:P\to N$ correspond to the different markings.

Consider a Petri net PN and a transition $t\in T$. The set of input places of t as denoted by $^\bullet t$ and called preset of t is

$$^\bullet t = \{p_i|(p_i,t)\in F\} \tag{7.4}$$

According to Equation (7.4), set $^\bullet t$ contains each place p_i being in the flow relation F with the given transition t. The set of output places belonging to t, called post-set of t, is

$$t^\bullet = \{p_i|(t,p_i)\in F\} \tag{7.5}$$

Similarly, the set of the input and output transitions pertaining to a given place p is

$$^\bullet p = \left\{t_j \middle| \left(t_j, p\right) \in F\right\}$$ (7.6)

$$p^\bullet = \left\{t_j \middle| \left(p, t_j\right) \in F\right\}$$ (7.7)

Input and output places of a transition are also named pre- and post-places. Similarly, input and output transitions of a place are also named pre- and post-transitions.

The next two definitions define the transition enabling and firing rules-they are together also called the execution rules of a Petri net.

Definition 7.4. (Enabling rule) A transition t in a given Petri net is called fireable or enabled by a marking M if and only if (iff for short):

    a. For each pre-place of t, its marking is equal or greater than the weight of the arc from it to t, or

    b. $^\bullet t$ has no pre-place.

Mathematically, a transition t is fireable iff

$$\forall p \in {}^\bullet t: \quad M(p) \geq W(p, t)$$ (7.8)

or

$$^\bullet t = \varnothing$$ (7.9)

Equation (7.9) means that a transition without any pre-place is fireable by any marking. Such a transition is called a source transition. The notation $W(p, t)$ means the value of the function W for the ordered pair (p, t). Strictly writing, it is $W\big((p, t)\big)$.

Definition 7.5. (Firing rules) Consider a Petri net and marking $M(p)$, $p \in \{p_1, p_2, ..., p_n\}$. Assume that transition t is fireable. Then, the marking after t's firing is

$$M'(p) = \begin{cases} M(p) - W(p, t) + W(t, p) & \text{if } \left(p \in {}^\bullet t\right) \wedge \left(p \in t^\bullet\right), \\ \quad \text{where} \wedge \text{ stands for the conjunction of logical expressions} \\ M(p) - W(p, t) & \text{if } \left(p \in {}^\bullet t\right) \wedge \left(p \notin t^\bullet\right) \\ M(p) + W(t, p) & \text{if } \left(p \notin {}^\bullet t\right) \wedge \left(p \in t^\bullet\right) \\ M(p) & \text{if } \left(p \notin {}^\bullet t\right) \wedge \left(p \notin t^\bullet\right) \end{cases}$$

(7.10)

M' is called immediately reachable from M. M is reachable from M_0 if firing a sequence of enabled transitions leads M_0 to M. All markings reachable from M_0 form a set called the reachability set. An alternative definition of a Petri net is often used as follows.

Definition 7.6. The Petri net PN is defined by a quintuple

$$PN = (P, T, I, O, M_0) \tag{7.11}$$

where

$P = \{p_1, p_2, ..., p_n\}$ is a finite non-empty set of elements called places,
$T = \{t_1, t_2, ..., t_m\}$ is a finite non-empty set of elements called transitions and $P \cap T = \varnothing$,
I is a function $I : P \times T \to N$ called input function,
O is a function $O : P \times T \to N$ called output function, and
M_0 is a function $M_0 : P \to N$ called the initial marking.

Definition 7.6 can define an equivalent Petri net with that defined by Definition 7.2. Let us formulate it as a theorem.

Theorem 7.1. A Petri net defined by Definition 7.2 is equivalent with that defined by Definition 7.6 iff the following conditions hold for F, W, I, and O:

$$(p, t) \in F \quad \Rightarrow \quad I(p, t) = W(p, t) \tag{7.12}$$
$$(p, t) \notin F \quad \Rightarrow \quad I(p, t) = 0 \tag{7.13}$$
$$(t, p) \in F \quad \Rightarrow \quad O(p, t) = W(t, p) \tag{7.14}$$
$$(t, p) \notin F \quad \Rightarrow \quad O(p, t) = 0 \tag{7.15}$$

and inversely

$$I(p, t) \neq 0 \quad \Rightarrow \quad (p, t) \in F \quad and \quad W(p, t) = I(p, t) \tag{7.16}$$
$$I(p, t) = 0 \quad \Rightarrow \quad (p, t) \notin F \quad and \quad W(p, t) \text{ is not defined} \tag{7.17}$$
$$O(p, t) \neq 0 \quad \Rightarrow \quad (t, p) \in F \quad and \quad W(t, p) = O(p, t) \tag{7.18}$$
$$O(p, t) = 0 \quad \Rightarrow \quad (t, p) \notin F \quad and \quad W(t, p) \text{ is not defined} \tag{7.19}$$

This theorem follows directly from Definitions 7.2 and 7.6.

Equation (7.10) can be written in a simple way using Definition 7.6. Firing an enabled (firable) transition t at M leads to a new marking M' such that

$$\forall p \in P, \ M'(p) = M(p) - I(p, t) + O(p, t) \tag{7.20}$$

From the graph-theoretical viewpoint, the Petri net defined by Definition 7.1 is a directed bi-partite labeled simple graph (not a multi-graph). The set of Petri net nodes consists of two disjunctive sets, namely P and T. The flow relation F corresponds to the set of directed (oriented) graph arcs. An arc is given by the relevant ordered pair. Direction of the arc is determined by the order of the pair elements. The weights defined by function W are the arc labels.

The same graph is defined according to Definition 7.6. If $I(p,t)=k, k \neq 0$ $\left(O(p,t)= k, k \neq 0\right)$ then there exists the directed arc from p to t (from t to p) with the weight equal k.

7.3 Vector and Matrix Representation of Petri Nets

In a given Petri net $PN = (P,T,F,W,M_0)$ for each transition t of PN and always for all places $p \in P$ we create the functions t^+, t^- and Δt as follows:

$$t^+(p)=\begin{cases} W(t,p) & \text{if } p \in t^{\bullet} \\ 0 & \text{otherwise} \end{cases} \tag{7.21}$$

$$t^-(p)=\begin{cases} W(p,t) & \text{if } p \in {}^{\bullet}t \\ 0 & \text{otherwise} \end{cases} \tag{7.22}$$

$$\Delta t(p)=t^+(p)-t^-(p) \tag{7.23}$$

Values of the functions t^+, t^-, and Δt along with their argument p can be represented as vectors $\mathbf{t}^+, \mathbf{t}^-$, and $\Delta \mathbf{t}$ having the dimension equal to the number n of the Petri net places. It is assumed that the first entry of each vector corresponds to place p_1, the second one to p_2, *etc.*, up to the n-th entry corresponding to p_n. By that construction places and transitions are numerated or indexed ordinarily according to Definition 7.1, namely, $P=\{p_1, p_2..., p_n\}$ and $T=\{t_1, t_2, ..., t_m\}$. Then

$$\mathbf{t}^+ = \begin{pmatrix} t^+(p_1) \\ t^+(p_2) \\ \vdots \\ t^+(p_n) \end{pmatrix}, \quad \mathbf{t}^- = \begin{pmatrix} t^-(p_1) \\ t^-(p_2) \\ \vdots \\ t^-(p_n) \end{pmatrix}, \quad \Delta \mathbf{t} = \begin{pmatrix} \Delta t(p_1) \\ \Delta t(p_2) \\ \vdots \\ \Delta t(p_n) \end{pmatrix} \tag{7.24}$$

Arithmetic operations of addition and subtraction are performed by vector entries, *e.g.*, the sum of two vectors $\Delta \mathbf{t}_2$ and $\Delta \mathbf{t}_4$ is

$$\Delta \mathbf{t}_2 + \Delta \mathbf{t}_4 = \begin{pmatrix} \Delta t_2(p_1) \\ \Delta t_2(p_2) \\ \vdots \\ \Delta t_2(p_n) \end{pmatrix} + \begin{pmatrix} \Delta t_4(p_1) \\ \Delta t_4(p_2) \\ \vdots \\ \Delta t_4(p_n) \end{pmatrix} = \begin{pmatrix} \Delta t_2(p_1) + \Delta t_4(p_1) \\ \Delta t_2(p_2) + \Delta t_4(p_2) \\ \vdots \\ \Delta t_2(p_n) + \Delta t_4(p_n) \end{pmatrix} \quad (7.25)$$

Another variable useful to be represented as a vector is the Petri net marking M. Similarly under the above assumptions we have

$$\mathbf{m} = \begin{pmatrix} M(p_1) \\ M(p_2) \\ \vdots \\ M(p_n) \end{pmatrix} \quad (7.26)$$

Consider two markings $\mathbf{m}$ and $\mathbf{m}'$ given for a Petri net. If $\mathbf{m}(p) \le \mathbf{m}'(p)$ for $\forall p \in P$, then $\mathbf{m} \le \mathbf{m}'$, where $\mathbf{m}(p)$ is a vector entry equal $M(p), p \in P$ according to Equation (7.26). Further if $\mathbf{m} \le \mathbf{m}'$ and $\mathbf{m} \ne \mathbf{m}'$, then $\mathbf{m} < \mathbf{m}'$. The inequality $\mathbf{m} \ne \mathbf{m}'$ is satisfied when at least for one corresponding pair of the vector entries holds the inequality $M(p) \ne M'(p)$.

Theorem 7.2. Consider a Petri net. Then a transition $t \in T$ is fireable by marking $\mathbf{m}$ iff $\mathbf{t}^- \le \mathbf{m}$. If t is fireable by marking $\mathbf{m}$ then marking $\mathbf{m}' = \mathbf{m} + \Delta \mathbf{t}$ is obtained by firing of t. The notation used for that event is $\mathbf{m}[t > \mathbf{m}'$. Expressed in another way, $\mathbf{m}[t > \mathbf{m}'$ means that firing of t leads from markings $\mathbf{m}$ to $\mathbf{m}'$.

Proof. The proof refers to Definition 7.4 dealing with the fireability of Petri net transitions, and to Equation (7.22). Consider a transition t and a marking $\mathbf{m}$. Let t be fireable by $\mathbf{m}$. If some place p is not a pre-place of the transition then the corresponding entry of the vector $\mathbf{t}^-$ is zero, which is less or equal to the corresponding entry of $\mathbf{m}$ for any value of the marking $\mathbf{m}$. If the considered place is a pre-place of t, the entry of $\mathbf{t}^-$ is the weight $W(p,t)$. According to Equation (7.8) it should be less or equal to $M(p)$ and this is just the condition contained in the corresponding vector entry of the expression $\mathbf{t}^- \le \mathbf{m}$. Thus the necessary condition (t is fireable $\Rightarrow$ $\mathbf{t}^- \le \mathbf{m}$) is proved. The sufficient condition ($\mathbf{t}^- \le \mathbf{m} \Rightarrow t$ is fireable) follows directly from the fireability definition. The validity of the expression $\mathbf{m}' = \mathbf{m} + \Delta \mathbf{t}$ results directly from the definition of the vector $\Delta \mathbf{t}$.

Dynamic behavior of the of a system represented by the Petri net can be expressed using the Petri net incidence matrix $\mathbf{A} = \left(a_{ij}\right)$. $\mathbf{A}$ is an $n \times m$ matrix with entries given by

$$a_{ij} = O\left(p_i, t_j\right) - I\left(p_i, t_j\right), \quad p_i \in P, i = 1, 2, \ldots, n, \quad t_j \in T, j = 1, 2, \ldots, m \tag{7.27}$$

The incidence matrix $\mathbf{A}$ can be given in terms of the vectors $\Delta\mathbf{t}_j$, $j = 1, 2, \ldots, m$, as

$$\mathbf{A} = \left(\Delta\mathbf{t}_1, \Delta\mathbf{t}_2, \ldots, \Delta\mathbf{t}_m\right) \tag{7.28}$$

Now, a new marking $\mathbf{m}'$ obtained from the marking $\mathbf{m}$ by firing the fireable transition t_j can be expressed using the incidence matrix as

$$\mathbf{m}' = \mathbf{m} + \mathbf{A}\,\mathbf{x} \tag{7.29}$$

where

$$\mathbf{x} = \quad j - th \begin{pmatrix} 0 \\ 0 \\ \cdot \\ 0 \\ 1 \\ 0 \\ \cdot \\ \cdot \\ 0 \end{pmatrix} \tag{7.30}$$

is an m-dimensional vector whose j-th vector entry equals 1 while the other entries equal zero.

Example 7.1. The behavior of the flexible manufacturing system with one robot was described in Example 5.2, Figure 5.2, Chapter 5, using the finite automaton (Figure 5.3). We will represent the system by the Petri net shown in Figure 7.6, in order to compare both ways of the DEDS modeling.

A token in the place p_{A0} represents a situation when the conveyor A is empty. The event "a workpiece occurred at the system input" changes the input state. Afterwards the workpiece is prepared for manufacturing. The event is represented by firing t_A. Transition t_A is fireable and on its firing the token from p_{A0} is removed and is placed into the place p_{A1}. In the Petri net it is not specified when the event happens. Places p_{B0} and p_{B1} have an analogous meaning for the input

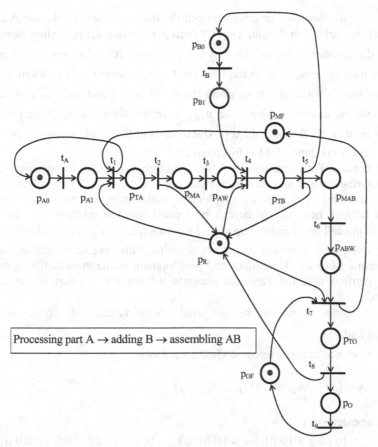

Figure 7.6. Petri net model of the flexible manufacturing system from Figure 5.2

workpieces of kind B. If the robot is available (a token is in p_R), a workpiece A is available (a token in p_{A1}), and the previous machine process is completed (a token in p_{MF}) then the robotic transfer of the workpiece A can start (firing transition t_1). Place p_{TA} is occupied by a token during the transport. The end of the transport and loading the workpiece into machine M is specified by transition t_2. After firing t_2 a token is placed in p_{MA}. End of processing of the workpiece A is specified by transition t_3 and the start of the transport of B into the machine by t_4. Then a token comes to the place p_{TB}, which indicates that the transfer of the workpiece B is in progress; t_5 denotes end of the transfer and start of the assembly operation; t_6 specifies end of the assembly operation; t_7 specifies start of the product transfer from M onto the conveyor C3 (t_7 is fired and a token appears in p_{TO}). After the transfer of the product AB on C3 and its leaving out the cell, the token moves to the p_{OF} place. The place p_{MF} ensures that the next part A is loaded into machine

M only after the assembly process has been finished. The next workpiece A can be loaded at the earliest on the conveyor C1 (firing t_A) when the preceding workpiece A is in the transfer to the machine M (p_{TA} marked). Considering it we have: t_1 fires and then t_A fires. First, A has to arrive to be processed in M (a token in p_{MA}) and after that it waits in the machine (token in p_{AW}) and only then B can be loaded into the machine. The place p_{ABW} corresponds to the machine output. A token is in p_{ABW} if the assembled product AB is at the machine output. After the product leaves machine M, M is free again.

The created Petri net specifies the structure and behavior of the flexible manufacturing system in a transparent and concise way. The relation between the system states and its dynamics is graphically visible using the properties and rules of Petri nets. A new workpiece can be loaded into the machine M when it is present at the cell input and the machine M is free (place p_{MF} has a token).

Moreover, the obtained Petri net specifies the cyclic repetition of the manufacturing process. Conditions and development of the manufacturing process can be verified thus the Petri net presents a basis for developing the control program.

The incidence matrix can be computed *via* the vectors Δt_j or *via* functions $O(p_i, t_j)$ and $I(p_i, t_j)$.

The incidence matrix expressed via the vectors is

$$\mathbf{A} = (\Delta \mathbf{t}_A, \Delta \mathbf{t}_B, \Delta \mathbf{t}_1, \Delta \mathbf{t}_2, ..., \Delta \mathbf{t}_8, \Delta \mathbf{t}_9) \qquad (7.31)$$

where, for example,

$$\mathbf{t}_A^+ = (0\,1\,0\,0\,0\,0\,0\,0\,0\,0\,0\,0\,0\,0)^T, \ \mathbf{t}_A^- = (1\,0\,0\,0\,0\,0\,0\,0\,0\,0\,0\,0\,0\,0)^T,$$

$$\Delta \mathbf{t}_A = (-1\,1\,0\,0\,0\,0\,0\,0\,0\,0\,0\,0\,0\,0)^T \text{ in which the order of the places is}$$

$$p_{A0}, p_{A1}, p_{B0}, p_{B1}, p_{TA}, \ p_{MA}, p_{AW}, p_{TB}, p_{MAB}, p_{AB}, p_{MF}, p_{TO}, p_O, p_{OF}, p_R ;$$

$$\mathbf{t}_B^+ = (0\,0\,0\,1\,0\,0\,0\,0\,0\,0\,0\,0\,0\,0)^T, \ \mathbf{t}_B^- = (0\,0\,1\,0\,0\,0\,0\,0\,0\,0\,0\,0\,0\,0)^T,$$

$$\Delta \mathbf{t}_B = (0\,0\,-1\,1\,0\,0\,0\,0\,0\,0\,0\,0\,0\,0)^T \ etc.$$

The functions $O(p_i, t_j)$ and $I(p_i, t_j)$ are calculated as follows:

$$O(p_{A0}, t_A) = 0, \ O(p_{A0}, t_B) = 0, \ O(p_{A0}, t_1) = 1, \ ..., O(p_{A0}, t_9) = 0,$$

$$O(p_{A1}, t_A) = 1, \ O(p_{A1}, t_B) = 0, \ O(p_{A1}, t_1) = 0, \ ..., O(p_{A1}, t_9) = 0, \ etc.$$

$$I(p_{A0}, t_A) = 1, \ I(p_{A0}, t_B) = 0, \ I(p_{A0}, t_1) = 0, \ ..., I(p_{A0}, t_9) = 0,$$

$$I(p_{A1}, t_A) = 0, \ I(p_{A1}, t_B) = 0, \ I(p_{A1}, t_1) = 1, \ ..., I(p_{A1}, t_9) = 0,$$

etc.

The incidence matrix using Equation (7.27) or (7.28) is

$$
\mathbf{A} = \begin{array}{c}
p_{A0} \\ p_{A1} \\ p_{B0} \\ p_{B1} \\ p_{TA} \\ p_{MA} \\ p_{AW} \\ p_{TB} \\ p_{MAB} \\ p_{AB} \\ p_{MF} \\ p_{TO} \\ p_{O} \\ p_{OF} \\ p_{R}
\end{array}
\left(\begin{array}{rrrrrrrrrrr}
-1 & 0 & 1 & 0 & 0 & 0 & 0 & 0 & 0 & 0 & 0 \\
1 & 0 & -1 & 0 & 0 & 0 & 0 & 0 & 0 & 0 & 0 \\
0 & -1 & 0 & 0 & 0 & 0 & 1 & 0 & 0 & 0 & 0 \\
0 & 1 & 0 & 0 & 0 & -1 & 0 & 0 & 0 & 0 & 0 \\
0 & 0 & 1 & -1 & 0 & 0 & 0 & 0 & 0 & 0 & 0 \\
0 & 0 & 0 & 1 & -1 & 0 & 0 & 0 & 0 & 0 & 0 \\
0 & 0 & 0 & 0 & 1 & -1 & 0 & 0 & 0 & 0 & 0 \\
0 & 0 & 0 & 0 & 0 & 1 & -1 & 0 & 0 & 0 & 0 \\
0 & 0 & 0 & 0 & 0 & 0 & 1 & -1 & 0 & 0 & 0 \\
0 & 0 & 0 & 0 & 0 & 0 & 0 & 1 & -1 & 0 & 0 \\
0 & 0 & -1 & 0 & 0 & 0 & 0 & 0 & 1 & 0 & 0 \\
0 & 0 & 0 & 0 & 0 & 0 & 0 & 0 & 1 & -1 & 0 \\
0 & 0 & 0 & 0 & 0 & 0 & 0 & 0 & 0 & 1 & -1 \\
0 & 0 & 0 & 0 & 0 & 0 & 0 & 0 & -1 & 0 & 1 \\
0 & 0 & -1 & 1 & 0 & -1 & 1 & 0 & -1 & 1 & 0
\end{array}\right) \qquad (7.32)
$$

The matrix columns from the 1st up to 11th one correspond to the transitions $t_A, t_B, t_1, t_2, \ldots, t_9$, respectively. The initial marking is

$$
\mathbf{m}_0 = (1,0,1,0,0,0,0,0,0,1,0,0,1,1)^T \qquad (7.33)
$$

The next marking after the initial one is obtained through firing t_A, i.e.,

$$
\mathbf{m}' = \mathbf{m}_0 + \mathbf{A}(1,0,0,\ldots,0)^T = (0,1,1,0,0,0,0,0,0,1,0,0,1,1)^T \qquad (7.34)
$$

The matrix representation of a Petri net can be simplified for a Petri net class called pure Petri nets. Figure 7.7 shows the case when the given Petri net includes a direct loop. A Petri net without the direct loops is called the pure Petri net. The limitation due to the structural purity condition is relatively small. A direct loop can be easily removed by the rearrangement shown in Figure 7.8, which practically does not bring about any discrepancy with the real system.

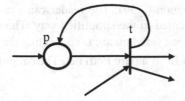

Figure 7.7. A direct loop in a Petri net

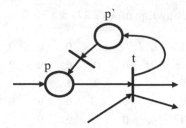

Figure 7.8. Removal of a direct loop from a Petri net

Formally, a Petri net is pure iff the following implication holds:

$$(p,t) \in F \Rightarrow (t,p) \notin F \tag{7.35}$$

Now let us return to the matrix simplification question. Consider a pure Petri net. A vector $\mathbf{t}_i$ can be associated with $t_i \in T$ in a different way than using Equation (7.24), in particular,

$$\mathbf{t}_i = \begin{pmatrix} t_{1i} \\ t_{2i} \\ \cdot \\ t_{ki} \\ \cdot \\ t_{|P|i} \end{pmatrix}, \quad t_{ki} = \begin{cases} -W(p_k, t_i) & \text{if } (p_k, t_i) \in F \\ +W(t_i, p_k) & \text{if } (t_i, p_k) \in F \\ 0 & \text{otherwise} \end{cases} \tag{7.36}$$

The number of the vector $\mathbf{t}_i$ entries is equal to the number of places in set P, *i.e.*, $|P|$. In the simplified case is the incidence matrix

$$\mathbf{A} = (\mathbf{t}_1, \mathbf{t}_2, \ldots, \mathbf{t}_m) \tag{7.37}$$

As before the new marking is given by

$$\mathbf{m}' = \mathbf{m} + \mathbf{A}\mathbf{x} \tag{7.38}$$

For the vector $\mathbf{x}$ see Equation (7.30). The reader can see why a direct loop in a Petri net cannot be represented in the simplified way. The direct loop case cannot be distinguished from the one when no arc is connecting the place p_k with t_i.

The Petri net in Figure 7.6 is a pure Petri net. The simplified incidence matrix is

$$
\mathbf{A} = \begin{pmatrix}
-1 & 0 & 1 & 0 & 0 & 0 & 0 & 0 & 0 & 0 & 0 \\
1 & 0 & -1 & 0 & 0 & 0 & 0 & 0 & 0 & 0 & 0 \\
0 & -1 & 0 & 0 & 0 & 0 & 1 & 0 & 0 & 0 & 0 \\
0 & 1 & 0 & 0 & 0 & -1 & 0 & 0 & 0 & 0 & 0 \\
0 & 0 & 1 & -1 & 0 & 0 & 0 & 0 & 0 & 0 & 0 \\
0 & 0 & 0 & 1 & -1 & 0 & 0 & 0 & 0 & 0 & 0 \\
0 & 0 & 0 & 0 & 1 & -1 & 0 & 0 & 0 & 0 & 0 \\
0 & 0 & 0 & 0 & 0 & 1 & -1 & 0 & 0 & 0 & 0 \\
0 & 0 & 0 & 0 & 0 & 0 & 1 & -1 & 0 & 0 & 0 \\
0 & 0 & 0 & 0 & 0 & 0 & 0 & 1 & -1 & 0 & 0 \\
0 & 0 & -1 & 0 & 0 & 0 & 0 & 0 & 1 & 0 & 0 \\
0 & 0 & 0 & 0 & 0 & 0 & 0 & 0 & 1 & -1 & 0 \\
0 & 0 & 0 & 0 & 0 & 0 & 0 & 0 & 0 & 1 & -1 \\
0 & 0 & 0 & 0 & 0 & 0 & 0 & 0 & -1 & 0 & 1 \\
0 & 0 & -1 & 1 & 0 & -1 & 1 & 0 & -1 & 1 & 0
\end{pmatrix}
\qquad (7.39)
$$

The modeling power of Petri nets can be increased by adding inhibitors and/or incidentors as a new kind of oriented arcs. They are defined next.

Definition 7.7. The Petri net with inhibitors and incidentors is defined by

$$
PI = (PN, INHD, \lambda_{INHD}, L_{INHD}) \qquad (7.40)
$$

where

PN is the Petri net according to Definition 7.2;

$INHD \subseteq P \times T \times \{0,1\}$, and the set of triples $(p, t, 0)$ for $p \in P$ and $t \in T$ is called the set of inhibitors, and the set of triples $(p,t,1)$ is called the set of incidentors;

$\lambda_{INHD} : INHD \to L_{INHD}$ is the function mapping the inhibitors or incidentors onto a set of logical assertions specifying the function of these special group of arcs. The assertions are related to markings of place p in $(p, t, 0)$ or $(p, t, 1)$. Such an assertion specifies additional conditions for the fireability of a transition pointed by the arc. For an inhibitor, a condition is defined stating when a transition is not fireable. Inversely, for an incidentor, a condition is stating when the corresponding transition is fireable.

For example, the following assertion is mapped to an inhibitor: a transition is not fireable if its inhibitor pre-place has one or more tokens. Another is that it is fireable if its incidentor pre-place has exactly three tokens. All the other rules for the net are in effect. Inhibitors and incidentors bring about additional firing conditions about transitions. However, they do not affect the token flow during any transition firing in a sense that no tokens "flow" through them.

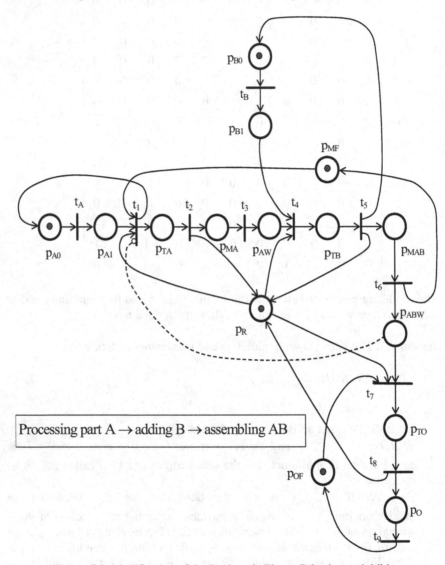

Figure 7.9. Modification of the Petri net in Figure 7.6 using an inhibitor

Example 7.2. Usage of inhibitors will be illustrated by means of a fine structural change in the Petri net from Figure 7.6. After substituting the arc (t_6, p_{MF}) for (t_7, p_{MF}) we have the modification; see Figure 7.9. The new structure indicates that after firing t_6 the machine M is again free. As assumed and modeled in Figure 7.6, it is not possible to enter any further workpiece into M for some technological reasons. Another way to express the situation is to use an inhibitor $(p_{ABW}, t_1, 0)$ according to Figure 7.9.

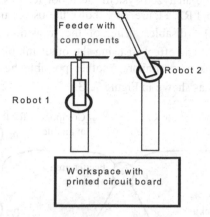

Figure 7.10. Flexible and automatic assembly of printed circuit board

Example 7.3. Another flexible manufacturing system will be considered in this example (Zhou and Leu 1991). It is a two robot flexible system for the automatic assembly of printed circuit boards. The block scheme of the system is in Figure 7.10. The two robot system cares for automatic picking and inserting electronic components onto a printed circuit board (PCB). The components are supplied from a feeder. The sequence of operations to be realized are: picking up a component from the feeder by a robot, pulling back the robot arm, moving to the workspace, inserting the component, pulling back the robot arm, moving back to the feeder *etc.*, cyclically.

Only one robot can be in the feeder and workspace, respectively. Therefore the system control should avoid a collision of robots in those areas. The robot arm is pulled back after component picking and inserting. Both robots with their arms pulled back can move in the space between the feeder and the PCB. The Petri net describing the system behavior is in Figure 7.11. The meaning of places and transitions with respect to the real system are inscribed in the Petri net. The places p_3 and p_4 ensure that picking or inserting is allowed only for one of the robots. The initial marking is given in Figure 7.11. In the initial state, both robots are available. At the beginning it is assumed that the robots are close to the feeder area. Both the feeder area and the workspace are free at the beginning. The Petri net shows possible continuations of the system behavior. Either of robots R1 or R2 picks up the electronic component from the feeder. Components of the Petri net show that either transition t_{11} or t_{21} can fire separately, but not together. It is not specified in the Petri net which transition fires, and when. The next required step (or in the Petri net terms: the next transition firing) has to be activated by a system control. We will deal with this question in the following sections. In what follows an attention will be focused on a kind of indeterminism contained in Petri nets. A potential possibility of an indeterministic situation arises when there is more than one arc coming out of a place. It is the case of the places p_3 and p_4 in Figure 7.11. Using inhibitors and incidentors is one possible way of avoiding this indeterminism. Let in the analyzed system the robot R1 has always a preference (priority) before the robot R2. Figure 7.12 shows the use of the inhibitors for this. An inhibitor $(p_{11}, t_{21}, 0)$ disables firing of the transition t_{21} by the marking depicted in Figure 7.12. The effect of using the other inhibitor when both places p_{14} and p_{24} have a token is similar. Another possible behavior would be the alternation of the robots as shown in Figure 7.13.

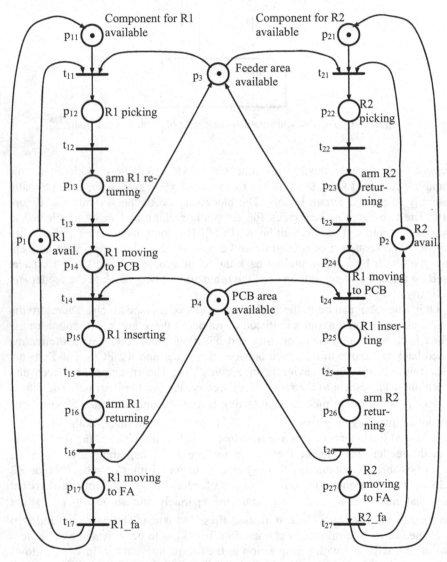

Figure 7.11. Petri net for the printed circuit board (PCB) assembly

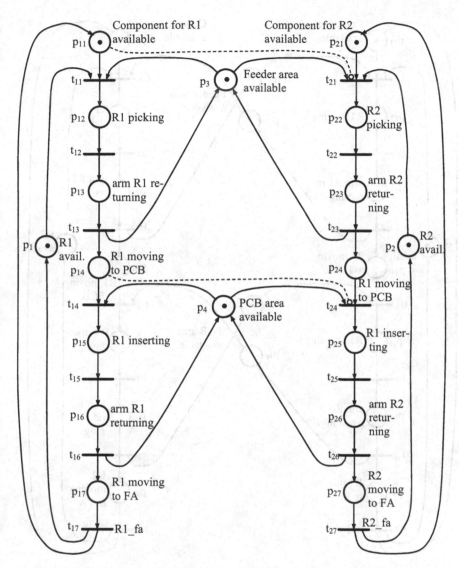

Figure 7.12. Petri net with inhibitors for the printed circuit board (PCB) assembly

7.4 Petri Net Classes

In Sections 7.1–7.3 we dealt with the most used standard Petri net definitions. They serve as basic or reference Petri net models. There are many modifications to the basic models. According to David and Alla (1994), they can be classified as abbreviations or extensions.

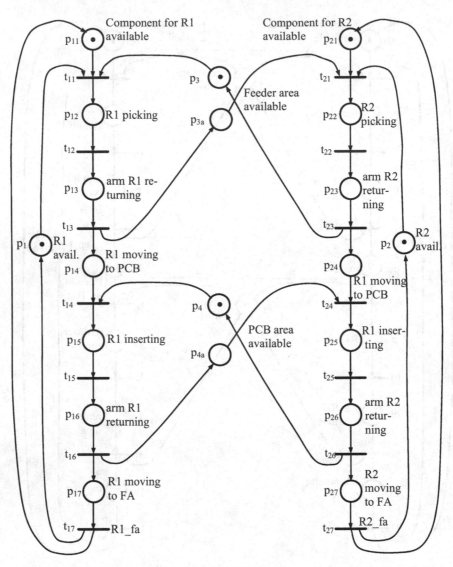

Figure 7.13. Petri net for printed circuit board (PCB) assembly with alternation of robots

Abbreviations of the basic Petri net model enable one to represent a DEDS in a simplified way, for example, colored Petri nets, Petri nets with capacities and others. An abbreviated representation can always be converted to a basic Petri net model though the latter may be much larger and less transparent.

On the other hand, the extensions are actually Petri net models with additional functional rules to those defined for the basic model. Extensions arise, for example, when special arcs called inhibitors and incidentors are added to the arc set or when further firing conditions are added to the transitions. The additional conditions may be deterministic, stochastic, timed *etc*. In such a case the transition firing is bound to external or internal states or events – a kind of model synchronization. Extensions are capable of representing many reactive control functions and therefore, the extension class is called the Petri nets interpreted for control. The chosen abbreviation and extension classes will be treated in the sequel.

Petri net models within each class can further be classified into sub-classes, with respect to the structural properties. Consider the class of standard Petri nets specified by Definition 7.2. The following sub-classes can be distinguished

1. Binary Petri nets.
   A Petri net is called binary or ordinary if all its weights are 1s, *i.e.,*

   $$W : F \rightarrow \{1\} \tag{7.41}$$
   An example of a binary Petri net is in Figure 7.3.

2. Petri net state machines.
   A Petri net state machine is a binary Petri net such that each transition has exactly one input place (pre-place) and exactly one output place (post-place). Given formally:

   $$\forall t \in T : |\bullet t| = |t \bullet| = 1 \tag{7.42}$$

   The name of this sub-class suggests that it is very close to the finite automata models. Figure 7.14 shows an example of the vending machine model. The machine accepts only 5 cent and 10 cent coins and it vends a bottle of coke for 20 c or candy for 15 c. Arrival of a token in places p_3, p_4 starts counting the time interval "int". It starts a counting renewal if the interval has not expired. If the required amount of money has not been accepted within the interval "int" the machine returns coins and waits.

3. Marked graphs.
   A binary Petri net is called a marked graph or event graph if each place has exactly one pre-transition and exactly one post-transition. Formally,

   $$\forall p \in P : |\bullet p| = |p \bullet| = 1 \tag{7.43}$$

   Figure 7.13 is an example of a marked graph.

4. Free-choice nets.
   A free-choice net is a binary Petri net such that every arc going out of a place is either (a) unique arc incoming into a transition and no other arcs go out of the place or (b) there are more arcs, but each of them is unique arc going into the transition. Figure 7.15 shows elementary structures characterizing the free-choice sub-class. Formal description of the sub-class is

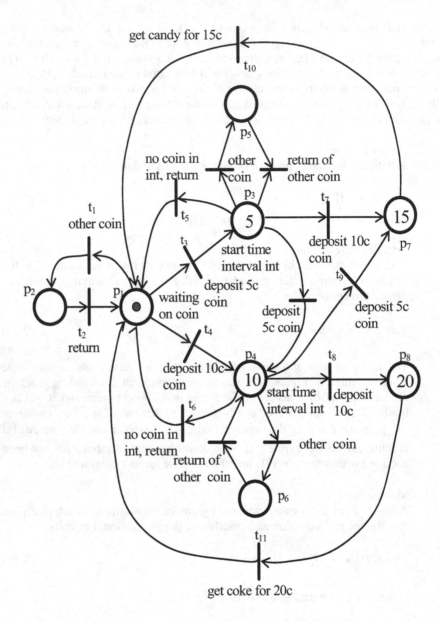

Figure 7.14. Petri net state machine model of a vending machine

$$\forall p \in P : |p \bullet| \le 1 \quad \text{or} \quad \bullet(p \bullet) = \{p\} \tag{7.44}$$

where $(p \bullet)$ are all post-transitions of place p, and $\bullet(p \bullet)$ means the set of pre-places of all transitions of the set $(p \bullet)$.

Note that in Figure 7.15, the first statement of Equation (7.44) holds for p_1, p_2, p_4, and p_5: $|p_1 \bullet| = 1$, $|p_2 \bullet| = 1$, $|p_4 \bullet| = 0$, $|p_5 \bullet| = 0$, and the second one holds for p_3: $(p_3 \bullet) = \{t_2, t_3\}$, $\bullet(p_3 \bullet) = \bullet\{t_2, t_3\} = \{p_3\}$.

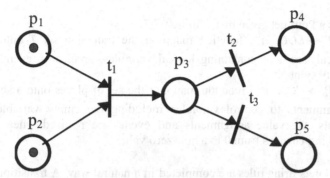

Figure 7.15. A free-choice net example

5. Safe Petri nets.
    A Petri net is safe if for all markings reachable from M_0:

$$\forall p \in P, \quad M(p_i) \le 1 \tag{7.45}$$

    For example the Petri net in Figure 7.3 is safe.

Each sub-class is specific as to various Petri net properties, which will be analyzed in the next chapter.

7.5 Petri Nets Interpreted for Control

In preceding sections Petri nets were used as a tool for discrete event system description from the observer's point of view. Internal structural relations in the system are expressed according to the Petri net components and rules. The relations determine the Petri net behavior so that the particular Petri net specifies the actual behavior and function of the represented system. The question of how to achieve the required behavior using Petri nets has not been treated yet.

DEDS control problems were considered in Chapter 4. It was shown that the system SYST in Figure 4.1 can be modeled as a Petri net. This was the main approach in the previous parts of this textbook. Now, we are interested in the DEDS control design. The control system C in Figure 4.1 can be represented as a Petri net as well. For that purpose the Petri net introduced in Definition 7.2 should be augmented. This new class of the modeling tools will be called the Petri nets interpreted for control.

Definition 7.8. A Petri net interpreted for the DEDS control is given by the quintuple

$$PC = (PI, \psi, LOG, \zeta, COM) \tag{7.46}$$

where

PI is a Petri net given by Definition 7.7;

$\psi : T \to LOG$ is a function mapping the transition set T onto a set of logical assertions containing logical variables, predicates, events, and the empty symbol;

$\zeta : P \to COM$ is a function mapping the set of places onto a set of value assignments to control variables including the empty variable, and of events, the value assignments and events are realized when the place marking changes from 0 to a non-zero value.

The Petri nets firing rules are completed in a natural way. A transition $t_i \in T$ of a Petri net interpreted for control (PC) is fireable if all fireability conditions for Petri nets with inhibitors and incidentors are fulfilled and all logical assertions mapped from the transition t_i are true. Logical assertions are logical expressions and predicates consisting of variables aggregated into vector $\mathbf{u}(t)$; See equation (4.1) in Chapter 4. These variables are outputs of the controlled system and inputs to the control system (as shown in Figure 4.1).

COM serves as a set of control commands. The commands are disabled when in a considered place to which they are mapped there is no token, and they are enabled when a token or tokens arrive in the place. Nature of the commands can be twofold: level commands (corresponding to setting of the control variable values) or impulse commands (corresponding to event forcing). This issue will be treated in detail later in connection with Grafcet. It is quite understandable that the commands correspond to the variable $\mathbf{w}(t)$ in Equation (4.7) related to the feedback structure of Figure 4.1.

As mentioned in the preceding section, Petri nets interpreted for control are sometimes called synchronized Petri nets (David and Alla 1994) because the transition firings are conditioned and synchronized by external variables.

Now we are able to generate a Petri net interpreted for control for the flexible manufacturing system depicted in Figure 5.2. This task is simplified with help of the Petri net in Figure 7.9. This problem was discussed in Chapter 4. The Petri net of the controlled system is always structurally very close to the Petri net interpreted for control of that system. Usually it is not the same. The Petri net interpreted for control PC is in Figure 7.16. Meanings of logical expressions mapped to transitions of the Petri net in Figure 7.16 are described in Table 7.3. Notation of control variables is given in Table 7.4.

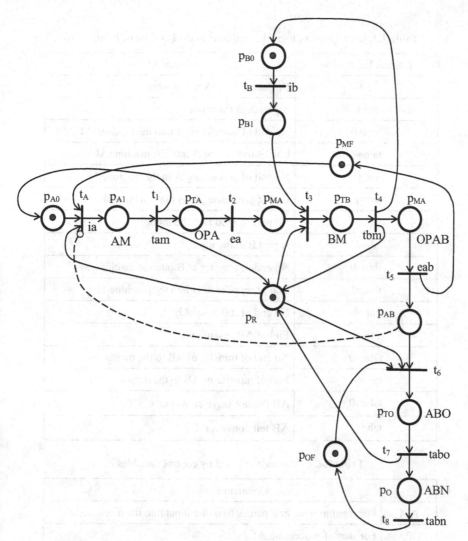

Figure 7.16. Control: processing of part A → adding B → assembling AB

Table 7.3. Description of logical conditions in the Petri net of Figure 7.16

Logical Expression	Meaning
ia=0	No part of kind A at the input
ia=1	A part A at the input
tam=0	No end of transfer of A into the machine M
tam=1	End of transfer of A into the machine M
ea=0	No end of processing A in the machine M
ea=1	End of processing A in the machine M
ib=0	No part of kind B at the input
ib=1	A part B at the input
tbm=0	No end of transfer of B into the machine M
tbm=1	End of transfer of B into the machine M
eab=0	No end of AB assembly
eab=1	End of AB assembly
tabo=0	No end of transfer of AB to the output
tabo=1	End of transfer of AB to the output
tabn=0	AB did not leave conveyor $C3$
tabn=1	AB left conveyor $C3$

Table 7.4. Commands realized by control variables

	Commands
AM	For robot to transfer a part A from the input into the machine M
OPA	For start of processing A
BM	For robot to transfer a part B from the input into the machine M
OPAB	To start the assembly of AB
ABO	For transfer AB on the output
ABN	To move AB out of conveyor $C3$

7.6 Petri Nets with Capacities

The modeling convenience of Petri nets can be enhanced by introducing the place capacity. The capacity restricts the number of tokens that can be located in a place as the following definition formally states.

Definition 7.9. The Petri net with capacities is given by a couple

$$PCA = (PN, \text{K}) \tag{7.47}$$

where *PN* is a standard Petri net by Definition 7.2 and **K** is the function

$$\text{K} : P \rightarrow N^+ \tag{7.48}$$

The firing rules for Petri nets with capacities are

$$\left(\mathbf{t}^- \le \mathbf{m}_i \right) \wedge \left(\text{K} \ge \mathbf{m}_i + \Delta\mathbf{t} \right) \Leftrightarrow \mathbf{m}_i \left[t > \mathbf{m}_j \right. \tag{7.49}$$

where capacities are expressed by a vector

$$\text{K} = \begin{pmatrix} \text{K}(p_1) \\ \text{K}(p_2) \\ \vdots \\ \text{K}(p_n) \end{pmatrix} \tag{7.50}$$

According to the firing rules it is not allowed to put into a place of a Petri net with capacities more tokens than its capacity. Such nets are also called finite-capacity Petri nets. The next manufacturing system example illustrates some essential differences between a traditional modeling method using Petri nets, termed as process-oriented modeling, and a resource-oriented modeling method using finite-capacity Petri nets. The latter was pioneered by Wu (1999) and later developed for various applications (Wu and Zhou 2001, 2004, 2005).

Example 7.4. An automated manufacturing system is shown in Figure 7.17. It contains machines M_1 and M_2 that can concurrently handle two types of parts, A and B. An A-part has two operations 1 and 2 to be processed by M_1 and M_2, respectively while a B-part requires M_2 to process first, and then M_1. We assume that raw materials are continuously supplied while the produced parts are shipped away when they are ready, *i.e.,* neither starving nor blocking exists. Each machine can process a part at a time.

By modeling each part's process using two places and two types of raw material availability using two separate places, we can derive the model as shown in Figure 7.18.

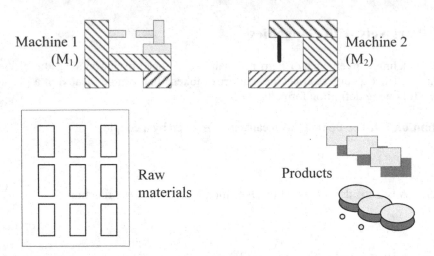

Figure 7.17. An automated manufacturing system producing A- and B-type parts

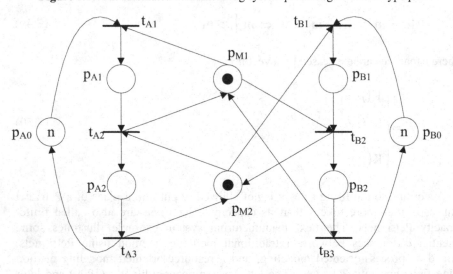

Figure 7.18. Petri net model resulting from a process-oriented modeling method

For $i=0$, 1 and 2, places p_{Ai} models raw pieces for part A are available, A-part's operations 1 and 2, respectively. For $i=1$, 2 and 3, transitions t_{Ai} models starting the first operation, second operation of a raw piece for part A and completion of an A-part, respectively. The explanations hold true for p_{Bi}, t_{Bi} and B-parts. Places p_{M1} and p_{M2} models the availability of machines 1 and 2. The arcs are added according to the operational requirements in order to produce A- and B-parts.

A resource-oriented modeling method models each resource as a finite-capacity place whose capacity is the resource's processing capability. In this system, since machine i has single capacity for $i=1$ and 2, place p_i has capacity $K(p_i)=1$. Place p_0 models the raw material supply and product take-away and, hence, $K(p_0)=n$ can be viewed as infinity. From the resource-oriented modeling viewpoint, an A-raw piece visits each resource in a pre-defined order starting from p_0, to p_1 and p_2, and then back to p_0. Similarly, a B-raw piece visits each resource in a pre-defined order starting from p_0, to p_2 and p_1, and then back to p_0. They can be built separately as shown in Figure 7.19a, b. Their union (by sharing those same places and transitions if applicable) leads to the Petri net model as shown in Figure 19c. Transition t_{ij} means an A- or B-piece is being transferred from resources i to j. For example, t_{01} means an A-rawpiece being transferred from raw material supply modeled by p_0 to machine 1 modeled by p_1. Transition t_{21} means a B-semi-finished piece being transferred from machine 2 modeled by p_2 to machine 1 modeled by p_1.

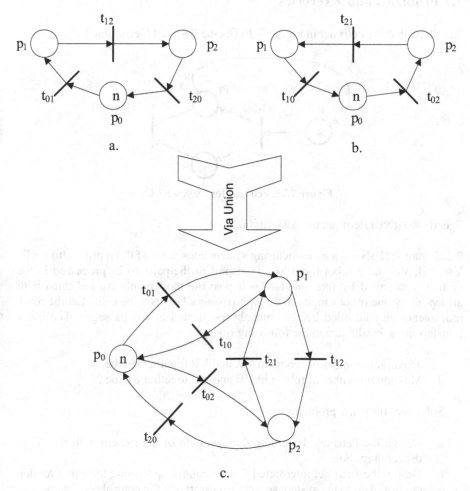

Figure 7.19. A finite-capacity Petri net model resulting from a resource-oriented modeling method

It is clear that the resource-oriented modeling method can lead to a much simpler Petri net model and thus bring certain advantages in the system analysis and deadlock control. On the other hand, a process-oriented Petri net modeling method is more generic as it can model the details and more complex resource requirements, *e.g.,* an operation requiring multiple resources.

According to Murata (1989), each finite-capacity Petri net can be transferred into an equally functioning standard Petri net without capacities called the complementary Petri net. We leave this to the reader as an exercise problem.

7.7 Problems and Exercises

7.1. Show that the Petri net in Figure 7.20 fits the Petri net Definition 7.2.

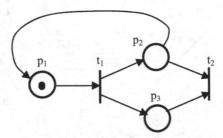

Figure 7.20. Petri net for Exercise 7.1

Specify the given Petri net using Definition 7.6.

7.2. Figure 7.21 shows a manufacturing system consisting of three production cells VA, VB, VC and a robot room SR. Transport of the parts to be processed in the system is executed by four robots. Their possible movements are indicated with arrows. Only one robot can move through passages between the cells. Let the robot movements be controlled by the semaphores located at the passages. The robot transfers are subordinated to the following rules:

1. Maximum number of robots in VA and VB together can be 3.
2. Maximum number of robots in VB and VC together can be 2.

Solve the following problems:

a. Design the Petri net describing the operation of the system with respect to the robot motion.
b. Design the Petri net interpreted for the control specifying the robot motion control. Add to the system sensors necessary for the control.

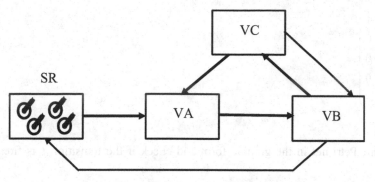

Figure 7.21. Manufacturing system with four robots

7.3. A Petri net is given in Figure 7.22. The initial marking is $\mathbf{m}_0 = (1\ 0\ 0)^{\mathrm{T}}$.

   a.   Compose vectors $\mathbf{t}_i^+, \mathbf{t}_i^-, \Delta\mathbf{t}_i, i = 1,2,3$.

   b.   Consider sequential firing of transitions t_1, t_2, t_3. Prove validity of the expression $\mathbf{m}_0[t_1 > \mathbf{m}_1[t_2 > \mathbf{m}_2[t_3 > \mathbf{m}_3$ and find values of $\mathbf{m}_1, \mathbf{m}_2, \mathbf{m}_3$.

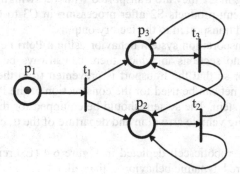

Figure 7.22. Petri net for Exercise 7.3

7.4. A pure Petri net is given by the following vectors:

$$\Delta\mathbf{t}_1 = \begin{pmatrix} -1 \\ 1 \\ 1 \\ 0 \\ 0 \end{pmatrix}, \Delta\mathbf{t}_2 = \begin{pmatrix} 0 \\ -1 \\ 0 \\ 1 \\ 0 \end{pmatrix}, \Delta\mathbf{t}_3 = \begin{pmatrix} 0 \\ 0 \\ -1 \\ 0 \\ 1 \end{pmatrix}, \Delta\mathbf{t}_4 = \begin{pmatrix} 1 \\ 1 \\ 0 \\ -1 \\ -1 \end{pmatrix}$$

The vectors correspond to transitions t_1–t_4. The initial marking is

$$\mathbf{m}_0 = \begin{pmatrix} 1 \\ 0 \\ 0 \\ 0 \\ 0 \end{pmatrix}.$$

Draw the Petri net in the graphic form and check if the transition t_1 is fireable at $\mathbf{m}_0$.

7.5. A factory transportation system using automatic guided vehicles is schematically depicted in Figure 7.23. Semi-products S1 are unloaded from the belt conveyor B1 and transported *via* the track T1 with vehicle V1 to the processing center C3. The emptied vehicle V1 returns back along the same track to load another semi-product from B1. Semi-products S2 are transported to the center C1. After processing in C1 they are transported with V3 swinging between C1 and C2. V4 transports semi-products S2 after processing in C3 to C2. S1 and S2 are assembled in C2 and transported to the factory output.

  Describe the transportation system behavior using a Petri net. For that purpose divide the tracks into sections in which there can always be only one vehicle. Specify the behavior so that the transport is prevented from the vehicle collision. Show how the Petri net can be used for the construction of the Petri net interpreted for control of the system. The system should be equipped for the control purposes with sensors detecting vehicle arrival in and departure of the track sections.

7.6. Propose for the robotic cell depicted in Figure 6.4 (Exercise 6.2) a Petri net specifying the required dynamic behavior of the cell.

7.7. Using the Petri net proposed in the previous Exercise 7.6, create a Petri net interpreted for control, which represents the control function ensured by a control computer such that the required function of the cell is achieved.

7.8. Convert the finite-capacity Petri net model in Figure 19c into an equivalent complementary Petri net. Suppose Machine M1's processing capacity becomes 2 and M2's becomes 3. What is the new equivalent complementary Petri net?

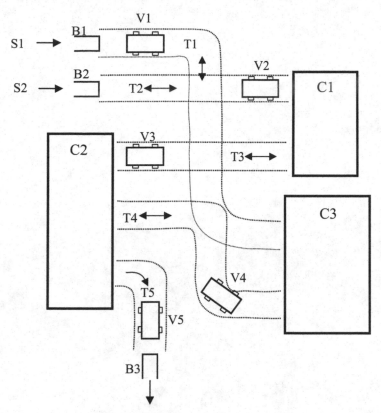

Figure 7.23. A factory AGV transportation system

8

Properties of Petri Nets

8.1 Marking Reachability

Many important properties of Petri nets can be analyzed by means of Petri net reachability and coverability graphs. The reachability graph relates to the Petri net marking reachability. A frequently asked question is whether a given marking $\mathbf{m}$ is reachable by a transition firing sequence. First let us define the transition firing sequence. The starting marking is important in that question.

Definition 8.1. Consider Petri net PN and its marking $\mathbf{m}$. The sequence

$$\tilde{\sigma} = t_{i_1}\, t_{i_2}\, \dots t_{i_k} \tag{8.1}$$

is called the transition firing sequence starting from marking $\mathbf{m}$ iff

$$t_{i_1} \in T,\quad t_{i_2} \in T,\quad \dots\,,\quad t_{i_k} \in T \tag{8.2}$$

$$\mathbf{t}_{i_1}^- \le \mathbf{m},\quad \mathbf{m}_1 = \mathbf{m} + \Delta \mathbf{t}_{i_1} \tag{8.3}$$

$$\mathbf{t}_{i_2}^- \le \mathbf{m}_1,\quad \mathbf{m}_2 = \mathbf{m}_1 + \Delta \mathbf{t}_{i_2} \tag{8.4}$$

$$\mathbf{t}_{i_k}^- \le \mathbf{m}_{k-1},\quad \mathbf{m}_k = \mathbf{m}_{k-1} + \Delta \mathbf{t}_{i_k} \tag{8.5}$$

The meaning of the notation in Equations (8.2)–(8.5) has been described in Section 7.3. Marking $\mathbf{m}_k$ obtained *via* a transition firing sequence starting from marking $\mathbf{m}$ is called the reachable marking in PN from $\mathbf{m}$. For handling convenience, assume that $\mathbf{m}$ is reachable from $\mathbf{m}$.

Substituting Equation (8.3) in Equation (8.4) yields

$$\mathbf{m}_2 = \mathbf{m} + \Delta \mathbf{t}_{i_1} + \Delta \mathbf{t}_{i_2} \tag{8.6}$$

By analogy we obtain

$$\mathbf{m}_k = \mathbf{m} + \sum_{s=1}^{k} \Delta \mathbf{t}_{i_s} \qquad (8.7)$$

The validity of Equation (8.7) alone is the necessary condition for the sequence $\tilde{\sigma} = t_{i_1} t_{i_2} \ldots t_{i_k}$ to be the transition firing sequence starting from $\mathbf{m}$. If Equation (8.7) does not hold, $\tilde{\sigma}$ is not a transition firing sequence starting from $\mathbf{m}$. The sufficiency can be formulated as follows: $\tilde{\sigma}$ is a transition firing sequence starting from $\mathbf{m}$ if

$$\mathbf{m}_r = \mathbf{m} + \sum_{s=1}^{r} \Delta \mathbf{t}_{i_s}, \; \forall r \in \{1,2,...,k\} \qquad (8.8)$$

and

$$\mathbf{t}_{i_1}^{-} \le \mathbf{m} \text{ and } \mathbf{t}_{i_r}^{-} \le \mathbf{m}_{r-1}, \; \forall r \in \{2,3,...,k\} \qquad (8.9)$$

The fireability of t_{i_s} at $\mathbf{m}_{s-1}$ for pure Petri nets can be formulated in the following way: t_{i_s} is fireable at $\mathbf{m}_{s-1}$ iff

$$\mathbf{m}_{s-1} + \Delta \mathbf{t}_{i_s} \ge \mathbf{0}. \qquad (8.10)$$

Considering Equation (8.10) in a different formulation: $\tilde{\sigma} = t_{i_1} t_{i_2} \ldots t_{i_k}$ is a transition firing sequence starting in marking $\mathbf{m}$ for the class of pure Petri nets iff

$$0 \le \mathbf{m} + \sum_{s=1}^{r} \Delta \mathbf{t}_{i_s}, \; \forall r \in \{1,2,...,k\} \qquad (8.11)$$

To provide a detailed explanation of the above reasoning, consider a non-pure Petri net. Let a transition has one direct loop consisting of one arc with weight $W(t_j, p_i)$ going back from transition t_j to its pre-place p_i, and another arc going from p_i to t_j with weight $W(p_i, t_j)$. If $W(p_i, t_j) > M(p_i)$ and $W(p_i, t_j) < W(t_j, p_i)$ then the condition at Equation (8.10) is fulfilled. However, in spite of it, t_j cannot be fired. Consider the non-sufficiency of Equation (8.7) for the following case. Let $\tilde{\sigma} = t_j$, $M(p_i)=1$, $W(p_i, t_j)=2$ and $W(t_j, p_i)=3$. Given $M_k(p_i)=2$, we have $M_k(p_i)=M(p_i)+\Delta t_j$. However, $M_k(p_i)$ is not reachable from $M(p_i)$ and $\tilde{\sigma}$ is not a fireable sequence at $M(p_i)$.

Equation (8.10) is a necessary and sufficient condition for the fireability of t_{i_s} at $\mathbf{m}_{s-1}$ in pure Petri nets, which can be explained as follows. Entries of vector $\mathbf{m}_{s-1}$

are $\mathbf{m}_{s-1}(p_i)$, $\forall p_i \in P$, where p_i is either a pre-place or post-place of t_{i_s}. In pure Petri nets it is impossible that both cases occur simultaneously. In the former case $t_{i_s}^+(p_i) = 0$ and $t_{i_s}^-(p_i) \ne 0$ so that from inequality $\mathbf{m}_{s-1} + \Delta t_{i_s} \ge 0$ we have $\mathbf{m}_{s-1}(p_i) + t_{i_s}^+(p_i) - t_{i_s}^-(p_i) = \mathbf{m}_{s-1}(p_i) - t_{i_s}^-(p_i) \ge 0$, which is the fireability condition by definition. In the latter case $t_{i_s}^+(p_i) \ne 0$ and $t_{i_s}^-(p_i) = 0$ yields $\mathbf{m}_{s-1}(p_i) \ge t_{i_s}^-(p_i) = 0$ from Equation (8.10), which is always true because $\mathbf{m}_{s-1}(p_i) \ge 0$. To conclude: in pure Petri nets $\mathbf{m}_{s-1} + \Delta t_{i_s} \ge 0$ is equivalent to $\mathbf{m}_{s-1}(p_i) \ge t_{i_s}^-(p_i), \forall p_i \in P$, or for vectors $\mathbf{m}_{s-1} \ge t_{i_s}^-$.

The most important reachability property in Petri nets is related to the initial marking value $\mathbf{m}_0$. In PN the set of all reachable markings from $\mathbf{m}_0$ is simply called the reachability set of PN. From that we have the following definition.

Definition 8.2. In PN, marking $\mathbf{m}_k$ is reachable iff a transition firing sequence starting in $\mathbf{m}_0$ exists for which $\mathbf{m}_k = \mathbf{m}_0 + \sum_{s=1}^{k} \Delta t_{i_s}$. The set of all reachable markings is called the reachability set of PN and is given by

$$R_{PN}(\mathbf{m}_0) = \{\mathbf{m}_k \mid \mathbf{m}_k \text{ is reachable}\} \tag{8.12}$$

$\mathbf{m}_0$ is defined as reachable: $\mathbf{m}_0 \in R_{PN}(\mathbf{m}_0)$ as introduced after Definition 8.1.

Sometimes it is useful to have an analogously defined set of markings reachable from other markings $\mathbf{m} \ne \mathbf{m}_0$. Such a set is called the reachability set from $\mathbf{m}$ and denoted as $R_{PN}(\mathbf{m})$.

8.2. Reachability Graph

The reachability set and firing sequences for a given Petri net can be graphically visualized using an oriented graph called the reachability graph.

Definition 8.3. Consider a Petri net $PN = (P, T, F, W, M_0)$ having a finite reachability set. The simple labeled directed mathematical graph (see Definition 2.2)

$$RG_{PN} = (R_{PN}(\mathbf{m}_0), REL, f_1, f_2, \varnothing, T) \tag{8.13}$$

is the reachability graph for PN, where

$R_{PN}(\mathbf{m}_0)$ is the set of nodes that is the reachability set of PN, and the node associated to M_0 or $\mathbf{m}_0$ is called the root of the graph;

$REL \subseteq R_{PN}(\mathbf{m}_0) \times R_{PN}(\mathbf{m}_0)$ is the relation defining the set of oriented arcs connecting the marking pairs (M_i, M_k), $M_i, M_k \in R_{PN}(\mathbf{m}_0)$, such that the marking M_k is obtained by firing a transition t_j at M_i;

f_1 is the function mapping P into the empty set $S_1 = \varnothing$ (the function is empty);

$f_2 : REL \to T$ is a function defining the arc label for each arc (M_i, M_k), $M_i, M_k \in R_{PN}(\mathbf{m}_0)$, where the label is transition t_j whose firing changes M_i into M_k, and the arc including its label can be denoted as a triple (M_i, t_j, M_k). Then, the relation REL including labels becomes

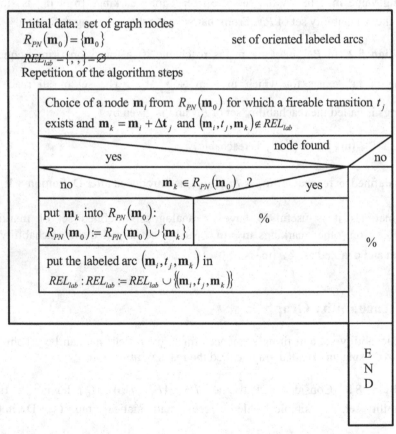

Figure 8.1. Nassi-Schneiderman structogramme of the reachability graph construction algorithm

$$REL_{lab} = \left\{ \left(M_i, t_j, M_k\right) \middle\| M_i, M_k \in R_{PN}\left(\mathbf{m}_0\right) \wedge t_j \in T \wedge M_i\left[t_j > M_k\right] \right\}$$

(8.14)

Instead of functions M_0, M_i, and M_k, the corresponding vectors $\mathbf{m}_0$, $\mathbf{m}_i$, and $\mathbf{m}_k$ can be used for the marking representation as explained before. The vector version of the algorithm for the reachability graph construction using Nassi-Schneiderman's structogramme is shown in Figure 8.1. The structogramme form of the algorithm description supports structured programming without GOTO instructions and lessens programming errors. Graphical elements of the structogramme are self-explanatory. Decision blocks are depicted as triangles. An action always continues through neighboring horizontal edges of the structogramme. An exception is the repetition block where the continuation edge after the bottom edge is the top edge of the whole repeated action group. "%" denotes the empty action, *i.e.*, a skip to the next edge.

More about reachability analysis can be found in Hudák (1999), Jeng and Peng (1999) and Wang *et al.* (2004).

Example 8.1. The reachability graph construction will be illustrated on a case study of a filling and mixing system with two tanks in Figure 8.2. V_{11}, V_{12}, V_{21}, and V_{22} are valves. The required volumes of two liquids to be mixed are prepared using tanks T_1 and T_2. The process start is initiated by the push-button S. Valves V_{12} and V_{22} are opened simultaneously when the liquid levels in both tanks achieve their maximum detected by the corresponding sensors L_{1max} and L_{2max}, respectively. Valves V_{12} and V_{22} are closed separately (independently) when the level L_{imin} in the corresponding tank is detected and the tank is immediately filled up: valve V_{i1} is opened. After reaching L_{imax}, valve V_{i1} is closed and, if the other tank is not yet full, it is necessary to wait. Then again both tanks are emptied together. At any time the filling can be stopped by the emergency button E. Then regardless of the system state both valves V_{11} and V_{21} are closed and both valves V_{12} and V_{22} are opened. The system returns to the ready state when both tanks are empty.

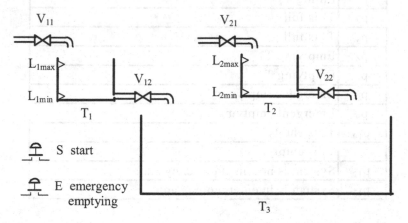

Figure 8.2. A two-tank filling system

Now, the Petri net describing the system behavior (Figure 8.3) is analyzed. Descriptions of places and transitions are in Tables 8.1 and 8.2. The inhibitors and incidentor are used in the net. For the sake of brevity, the inhibitors going out from p_{12} and p_{13} are branched from a common arc.

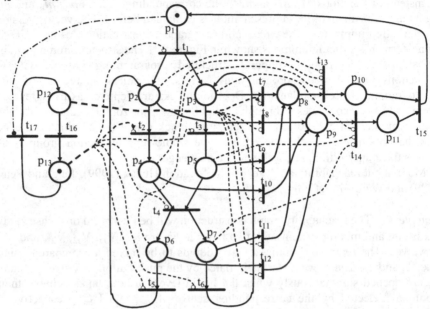

Figure 8.3. Petri net for the two tank system

Table 8.1. Description of the Petri net places

p_1	Ready state of the system, both tanks are empty
p_2	Filling T_1
p_3	Filling T_2
p_4	T_1 is full
p_5	T_2 is full
p_6	Emptying T_1
p_7	Emptying T_2
p_8	Emergent emptying T_2
p_9	Emergent emptying T_1
p_{10}	T_2 is empty
p_{11}	T_1 is empty
p_{12}	System is not in emergency state
p_{13}	System is in emergency state

Table 8.2. Description of the Petri net transitions

t_1	Start of filling
t_2	T_1 achieved its maximum level
t_3	T_2 achieved its maximum level
t_4	Start of both tanks emptying
t_5	T_1 achieved its minimum level
t_6	T_2 achieved its minimum level
t_7	Interrupt of T_2 filling and start emptying by emergency stop
t_8	Interrupt of T_1 filling and start emptying by emergency stop
t_9	Start of T_2 emptying by emergency stop
t_{10}	Start of $T_{1\,emptying}$ by emergency stop
t_{11}	Confirmation of T_2 emptying by emergency stop
t_{12}	Confirmation of T_1 emptying by emergency stop
t_{13}	T2 achieved its minimum level
t_{14}	T1 achieved its minimum level
t_{15}	Transition to the ready state
t_{16}	The emergency push-button activated
t_{17}	Start or transition to the non-emergency state when system is in ready state

As already discussed, it is assumed that only one transition can fire in a time point. For example an "immediate" parallel emptying of the tanks in case of the emergency stop is in fact decomposed into an event sequence. Consider for example that a token is in p_3 and another in p_4, i.e., V_{21} is open and V_{22} closed, tank T_2 is being filled, T_1 is full, and V_{11} and V_{12} are closed. If in such a situation the emergency switch is pressed, the token from p_{12} goes into p_{13}. Transition t_{16} corresponds to this event. Transitions t_3 and t_4 cannot fire but t_7 or t_{10} can. A possible event sequence can be as follows: firing of t_7, then t_{10}. Transition t_{13} cannot fire due to the inhibitor arc (p_4,t_{13}). A next continuation can be t_{14} and t_{13}, so that we obtain the sequence $\tilde{\sigma} = t_7\,t_{10}\,t_{14}\,t_{13}\,t_{15}$. Note that a verbal description of the tank system can hardly be as exact as the Petri net description.

The particular event sequence depends on the chosen control policy. There is no special priority requirement in our example. Anyhow, a decision should be made in the control process. From the point of view of the required system function, the inhibitors going into t_{13} and t_{14} are redundant, but they help to reduce the reachability graph dimension (Figure 8.4). The reachability graph is in Figure 8.4. It has been constructed according to the algorithm described above. By using inhibitors, the number of nodes has been reduced as indicated above.

According to Definition 8.3, the reachability graph is a simple labeled directed graph (see also Definition 2.2). Labels of the graph are the underlying Petri net transitions causing change from one marking to another. The reachability graph is always a connected one. It follows from the connectivity according to Definition 2.6. The marking reachability can be determined using the reachability graph. In

PN, marking $\mathbf{m}_k$ is reachable from $\mathbf{m}$ if there exists a directed path $\tilde{a}$ from $\mathbf{m}$ to $\mathbf{m}_k$ in the reachability graph. The reachability set $R_{PN}(\mathbf{m}_0)$ corresponds to the set of all nodes of the reachability graph constructed for *PN*.

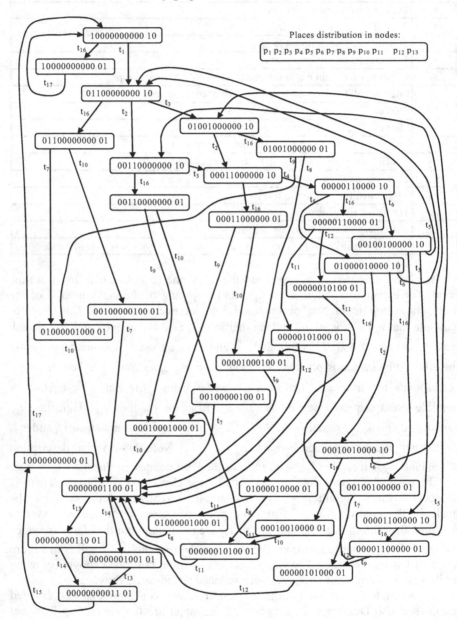

Figure 8.4. Reachability graph of the Petri net

A reachable marking **m** for which there is no fireable transition is called the dead marking. It represents the system deadlock. The system stops in the state corresponding to **m** because no event is executable.

8.3 Boundedness

On specifying behavior/control of a DEDS, an important question arises about whether the set of states of the specified system is finite. This question is frequent in case of real systems. In Petri nets this question concerns boundedness.

Definition 8.4. Let p be a place of PN, i.e., $p \in P = \{p_1, p_2, ..., p_n\}$. It is bounded if for each $\mathbf{m}_k$, which is reachable in PN, i.e., $\mathbf{m}_k \in R_{PN}(\mathbf{m}_0)$, the following holds:

$$\mathbf{m}_k(p) \le j, j \in N^+ \tag{8.15}$$

If a place of PN is bounded by some j for which $\mathbf{m}_k(p) \le j$, we can state more exactly that it is j-bounded. PN is j-bounded if each place of it is j-bounded. PN is bounded if all its places are bounded. The relationship between PN's boundedness and finiteness of its reachability set is well-stated in the following theorem.

Theorem 8.1. A Petri net PN is bounded iff its reachability set is finite.

Proof. $\Leftarrow$: Any initial marking is given by finite natural numbers associated with places. Weights of a Petri net are finite positive integers. Due to the firing rules in Definition 7.5, the generated markings have vector components again including finite positive integers. This is due to finite reachability set generated using only a finite number of sums of finite integers. The Petri net is bounded.

$\Rightarrow$: Conversely, if PN is bounded, markings of the reachability set are always achieved by a finite number of firings so that the reachability set is finite.

Corollary 8.1. The reachability graph of a Petri net can be constructed if and only if the Petri net is bounded.

If a Petri net is not bounded, there are an infinite number of markings and correspondingly an infinite number of reachability graph nodes. Therefore, it is not possible to construct the entire graph.

Example 8.2. An unbounded Petri net is shown in Figure 8.5. A transition firing sequence

$$\tilde{\sigma} = t_1 \, t_2 \, t_1 \, t_2 \, t_1 \, t_2 \, t_1 \, t_2 \, ... \, etc., \tag{8.16}$$

generates the sequence of marking vectors as follows:

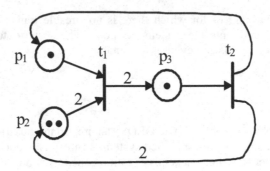

Figure 8.5. Unbounded Petri net

$$\tilde{\pi} = \begin{pmatrix} 1 \\ 2 \\ 1 \end{pmatrix}\begin{pmatrix} 0 \\ 0 \\ 3 \end{pmatrix}\begin{pmatrix} 1 \\ 2 \\ 2 \end{pmatrix}\begin{pmatrix} 0 \\ 0 \\ 4 \end{pmatrix}\begin{pmatrix} 1 \\ 2 \\ 3 \end{pmatrix}\begin{pmatrix} 0 \\ 0 \\ 5 \end{pmatrix} \dots \begin{pmatrix} 1 \\ 2 \\ \infty \end{pmatrix}\begin{pmatrix} 0 \\ 0 \\ \infty \end{pmatrix} \dots \qquad (8.17)$$

Another firing sequence, $t_2\, t_1\, t_2\, t_1 \dots$, can generate anthoer sequence of markings:

$$\pi' = \begin{pmatrix} 1 \\ 2 \\ 1 \end{pmatrix}\begin{pmatrix} 2 \\ 4 \\ 0 \end{pmatrix}\begin{pmatrix} 1 \\ 2 \\ 2 \end{pmatrix}\begin{pmatrix} 2 \\ 4 \\ 1 \end{pmatrix}\begin{pmatrix} 1 \\ 2 \\ 3 \end{pmatrix}\begin{pmatrix} 2 \\ 4 \\ 2 \end{pmatrix} \dots \begin{pmatrix} 1 \\ 2 \\ \infty \end{pmatrix}\begin{pmatrix} 2 \\ 4 \\ \infty \end{pmatrix} \dots \qquad (8.18)$$

Marking $M(p_3)$ of p_3 grows to infinity. The number of reachability graph nodes is unlimited. An application of the reachability graph construction algorithm in Figure 8.6 shows that the marking of p_3 can be infinite in combination with many other markings of p_1 and p_2. These markings are as follows

$$\begin{pmatrix} 1 \\ 2 \\ \infty \end{pmatrix}, \begin{pmatrix} 2 \\ 4 \\ \infty \end{pmatrix}, \begin{pmatrix} 3 \\ 6 \\ \infty \end{pmatrix}, \dots, \begin{pmatrix} i \\ 2i \\ \infty \end{pmatrix} \quad \text{where } i = 1,2,\dots,\infty \qquad (8.19)$$

As a result, every place can grow its marking to infinity.

8.4 Coverability

The reader could ask how to represent the marking state space for unbounded Petri nets in a finite way. It is possible using the coverability property and the so-called coverability graph.

Definition 8.5. Let $PN = (P, T, F, W, M_0)$ be a Petri net. A reachable marking **m** of the Petri net PN is said to be covered by a marking **m**$'$ iff

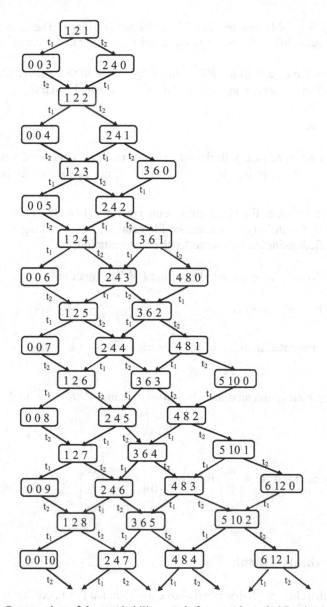

Figure 8.6. Construction of the reachability graph for an unbounded Petri net; to construct the complete graph is impossible

$$\mathbf{m'} \geq \mathbf{m} \qquad\qquad (8.20)$$

The inequality at Equation (8.20) means that for each pair of corresponding vector entries $m'(i) \geq m(i)$ for $i = 1,2,...,|P|$. Note that Definition 8.5 concerns any marking $\mathbf{m'}$ of *PN*, not only the reachable ones. Here, recall Definition 7.3

defining both reachable and unreachable Petri net markings. The next definition deals with coverability with respect to a reachable marking in the given Petri net.

Definition 8.6. Given a Petri net PN. A marking $\mathbf{m}$ is said to be coverable in PN if there is a reachable marking $\mathbf{m}'$ in PN, *i.e.*, $\mathbf{m}' \in R_{PN}(\mathbf{m}_0)$, such that

$$\mathbf{m}' \geq \mathbf{m} \tag{8.21}$$

Definition 8.6 specifically deals with the coverability connected with a given Petri net. The initial marking M_0 used in the Petri net definition can be replaced by vector $\mathbf{m}_0$.

In order to augment the coverability concept for unbounded Petri nets, ω, a special symbol for infinity, has been supplemented to the marking value set. A slightly modified definition will serve for the augmentation.

Definition 8.7. Given an unbounded Petri net PN. The function

$$M : P \rightarrow N^+ \cup \{0\} \cup \{\omega\} \tag{8.22}$$

is called the augmented marking of PN. For each $n \in N^+$: $\omega > n, \omega \pm n = \omega, \omega \geq \omega$.

For example an augmented marking $\mathbf{m} = \begin{pmatrix} 1 \\ 10 \\ 0 \\ \omega \end{pmatrix}$ in a Petri net with four places

is covered by marking $\mathbf{m}' = \begin{pmatrix} 2 \\ \omega \\ 0 \\ \omega \end{pmatrix}$ because $\begin{pmatrix} 2 \\ \omega \\ 0 \\ \omega \end{pmatrix} \geq \mathbf{m} = \begin{pmatrix} 1 \\ 10 \\ 0 \\ \omega \end{pmatrix}$.

8.5 Coverability Graph

Like the reachability graph described above for bounded Petri nets, the coverability graph is used for unbounded Petri nets. The coverability graph enables to represent in a finite form the marking state space so that the finite number of augmented markings cover all actual markings of the considered Petri net. The coverability graph construction is based on the following theorem.

Theorem 8.2. Given a Petri net $PN = (P, T, F, W, M_0)$ and two firing sequences $\tilde{\mu}$ and $\tilde{\nu}$, where $\tilde{\nu}$ is a non-empty firing sequence starting from marking $\mathbf{m}_0$, $\tilde{\mu}$ can also be the empty firing sequence. If $\tilde{\mu}$ is empty ($\tilde{\mu} = \tilde{\varepsilon}$), then $\mathbf{m}_0[\tilde{\varepsilon} > \mathbf{m}_0$. If

$$\mathbf{m}_0 [\tilde{\mu} > \mathbf{m}[\tilde{\nu} > \mathbf{m}' \quad \text{and} \quad \mathbf{m}' > \mathbf{m}$$ (8.23)

the Petri net *PN* is unbounded.

Proof. Inequality $\mathbf{m}' > \mathbf{m}$ means that, at least for one entry of those vectors,

$$m'(i) > m(i), \ i \in 1,2,...,|P|$$ (8.24)

while for the other entries for which Equation (8.24) is not valid, the equality holds. The vector entries considered in Equation (8.24) correspond to place p_i.

Let the first transition in the non-empty firing sequence $\tilde{\nu}$ be t_k. If t_k is fireable by $\mathbf{m}$ even more it is fireable by marking $\mathbf{m}'$. The structure and weights of the Petri net *PN* are fixed. Some tokens are removed by firing t_k from p_i and some are added in p_i according to the firing rules in Definition 7.5. If there are more tokens added to p_i than those removed from it, Equation (8.24) holds. Consider that the firing sequence $\tilde{\nu}$ is again applied starting from marking $\mathbf{m}'$ and $\mathbf{m}'[\tilde{\nu} > \mathbf{m}''$. The inequality $\mathbf{m}'' > \mathbf{m}'$ holds for the above-mentioned reason. Continuing the reasoning in the described way, it can be concluded that the number of tokens in p_i grows to infinity. A similar situation can occur for more than one place. In such a case, the number of tokens in such places increases beyond any integer limit.

The idea of the coverability graph is to construct a graph of markings using the augmented markings and to achieve a finite graphical representation so that all reachable markings even unlimited are covered by a finite number of augmented markings. Then the coverability graph has a finite number of graph nodes.

Definition 8.8. The coverability set for a given unbounded Petri net *PN* is a finite set of augmented markings covering all reachable markings in *PN*.

Now we can formulate the coverability graph definition similar to the one of reachability graph.

Definition 8.9. Consider a Petri net $PN = (P,T,F,W,M_0)$ be given. The simple labeled directed mathematical graph

$$CG_{PN} = (C_{PN}(\mathbf{m}_0), RELC, f_1, f_2, \varnothing, T)$$ (8.25)

is the coverability graph of *PN*, where
    $C_{PN}(\mathbf{m}_0)$ is the coverability set of *PN*
    $RELC \subseteq C_{PN}(\mathbf{m}_0) \times C_{PN}(\mathbf{m}_0)$ is the set of oriented arcs labeled by the
    function f_2

$f_1 : C_{PN}(\mathbf{m}_0) \to \varnothing$ is not defined or empty function

$f_2 : RELC \to T$ is a function mapping the arcs of $RELC$ into the set of transitions. The labeled arcs are represented *via* the triples (M_{Ai}, t_j, M_{Ak}). The labeled relation is $RELC_{lab}$. Transition t_j is fireable by all reachable markings (standard, *i.e.,* not augmented marking) covered by the augmented marking M_{Ai} and the firing brings about a marking covered by an augmented marking M_{Ak}, t_j is a label of the arc from M_{Ai} to M_{Ak}. In this way, all reachable markings of PN are covered by $C_{PN}(\mathbf{m}_0)$ and all possible transition firings are among the labels of the coverability graph.

A coverability graph node corresponding to an augmented marking may cover some subset of standard reachable markings in PN. For each possible transition firing in these reachable markings there exists an arc labeled with the transition and going into another augmented marking, possibly in the same augmented marking (feedback loop). Several reachable markings covered by an augmented marking can enable firing of the same transition represented by the same arc. Construction of a coverability graph according to Definition 8.9 is not an unambiguous task. There can be several coverability graphs for the same Petri net. Each node of the coverability graph represents a marking that covers a subset of the reachable marking of the given Petri net. Firing of a transition can change a marking that belongs to one covered subset to another marking belonging to another subset. Each subset is represented by an augmented marking that covers its subset. In a special case, a transition firing can return to the same subset.

A coverability graph of PN can be constructed according to the following verbal algorithm. The algorithm proceeds in consecutive steps if not otherwise specified.

STEP 1.   Define three empty sets denoted as "NOT_ANALYZED", "DEAD", and "ANALYZED".

STEP 2.   Put the initial marking represented by the vector $\mathbf{m}_0$ into NOT_ANALYZED.

STEP 3.   If NOT_ANALYZED is empty, go to STEP 11, otherwise go to STEP 4.

STEP 4.   Select and withdraw a marking from the set NOT_ANALYZED and denote it as $\mathbf{m}$. Draw a node of the coverability graph which corresponds to the marking $\mathbf{m}$.

STEP 5.   If no transition is fireable by $\mathbf{m}$ put marking $\mathbf{m}$ into DEAD and go to STEP 3, otherwise go to STEP 6.

STEP 6.   Define set $T_m \subseteq T$ of all fireable transitions by $\mathbf{m}$.

STEP 7.   If T_m is empty go to STEP 3, and otherwise go to STEP 8.

STEP 8.   Withdraw a transition t from T_m and by its firing in PN obtain marking $\mathbf{m}'$, that is $\mathbf{m}[t > \mathbf{m}'$.

STEP 9.   If there is a marking $\mathbf{m}_e$ equal to $\mathbf{m}'$ in ANALYZED or DEAD, draw an arc from node $\mathbf{m}$ to $\mathbf{m}_e$, label it with t and go to STEP 7, otherwise go to STEP 10.

STEP 10. If there is one or more markings $\mathbf{m}''$ on any possible directed path from $\mathbf{m}_0$ to $\mathbf{m}'$, for which $\mathbf{m}' > \mathbf{m}''$, then replace the number of tokens for each place $p \in P$ for which $\mathbf{m}'(p) > \mathbf{m}''(p)$ with symbol ω standing for the infinite number of tokens, i.e., the original marking is replaced by an appropriate augmented marking. Recall that ω does not strictly cover ω because it is defined that $\omega \geq \omega$, this case does not lead to a new covering marking. In the case that there is not the same marking as $\mathbf{m}'$ in ANALYZED or DEAD, add a new node corresponding to the modified augmented marking $\mathbf{m}'$ into the coverability graph and draw an arc from $\mathbf{m}$ to the modified marking $\mathbf{m}'$. Put marking $\mathbf{m}'$ into set NOT_ANALYZED, else draw only an arc. Go to STEP 7.

STEP 11. END.

It is evident that STEP 10 is directly implied by Theorem 8.2. If some newly generated marking "sharply" covers a marking somewhere on the backtracked path in the coverability graph, then Theorem 8.2 can be applied. The fact that in one place tokens will be cumulated without limits is interpreted by replacing the marking with infinity ω. From this moment when a transition t is fired by $\mathbf{m}'$, the number of tokens in place p is considered infinitely large. Recall that the addition of k tokens to p gives $\omega + k = \omega$ and similarly for any subtraction. For covering ω holds $\omega \geq \omega$.

A Nassi-Schneiderman structogramme is depicted in Figure 8.7. It specifies practically the same construction algorithm as verbally described earlier. The structogramme is more transparent and enables to program better the algorithm using a computer. The structogramme design principles have been explained when constructing the reachability graph.

Example 8.3. Construction of the coverability graph is illustrated using the Petri net from Example 8.2, which is depicted in Figure 8.5.

The initial marking is $\mathbf{m}_0 = \begin{pmatrix} 1 & 2 & 1 \end{pmatrix}^T$. There are two transitions fireable by $\mathbf{m}_0$: t_1 and t_2. Next markings are $\begin{pmatrix} 0 & 0 & 3 \end{pmatrix}^T$ and $\begin{pmatrix} 2 & 4 & 0 \end{pmatrix}^T$. None of these markings cover some of their ancestors, namely $\begin{pmatrix} 0 & 0 & 3 \end{pmatrix}^T \not> \begin{pmatrix} 1 & 2 & 1 \end{pmatrix}^T$ and $\begin{pmatrix} 2 & 4 & 0 \end{pmatrix}^T \not> \begin{pmatrix} 1 & 2 & 1 \end{pmatrix}^T$, either. Now there are two markings in the set NOT_ANALYZED. Considering first $\begin{pmatrix} 0 & 0 & 3 \end{pmatrix}^T$, only t_2 is fireable at $\begin{pmatrix} 0 & 0 & 3 \end{pmatrix}^T$. The next marking is $\begin{pmatrix} 1 & 2 & 2 \end{pmatrix}^T$ which sharply covers $\mathbf{m}_0 = \begin{pmatrix} 1 & 2 & 1 \end{pmatrix}^T$ as $\mathbf{m}' = \begin{pmatrix} 1 & 2 & 2 \end{pmatrix}^T > \begin{pmatrix} 1 & 2 & 1 \end{pmatrix}^T = \mathbf{m}'' = \mathbf{m}_0$. For p_3, since $\mathbf{m}'(p_3) > \mathbf{m}''(p_3)$, the marking at p_3 becomes ω. Since $\omega - 1 = \omega$, the next marking resulting from firing t_1 is

$(0\ 0\ \omega)^{\mathrm{T}}$. Continue the process according to the algorithm; we obtain the coverability graph in Figure 8.8. It has only six distinct nodes, *i.e.,*

$$\begin{pmatrix} 1 \\ 2 \\ 1 \end{pmatrix}, \begin{pmatrix} 0 \\ 0 \\ 3 \end{pmatrix}, \begin{pmatrix} 2 \\ 4 \\ 0 \end{pmatrix}, \begin{pmatrix} 1 \\ 2 \\ \omega \end{pmatrix}, \begin{pmatrix} 0 \\ 0 \\ \omega \end{pmatrix}, \text{ and } \begin{pmatrix} \omega \\ \omega \\ \omega \end{pmatrix}.$$

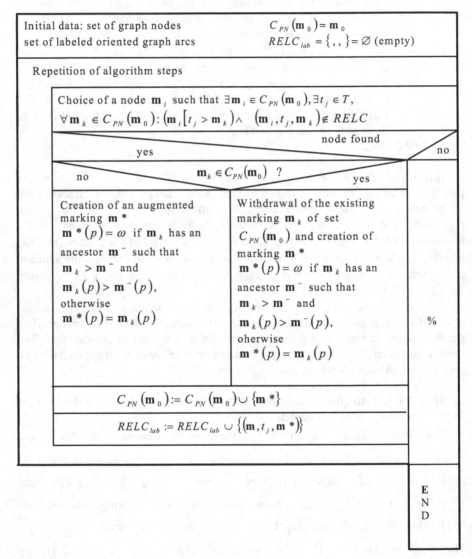

Figure 8.7. Nassi-Schneiderman structogramme for the coverability graph construction algorithm

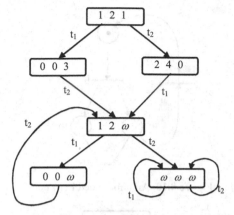

Figure 8.8. Coverability graph for the Petri net in Figure 8.5

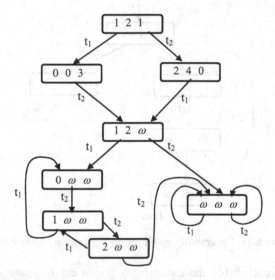

Figure 8.9. Coverability graph for the Petri net in Figure 8.5

The example allows one to demonstrate the ambiguity of the coverability graph construction. In Figure 8.9 there is a different coverability graph that complies with Definition 8.9 though it has not been designed using the above described algorithm.

With its augmented markings the coverability graph in Figure 8.9 covers all reachable markings of the net in Figure 8.5. Each possible firing has an associated arc from one augmented marking covering of the coverability graph to another of the next marking after the firing.

Example 8.4. Consider the net in Figure 8.10. It comprises a transition without any outgoing arc. Such a structure complies with Definition 7.2. Its coverability graph is shown in Figure 8.11 and has two dead markings. They are contained in the set DEAD filled up during the algorithm execution.

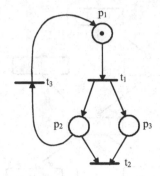

Figure 8.10. An unbounded Petri net

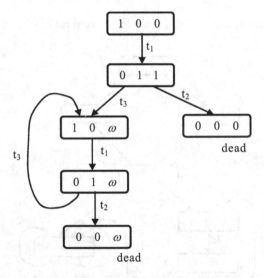

Figure 8.11. Coverability graph for the Petri net in Figure 8.10

Design of the reachability and coverability graphs can be analogously done for the Petri nets with inhibitors and incidentors *PI*, for the Petri nets with capacities *PCA*, and for the Petri nets interpreted for control *PC*. The *PI*'s and *PCA*'s firing of a transition is subordinated to the existence of inhibitors or incidentors, and to the place capacities, as well. For *PC* the arcs in the reachability and coverability graphs are labeled both with firing transitions and with additional firing conditions.

8.6 Liveness

The next property we intend to study is the liveness. It is related to the deadlock free operation of discrete event systems. Liveness can be analyzed in terms of marking and event.

Given a Petri net PN we say that a reachable marking $\mathbf{m}_d \in R_{PN}(\mathbf{m}_0)$ is dead if there is no transition fireable at $\mathbf{m}_d$. Put simply, if a Petri net is a correct model of a real system and reaches $\mathbf{m}_d$, then the system cannot continue its operation/function. No event is executable and the system is in a deadlock.

Moreover, we are interested in the system event history leading to a deadlock. The history can be expressed in terms of firing sequences. Firing sequences are a very basis for the definition of Petri net liveness.

Definition 8.10. Consider $PN = (P, T, F, W, M_0)$. A transition $t \in T$ is said to be

- Live at level $L0$ (or dead) if there is no firing sequence beginning at $\mathbf{m}_0$ and containing t,
- Live at level $L1$ if t can be fired at least once in a firing sequence beginning at $\mathbf{m}_0$,
- Live at level $L2$ if t can be fired at least k ($k \geq 1$) times in a firing sequence beginning at $\mathbf{m}_0$,
- Live at level $L3$ if t can be fired infinite times in a firing sequence beginning at $\mathbf{m}_0$,
- Live at level $L4$ (or live) if at each reachable marking of PN, there is at least one firing sequence in which t fires at least once.

It is possible to express $L4$ in terms of $L1$ as follows: transition t is live at level $L4$ if for each reachable marking $\mathbf{m} \in R_{PN}(\mathbf{m}_0)$ it is $L1$-live. In other words, starting from every reachable marking there should exist a firing sequence in which t fires at least once.

Liveness at level Lk is denoted as Lk-liveness for brevity.

The following implications can be easily understood:

$$L4\text{-liveness} \Rightarrow L3\text{-liveness} \Rightarrow L2\text{-liveness} \Rightarrow L1\text{-liveness}$$

t is said to be strictly Lk-live if it is Lk-live but at the same time it is not $L(k+1)$-live.

The minimum liveness property with respect to all transitions of a Petri net is transferred into the property of the whole Petri net, *i.e.,* if all transitions of PN are Lk-live, PN is Lk-live.

Many authors use the tag "live" for the $L4$-liveness. Consequently if t is not live then it may not be dead, it can be, *e.g.,* $L2$-live, and inverse of dead ($L0$-liveness) is not live but can be any Lk-liveness, k=1,2,3, and 4.

Example 8.5. Figure 8.12 shows a Petri net that serves for liveness study. Only t_1 is fireable at initial marking $\mathbf{m}_0 = (1\ 0\ 0\ 1\ 0)^T$. After firing t_1 the next marking is $(0\ 1\ 0\ 0\ 0)^T$ and then only t_3 is fireable. Firing it leads to $(1\ 0\ 1\ 0\ 0)^T$.

Afterwards, t_4 and t_5 can alternate their firing infinitely. Transition t_2 never fires. Therefore it is $L0$-live. Transitions t_1 and t_3 fire exactly once – they are strictly $L1$-live, and t_4 and t_5 can fire for infinite times – they are live at level $L3$. The Petri net as a whole is live at level $L0$, thus it is dead. The liveness situation is evident from the reachability graph in Figure 8.13. Obviously, in the loop $(1\,0\,1\,0\,0)^T \rightarrow (0\,0\,0\,0\,1)^T \rightarrow (1\,0\,1\,0\,0)^T \rightarrow ....$ transitions t_4 and t_5 can fire for an infinity number of times.

The analyzed Petri net is bounded.

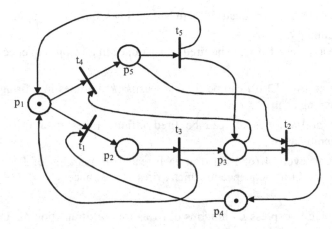

Figure 8.12. Petri net showing Lk liveness, k=0, 1, 2, 3, and 4

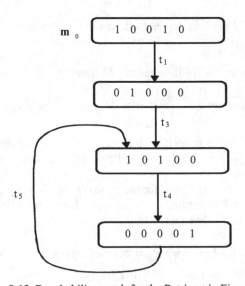

Figure 8.13. Reachability graph for the Petri net in Figure 8.12

8.7 Reversibility

Many processes in DEDS have a cyclic character. For example, production jobs in manufacturing systems run cyclically. Some product is manufactured out of input parts through a sequence of operations that are repeatedly executed during a work shift. The Petri net model of a DEDS reflects the cyclic character through a property called reversibility defined as follows.

Definition 8.11. A Petri net *PN* is reversible iff for each reachable marking $\mathbf{m} \in R_{PN}(\mathbf{m}_0)$ the initial marking $\mathbf{m}_0$ belongs to the reachability set from $\mathbf{m}$ in *PN*, i.e.,

$$\mathbf{m}_0 \in R_{PN}(\mathbf{m}) \tag{8.26}$$

Simply speaking, the reversibility property says that for every reachable marking there exists at least one firing sequence beginning at it and going back to the initial marking. From the point of view of the reachability graph, for every graph node there exists a directed path starting in the node and going back to the root. The reversibility is illustrated *via* the following example.

Example 8.6. Consider the Petri net in Figure 8.14. At the beginning, just t_2 is fireable, then, either t_1 or t_3 is fireable. After having fired t_1, transition t_2 can be fired. After repeating this sequence *k*-times, *k*-tokens are delivered into p_3. When afterwards transitions t_3 and t_4 are fired *k*-times, the initial marking $(0\ 1\ 0\ 1\ 0)^T$ is obtained.

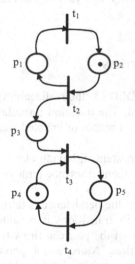

Figure 8.14. Petri net for the reversibility illustration

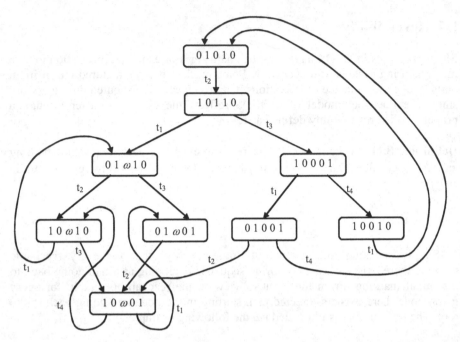

Figure 8.15. Coverability graph for the reversible Petri net in Figure 8.14

Evidently, the Petri net is a reversible one because from any reachable marking it is always possible to return to the initial marking through appropriate transition firings. The corresponding coverability graph is given in Figure 8.15. The graph will later be used in the summary analysis of the Petri net properties in Section 8.12.

8.8 Persistence and Fairness

An interesting property of the DEDS is the realizeability of an event with respect to the realization of another event. The question formulated in terms of Petri nets is whether a transition once being fireable or enabled can lose the fireability prior to its actually being fired.

A Petri net is said to be persistent if, having two or more transitions fireable in a state, firing a transition of them does not remove the fireability of the other transitions.

A property closely related to the persistence is fairness. A Petri net is said to be bounded fair if every pair of its transitions is bounded fair. A transition pair is bounded fair if every transition of the pair can fire k-times at maximum before the other transition in the pair fires. More about persistence in connection with concurrency and conflict will be written later in this textbook. Fairness is dealt with, for example, in the excellent survey paper of Murata (1989).

8.9 Conservativeness

Petri nets are often used to model the utilization of various system resources. For example, the utilization of computer processors, memory devices, input/output units *etc.*, requires modeling their availability or occupation in distributed computer systems. Frequently the number of resources in a system remains constant. In Petri nets the situation is reflected through conservativeness.

Definition 8.12. A Petri net *PN* is strictly conservative if in all reachable markings the total sum of tokens in the net is constant. Formally expressed

$$\sum_{i=1}^{|P|} \mathbf{m}(p_i) = \sum_{i=1}^{|P|} \mathbf{m}_0(p_i) \text{ for } \forall \mathbf{m} \in R_{PN}(\mathbf{m}_0) \qquad (8.27)$$

If two or more resources, *e.g.,* a part and robot, are represented each by a separate token and then the action "The robot picks up the part" by one token, then the number of tokens is changed from two to one. That kind of system situation can be coped with the weighted conservativeness, which is subject of the next definition.

Definition 8.13. A Petri net *PN* is conservative with respect to a weight vector **v** if

$$\mathbf{v}^T \mathbf{m} = \mathbf{v}^T \mathbf{m}_0 = \sum_{i=1}^{|P|} \mathbf{v}(p_i)\mathbf{m}(p_i) = \sum_{i=1}^{|P|} \mathbf{v}(p_i)\mathbf{m}_0(p_i) \qquad (8.28)$$

for $\forall \mathbf{m} \in R_{PN}(\mathbf{m}_0)$ and $\mathbf{v} \in N^{|P|}$

Example 8.7. Distributed computer networks are a frequently used discrete event systems. In the adapted example inspired by Starke (1990), the system comprises three processors communicating *via* two transmission channels. Conflict situations occur due to the use of only two communication resources by three competing users. The system layout with main data channels is shown in Figure 8.16.

Variables:

r_i represents the requirement of the processor Proc i to communicate data;

f_i announces disconnection from the used channel and end of the communication;

w_i is the communication start command for Proc i.

First, the system is specified as a deterministic finite automaton, and then as a Petri net. Thus the example is used to compare both approaches.

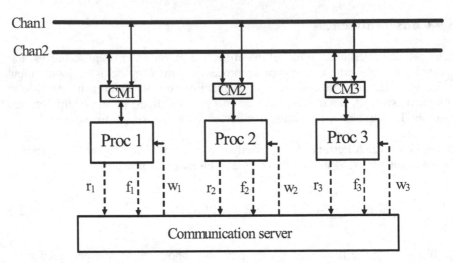

Figure 8.16. Data communication network

There are seven states characterizing the system situations and six events driving the system behavior. Only one processor can use a channel at a time. The state transition table of the automaton is shown in Table 8.3 and its graphical representation is in Figure 8.17. The abbreviation "Proc i" is used for the i-th system processor. The used communication channel is not specified, as it can be seen from the state transition table.

Important information is if one channel, regardless which one of two available channels, is busy or not. The Petri net of the described system is shown in Figure 8.18. The meaning of places and transitions is described in Tables 8.4 and 8.5.

The state transition table or graph of the finite automaton (Figure 8.17) provides much less information about structural and behavioral properties of the modeled system than the corresponding Petri net does. From the finite automaton specification it can be found that, if a state change occurs by event e_i, the inverse back leading change occurs by event e_{i+3}.

A finite automaton specifying the control of the communication is shown in Figure 8.19. It is a part of the complete automaton. Control commands w_i are set to value C_i in the respective states. The complete automaton would be rather complex.

Place p_7 in the Petri net represents free channels. The initial marking $\mathbf{m}_0(p_7) = 2$ says that there are two free channels available for processors at the beginning. From the net we know that two activities run in parallel at a reachable marking $\mathbf{m} = (1\ 0\ 1\ 0\ 1\ 0\ 0)^T$: processor 1 works using one channel and processor 3 works using the other channel. No other channel is available until one of the processors returns a channel it occupies.

Table 8.3. State transition table

	e_1	e_2	e_3	e_4	e_5	e_6	
q_0	q_1	q_2	q_3				Initial state
q_1		q_4	q_5	q_0			Proc1 uses one channel
q_2	q_4		q_6		q_0		Proc2 uses one channel
q_3	q_5	q_6				q_0	Proc3 uses one channel
q_4				q_2	q_1		Proc1 and Proc2 use both channels
q_5				q_3		q_1	Proc1 and Proc3 use both channels
q_6					q_3	q_2	Proc2 and Proc3 use both channels
	Proc1 starts using a channel	Proc2 starts using a channel	Proc3 starts using a channel	Proc1 ends using a channel	Proc2 ends using a channel	Proc3 ends using a channel	

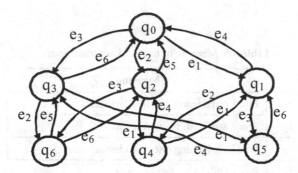

Figure 8.17. Graphical representation of the finite automaton

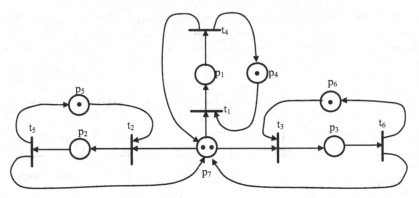

Figure 8.18. Petri net for a three-processor and two-channel system

Table 8.4. Meaning of the Petri net places

	Meaning
P_1	Proc1 uses one channel
P_2	Proc2 uses one channel
P_3	Proc3 uses one channel
P_4	Proc1 does not use a channel
P_5	Proc2 does not use a channel
P_6	Proc3 does not use a channel
P_7	Number of tokens in it indicates the number of free channels

Table 8.5. Meaning of the Petri net transitions

	Meaning
t_1	Proc1 starts using one free channel
t_2	Proc2 starts using one free channel
t_3	Proc3 starts using one free channel
t_4	Proc1 ends using one channel
t_5	Proc2 ends using one channel
t_6	Proc3 ends using one channel

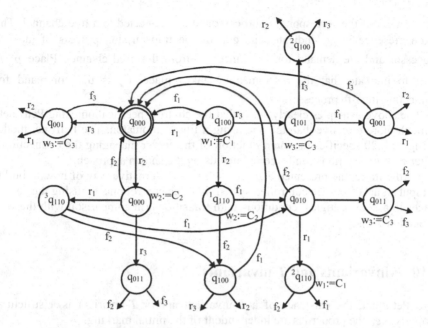

Figure 8.19. Partial finite automaton with outputs specifying control

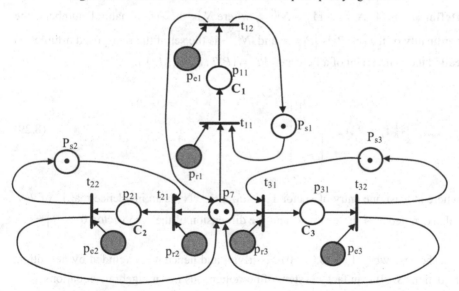

Figure 8.20. Petri net interpreted for control of a three-processor and two-channel system

The Petri net interpreted as a control model for the system of three processors and two channels is shown in Figure 8.20. A new element is used in the net, namely the source place. Source places are hatched in the figure. A token is generated in the source place when an associated requirement of the system resource occurs. Depositing a token in source places p_{r1}, p_{r2} and p_{r3} indicates the requirement of the corresponding processor to be connected to a free channel. The token presence in p_{ei} indicates the end of the transmission process of the i-th processor and the demand for disconnecting from the used channel. Place p_{si} is used to indicate that the disconnection takes place. C_i is the command for connecting the i-th processor.

The above examples clearly show that the control specification using Petri nets is much more effective than the one using a finite automaton. The Petri net model in Figure 8.20 specifies the main function of the server managing the connections. Other control functions can be completed using the shown approach.

Furthermore, the presented example shows an alternative way of how to build-in input signals from the controlled system into the control one, by using the source places to associate logical conditions with transitions (in comparison with the way in Section 7.4).

8.10 *P*-invariants and *T*-invariants

In a Petri net, the existence of a *P*-invariant and/or *T*-invariant is a structural property, *i.e.*, the properties are independent of the initial marking.

Definition 8.14. A vector $\mathbf{i}_p \in N^{|P|\mathrm{T}}$, where N is the set of natural numbers, the cardinality of the set P is $|P| = n$ and $N^{|P|\mathrm{T}}$ is the set of the transposed n-tuples, is called the *P*-invariant of a Petri net $PN = (P, T, F, W, M_0)$ if

$$\mathbf{N}_\Delta^{\mathrm{T}} \mathbf{i}_P = \mathbf{0}, \mathbf{i}_P = \begin{pmatrix} {}^P i_1 \\ {}^P i_2 \\ \vdots \\ {}^P i_n \end{pmatrix} \tag{8.29}$$

where at least one entry of vector $\mathbf{i}_P$ is nonzero. $\mathbf{N}_\Delta$ is an incidence matrix of PN (also denoted as A in Section 7) and the dimension of the zero vector $\mathbf{0}$ is $|T| = m$.

The case when $\mathbf{i}_P = (0, 0, \ldots, 0)^{\mathrm{T}}$ is trivial and hence it is excluded by definition. Equation (8.29) is in fact a system of homogeneous linear algebraic equations

$$\begin{aligned}
a_{11} {}^P i_1 + a_{12} {}^P i_2 + \ldots + a_{1n} {}^P i_n &= 0 \\
a_{21} {}^P i_1 + a_{22} {}^P i_2 + \ldots + a_{2n} {}^P i_n &= 0 \\
\vdots \\
a_{m1} {}^P i_1 + a_{m2} {}^P i_2 + \ldots + a_{mn} {}^P i_n &= 0
\end{aligned} \tag{8.30}$$

Obviously, its zero solution always exists. Generally, it is implied by the Frobenius theorem stating that if the rank of the coefficient matrix of the system at Equation (8.29) is equal to the rank of the augmented matrix, the system has a solution. The zero solution of Equation (8.30) is

$$^P i_1 = {}^P i_2 = \ldots = {}^P i_n = 0 \tag{8.31}$$

For a system of linear algebraic equations it is well known that when the rank r of its coefficient matrix is equal to the number of equations $r = n$, exactly one solution exists. For the case of the homogeneous system at Equation (8.30) there is the solution at Equation (8.31). Coefficients a_{ij} in the system at Equation (8.30) are integers as it follows from the construction of the Δ-incidence matrix. If the searched unknowns are allowed to be real numbers then there are infinitely many solutions for $r < n$. If only integers are allowed, the condition $r < n$ is necessary but not sufficient for the existence of solutions other than Equation (8.31). Sometimes, even a stronger restriction can be imposed on the solution space, namely that the unknowns should be natural numbers (non-negative integers). In this case, even more so the condition $r < n$ is not sufficient.

As mentioned above we are interested in nonzero P-invariants and consequently in nonzero solutions of Equation (8.30) in the integer space or even in the space of natural numbers.

In what follows we will show that the conservativeness property of a Petri net is a corollary of the Petri net P-invariant presence. Suppose a given Petri net has a P-invariant $\mathbf{i}_P$. Then owing to Definition 8.14

$$\mathbf{N}_\Delta^T \mathbf{i}_P = 0 \tag{8.32}$$

Recalling Equation (7.29), Definition 8.1 and Equation (8.7) we have for a reachable marking $\mathbf{m}_k$

$$\mathbf{m}_k = \mathbf{m}_0 + \sum_{s=1}^k \Delta \mathbf{t}_{i_s} \tag{8.33}$$

which can be given as

$$\mathbf{m}_k = \mathbf{m}_0 + \mathbf{N}_\Delta \mathbf{z} \tag{8.34}$$

where $\mathbf{z} = (z_1, z_2, \ldots, z_m)^T$, $z_i \in N$, $i \in (1, 2, \ldots, m)$ determines how many times the vector $\Delta \mathbf{t}_i$ occurs in a particular firing sequence.

From Equation (8.34) we have

$$\mathbf{m}_k - \mathbf{m}_0 = \mathbf{N}_\Delta \mathbf{z} \tag{8.35}$$

Transposing Equation (8.35) yields

$$\mathbf{m}_k^T - \mathbf{m}_0^T = \mathbf{z}^T \; \mathbf{N}_\Delta^T \tag{8.36}$$

and multiplying by $\mathbf{i}_P$ gives

$$\left(\mathbf{m}_k^T - \mathbf{m}_0^T\right)\mathbf{i}_P = \mathbf{z}^T \; \mathbf{N}_\Delta^T \; \mathbf{i}_P \tag{8.37}$$

Due to the assumption of P-invariant

$$\left(\mathbf{m}_k^T - \mathbf{m}_0^T\right)\mathbf{i}_P = \mathbf{z}^T \; \mathbf{0} = 0 \tag{8.38}$$

$$\mathbf{m}_k^T \; \mathbf{i}_P = \mathbf{m}_0^T \; \mathbf{i}_P \tag{8.39}$$

$$\left(m_{1k}, m_{2k}, ..., m_{|P|k}\right)\begin{pmatrix} ^P i_1 \\ ^P i_2 \\ \vdots \\ ^P i_n \end{pmatrix} = \left(m_{10}, m_{20}, ..., m_{|P|0}\right)\begin{pmatrix} ^P i_1 \\ ^P i_2 \\ \vdots \\ ^P i_n \end{pmatrix} \tag{8.40}$$

$$m_{1k} \; ^P i_1 + m_{2k} \; ^P i_2 + ... + m_{|P|k} \; ^P i_n = m_{10} \; ^P i_1 + m_{20} \; ^P i_2 + ... + m_{|P|0} \; ^P i_n \tag{8.41}$$

According to Equation (8.41) in a Petri net the sum of tokens distributed in places with respect to the initial marking, and weighted (multiplied) by the components of P-invariant $\mathbf{i}_P$ is constant   for all reachable markings $\mathbf{m}_k \in R_{PN}(\mathbf{m}_0)$. If some of the P-invariant components is zero, the number of tokens in the corresponding place is excluded from the sum (8.41). A term P-invariant support is used in this connection. The P-invariant support is the set of Petri net places corresponding to nonzero entries in $\mathbf{i}_P$. Using the concept of support, the weighted sum of tokens in the places of the P-invariant support is constant for all reachable markings.

It is possible to prove the inverse proposition to (8.32) $\Rightarrow$ (8.39), namely (8.39) $\Rightarrow$ (8.32) for L1-live Petri net. Assume for L1-live Petri net $\mathbf{m}_k^T \; \mathbf{i}_P = \mathbf{m}_0^T \; \mathbf{i}_P$ for all reachable markings $\mathbf{m}_k \in R_{PN}(\mathbf{m}_0)$ and for some $\mathbf{i}_P \in N^{|P|T}$. Then $\mathbf{m}_k^T \; \mathbf{i}_P - \mathbf{m}_0^T \; \mathbf{i}_P = 0$ and $\left(\mathbf{m}_k^T - \mathbf{m}_0^T\right)\mathbf{i}_P = 0$. Further $\mathbf{m}_k^T = \mathbf{m}_0^T + \mathbf{z}^T \; \mathbf{N}_\Delta^T$ so that $\left(\mathbf{m}_0^T + \mathbf{z}^T \; \mathbf{N}_\Delta^T - \mathbf{m}_0^T\right)\mathbf{i}_P = 0$ where $\mathbf{z}^T \neq \mathbf{0}$. We have $\left(\mathbf{z}^T \; \mathbf{N}_\Delta^T\right)\mathbf{i}_P = 0$ and $\mathbf{z}^T \; \mathbf{N}_\Delta^T \neq \mathbf{0}$ for entries corresponding $\mathbf{i}_P$ yielding $\mathbf{N}_\Delta^T \; \mathbf{i}_P = \mathbf{0}$.

The concept of P-invariants is useful for the solution to various problems in Petri nets and via them also in DEDS as we will see later on. A similar concept of T-invariants is introduced in the following definition.

Definition 8.15. A vector $\mathbf{i}_T \in N^{|T|\text{T}}$, where N is the set of natural numbers, the cardinality of set T is $|T| = m$ and $N^{|T|\text{T}}$ is the set of transposed m-tuples, is called the T-invariant of a Petri net $PN = (P, T, F, W, M_0)$ if

$$\mathbf{N}_\Delta\, \mathbf{i}_T = \mathbf{0}, \ \mathbf{i}_T = \begin{pmatrix} ^T i_1 \\ ^T i_2 \\ \vdots \\ ^T i_m \end{pmatrix} \tag{8.42}$$

where at least one entry of the vector $\mathbf{i}_T$ is nonzero. $\mathbf{N}_\Delta$ is an incidence matrix of the Petri net PN and the dimension of the zero vector $\mathbf{0}$ is $|P| = n$.

Note that the incidence matrix in the definition is not transposed. If $t_{s_1} t_{s_2} ... t_{s_k}$ is a transition firing sequence beginning in marking $\mathbf{m}_0$ and referring to (8.33) and (8.34) we have

$$\mathbf{m}_k = \mathbf{m}_0 + \sum_{s=1}^{k} \Delta t_{i_s} = \mathbf{m}_0 + \mathbf{N}_\Delta\, \mathbf{z} \tag{8.43}$$

If $\mathbf{z}$ is a T-invariant it follows

$$\mathbf{N}_\Delta\, \mathbf{z} = \mathbf{0} \quad \text{and} \quad \mathbf{m}_k = \mathbf{m}_0 \tag{8.44}$$

Theorem 8.3. The existence of a T-invariant is a necessary condition for a Petri net to be reversible.

Proof. The principle of the Petri net reversibility consists in that for each reachable marking $\mathbf{m}_k$ there is a transition firing sequence continuation from $\mathbf{m}_k$ reaching the marking $\mathbf{m}_0$, *i.e.*, in terms of Equation 8.43, $\mathbf{m}_0 = \mathbf{m}_0 + \mathbf{N}_\Delta\, \mathbf{z}$. Therefore $\mathbf{N}_\Delta \mathbf{z} = \mathbf{0}$ has to hold, hence $\mathbf{z}$ is a T-invariant.

Proposition of Theorem 8.3 can be expressed by the implication:

Petri net is reversible $\Rightarrow$ T-invariant exists

Using a counterexample it will be shown that the inverse implication:

T-invariant exists $\Rightarrow$ Petri net is reversible

does not hold. See Petri net in Figure 8.21. We have:

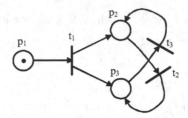

Figure 8.21. Petri net with a T-invariant

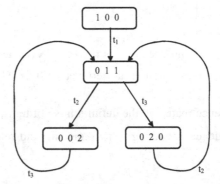

Figure 8.22. Reachability graph for the Petri net in Figure 8.21

$$\begin{pmatrix} -1 & 0 & 0 \\ 1 & -1 & 1 \\ 1 & 1 & -1 \end{pmatrix} \begin{pmatrix} 0 \\ 1 \\ 1 \end{pmatrix} = \begin{pmatrix} 0 \\ 0 \\ 0 \end{pmatrix}$$

A T-invariant exists, it is $\mathbf{i}_T = \begin{pmatrix} 0 \\ 1 \\ 1 \end{pmatrix}$, but the reachability graph indicates that the

Petri net is not reversible (Figure 8.22).

A way to find the solutions to the linear algebraic system of equations (8.32) in the space of integer numbers is to use the standard methods used for the real number space hoping that an integer solution could be obtained or extracted from a real number solution. The following example illustrates this solution way. An exact method yielding the solution directly is described later.

Example 8.8. A robotic manufacturing cell as adapted from Abel (1990) is depicted in Figure 8.23. Workpieces of type A are transported into the cell with conveyor C1 and workpieces B with C3. The Petri net specifying this cell's operation is shown in Figure 8.24. Presence of a token in place p_1 corresponds to the presence of a workpiece A ready for technological processing at the end of conveyor C1. The part of the net for B-workpieces is quite symmetrical with respect to that for

A-workpieces. Transition t_1 expresses the technological processing start with assistance of robot R1. Availability of R1 is expressed by a token in p_7. After t_1 fires the tokens are removed from p_1 and p_7 and a token is placed in p_2, which means that the processing is in progress in machine center MA1 using robot R1. An example is coachwork welding in the automobile industry. The end of this technological step and start of the processing in MA2 using R2 is marked by firing t_2. Firing t_3 means that the processing in MA2 has been completed and the next workpiece to be processed is available at the input on C1. The last event represented by t_3 may seem to be a little bit artificial: the end of the processing and arrival of an A-workpiece both represented by one transition means that the robot R2 is kept busy until the A-workpiece arrival.

It would be possible to add more places and transitions and to express the operation more precisely. The used representation illustrates a natural way of a Petri net construction – from a simpler and rough net to the extensions. This construction way is called top-down design. We will keep the Petri net in the form in Figure 8.24 just for the sake of simplicity of the following considerations.

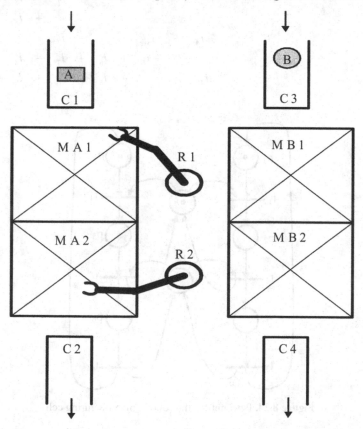

Figure 8.23. A two-robot manufacturing cell

In the example, the task is to find P-invariants for the Petri net describing the robotic cell. The transposed Δ-incidence matrix is

$$
\mathbf{N}_\Delta^T =
\begin{pmatrix}
-1 & 1 & 0 & 0 & 0 & 0 & -1 & 0 \\
0 & -1 & 1 & 0 & 0 & 0 & 1 & -1 \\
1 & 0 & -1 & 0 & 0 & 0 & 0 & 1 \\
0 & 0 & 0 & -1 & 1 & 0 & -1 & 0 \\
0 & 0 & 0 & 0 & -1 & 1 & 1 & -1 \\
0 & 0 & 0 & 1 & 0 & -1 & 0 & 1
\end{pmatrix}
\tag{8.45}
$$

From the equation $\mathbf{N}_\Delta^T \mathbf{i}_P = \mathbf{0}$, we get a system of equations to be solved (having omitted the index P for simplicity):

$$
\begin{aligned}
-i_1 + i_2 & & & & & & & & & = 0 \\
& -i_2 + i_3 & & & & & + i_7 & - i_8 & = 0 \\
i_1 & & - i_3 & & & & & + i_8 & = 0 \\
& & & -i_4 + i_5 & & & - i_7 & & = 0 \\
& & & & -i_5 + i_6 & + i_7 & - i_8 & & = 0 \\
& & & i_4 & & - i_6 & & + i_8 & = 0
\end{aligned}
\tag{8.46}
$$

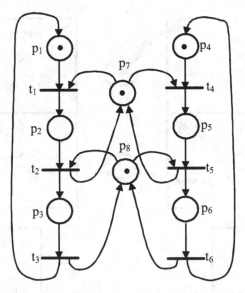

Figure 8.24. Petri net for the robotic manufacturing cell

The calculation of the coefficient matrix determinant and of the lower order determinants shows that the rank of the coefficient matrix is $r = 4$. It means that there are two equations in Equation (8.46), which linearly depend on the remaining four. Summing the first and second equations gives:

$$
\begin{array}{rcl}
-i_1 \; + \; i_2 \qquad\qquad\qquad\quad -\; i_7 \qquad\qquad & = & 0 \\
-i_2 \; + \; i_3 \qquad\quad +\; i_7 \; - \; i_8 & = & 0 \\
\hline
-i_1 \qquad\qquad +\; i_3 \qquad\qquad\quad -\; i_8 & = & 0
\end{array}
\tag{8.47}
$$

and multiplication by -1 gives the third equation in Equation (8.46). Similarly, the sum of the fourth and fifth equations multiplied by -1 gives the sixth equation. Hence, both the third and sixth equations can be omitted and we have finally

$$
\begin{array}{rcl}
-i_1 \; + \; i_2 \qquad\qquad\qquad\qquad\qquad -\; i_7 \qquad\qquad & = & 0 \\
-i_2 \; + \; i_3 \qquad\qquad\qquad +\; i_7 \; - \; i_8 & = & 0 \\
-i_4 \; + \; i_5 \qquad\qquad -\; i_7 \qquad\qquad & = & 0 \\
-i_5 \; + \; i_6 \quad +\; i_7 \; - \; i_8 & = & 0
\end{array}
\tag{8.48}
$$

The particular integer values of four unknowns can be chosen arbitrarily of and the remaining four unknowns can be calculated. Values of the four unknowns are chosen as follows:

$$
\begin{aligned}
i_1 &= \lambda_1 \\
i_4 &= \lambda_2 \\
i_7 &= \lambda_3 \\
i_8 &= \lambda_4
\end{aligned}
\tag{8.49}
$$

and for transparency we denote the remaining dependent unknowns as

$$
\begin{aligned}
x_1 &= i_2 \\
x_2 &= i_3 \\
x_3 &= i_5 \\
x_4 &= i_6
\end{aligned}
\tag{8.50}
$$

Now, the system is

$$
\begin{aligned}
x_1 & = \lambda_1 + \lambda_3 \\
-x_1 + x_2 & = \lambda_4 - \lambda_3 \\
x_3 & = \lambda_2 + \lambda_3 \\
-x_3 + x_4 & = \lambda_4 - \lambda_3
\end{aligned}
\tag{8.51}
$$

Using the Cramer's rule

$$
x_1 = \frac{\begin{vmatrix} \lambda_1+\lambda_3 & 0 & 0 & 0 \\ \lambda_4-\lambda_3 & 1 & 0 & 0 \\ \lambda_2+\lambda_3 & 0 & 1 & 0 \\ \lambda_4-\lambda_3 & 0 & -1 & 1 \end{vmatrix}}{\begin{vmatrix} 1 & 0 & 0 & 0 \\ -1 & 1 & 0 & 0 \\ 0 & 0 & 1 & 0 \\ 0 & 0 & -1 & 1 \end{vmatrix}} = \frac{(\lambda_1+\lambda_3)\begin{vmatrix} 1 & 0 & 0 \\ 0 & 1 & 0 \\ 0 & -1 & 1 \end{vmatrix}}{1} = \lambda_1+\lambda_3
\tag{8.52}
$$

$$
\begin{aligned}
x_2 &= \lambda_1 + \lambda_4 \\
x_3 &= \lambda_2 + \lambda_3 \\
x_4 &= \lambda_2 + \lambda_4
\end{aligned}
$$

Returning to the original notation the P-invariants are given by

$$
\begin{pmatrix} i_1 \\ i_2 \\ i_3 \\ i_4 \\ i_5 \\ i_6 \\ i_7 \\ i_8 \end{pmatrix} = \lambda_1 \begin{pmatrix} 1 \\ 1 \\ 1 \\ 0 \\ 0 \\ 0 \\ 0 \\ 0 \end{pmatrix} + \lambda_2 \begin{pmatrix} 0 \\ 0 \\ 0 \\ 1 \\ 0 \\ 1 \\ 0 \\ 0 \end{pmatrix} + \lambda_3 \begin{pmatrix} 0 \\ 1 \\ 0 \\ 0 \\ 1 \\ 0 \\ 1 \\ 0 \end{pmatrix} + \lambda_4 \begin{pmatrix} 0 \\ 0 \\ 1 \\ 0 \\ 0 \\ 1 \\ 0 \\ 1 \end{pmatrix}
\tag{8.53}
$$

This equation determines all P-invariants for the values of $\lambda_1, \lambda_2, \lambda_3, \lambda_4$; λ_1, λ_2, λ_3, and λ_4 are chosen from the set of integers. Obviously, the non-negative P-invariants are obtained choosing them from the set of natural numbers. First, consider the P-invariants obtained for their values from $\{0,1\}$ only. The obtained P-invariants and their corresponding supports are listed below:

$$\begin{pmatrix} 1 \\ 1 \\ 1 \\ 0 \\ 0 \\ 0 \\ 0 \\ 0 \end{pmatrix} \Rightarrow \{p_1, p_2, p_3\}, \quad \begin{pmatrix} 0 \\ 0 \\ 0 \\ 1 \\ 1 \\ 1 \\ 0 \\ 0 \end{pmatrix} \Rightarrow \{p_4, p_5, p_6\}, \quad \begin{pmatrix} 0 \\ 1 \\ 0 \\ 0 \\ 1 \\ 0 \\ 1 \\ 0 \end{pmatrix} \Rightarrow \{p_2, p_5, p_7\},$$

$$\begin{pmatrix} 0 \\ 0 \\ 1 \\ 0 \\ 0 \\ 1 \\ 0 \\ 1 \end{pmatrix} \Rightarrow \{p_3, p_6, p_8\}, \quad \begin{pmatrix} 1 \\ 1 \\ 1 \\ 1 \\ 1 \\ 1 \\ 0 \\ 0 \end{pmatrix} \Rightarrow \{p_1, ..., p_6\}, \quad \begin{pmatrix} 1 \\ 2 \\ 1 \\ 0 \\ 1 \\ 0 \\ 1 \\ 0 \end{pmatrix} \Rightarrow \{p_1, p_2, p_3, p_5, p_7\},$$

etc.

$$(8.54)$$

For example, according to the first P-invariant the number of tokens in places $\{p_1, p_2, p_3\}$ is equal to 1 for all reachable markings. Consider the last P-invariant shown above. The number of tokens in p_2 is weighted by 2. By inspection of the Petri net in Figure 8.24 the weighted sum of tokens in places $\{p_1, p_2, p_3.p_5, p_7\}$ is always 2.

The P-invariants express the grouping of the system components from the viewpoint of technology and operation. For example the first P-invariant in Equation (8.54) groups the places comprising the component route of the workpieces A. The second P-invariant has an analogous meaning for the workpieces B. On the other hand, the third P-invariant is associated with the operation state of robot R1. It works either with workpiece A or B or is waiting to start work.

The procedure shown above may not be successful in finding the existing integer solutions of the equation system at Equation (8.30). The main idea of a method always giving the answer about the system solvability and in the positive case the solutions themselves will be presented next.

For the sake of a transparent explanation, consider a very simple case

$$a_1 x_1 + a_2 x_2 = b \tag{8.55}$$

where the coefficients a_1, a_2 and b and unknowns x_1 and x_2 are supposed to be integers, *i.e.,*

$$a_1, a_2, b, x_1, x_2 \in I. \tag{8.56}$$

The greatest common divisor of the coefficients a_1, a_2 will be denoted $GDIV(a_1, a_2)$. It can always be extracted from the left-hand side of Equation (8.55)

$$GDIV(a_1, a_2)(a'_1 \ x_1 + a'_2 \ x) = b, \ a'_1, a'_2 \in I \qquad (8.57)$$

and

$$a'_1 \ x_1 + a'_2 \ x_2 = \frac{b}{GDIV(a_1, a_2)} \qquad (8.58)$$

The following theorem is based on the previous facts.

Theorem 8.4. Equation (8.55) under the condition at Equation (8.56) has a solution if and only if the right-hand side coefficient b is divisible by the greatest common divisor of a_1 and a_2.

Theorem 8.4 can be easily extended to the system

$$a_{11}x_1 + a_{12}x_2 + ... + a_{1j}x_j + ... + a_{1n}x_n = b_1$$
$$a_{21}x_1 + a_{22}x_2 + ... + a_{2j}x_j + ... + a_{2n}x_n = b_2$$
$$...$$
$$a_{i1}x_1 + a_{i2}x_2 + ... + a_{ij}x_j + ... + a_{in}x_n = b_i \qquad (8.59)$$
$$...$$
$$a_{m1}x_1 + a_{m2}x_2 + ... + a_{mj}x_j + ... + a_{mn}x_n = b_m$$

The divisibility of b_i through $GDIV(a_{i1}, a_{i2}, ..., a_{in})$ is a necessary and sufficient condition for the solution existence of the system at Equation (8.59) together with Frobenius condition.

Further consider that for the absolute values in Equation (8.55) the following holds:

$$|a_1| \geq |a_2| \qquad (8.60)$$

Using the Euclid algorithm we can write

$$GDIV(a_1, a_2) = GDIV(\mathbf{Mod}(a_1, a_2), a_2) \qquad (8.61)$$

where $\mathbf{Mod}(a_1, a_2)$ is the remainder after the integer division of a_1 by a_2. The result of the integer division is denoted by $\left\lfloor \dfrac{a_1}{a_2} \right\rfloor$. For example, for

$a_1 = 21, \quad a_2 = 18$, **Mod**$(21, 18)=3$, because $\left\lfloor \dfrac{21}{18} \right\rfloor = 1$ with the remainder 3. The

following inequality is always fulfilled:

$$|\mathbf{Mod}(a_1, a_2)| < |a_2| \qquad (8.62)$$

if Equation (8.60) holds. By substituting in Equation (8.55) the following unknowns,

$$
\begin{aligned}
x_1 &= \xi_1 \\
x_2 &= -\left\lfloor \frac{a_1}{a_2} \right\rfloor \xi_1 + \xi_2
\end{aligned}
\qquad (8.63)
$$

we obtain

$$a_1 \xi_1 + a_2 \left(\xi_2 - \left\lfloor \frac{a_1}{a_2} \right\rfloor \xi_1 \right) = b \qquad (8.64)$$

$$\left(a_1 - a_2 \left\lfloor \frac{a_1}{a_2} \right\rfloor \right) \xi_1 + a_2 \xi_2 = b \qquad (8.65)$$

$$\mathbf{Mod}(a_1, a_2)\xi_1 + a_2 \xi_2 = b \qquad (8.66)$$

Applying Theorem 8.4, the necessary and sufficient condition for an integer solution of Equation (8.66) is divisibility of b by $GDIV(\mathbf{Mod}(a_1, a_2)), a_2$, which is equivalent to the divisibility of b by $GDIV(a_1, a_2)$ due to (8.61). Therefore, if Equation (8.66) has an integer solution then Equation (8.55) does have, too. Between unknowns $\xi_1 (\xi_2)$ and $x_1 (x_2)$ there is an injection relation. Due to the important fact, namely $|\mathbf{Mod}(a_1, a_2)| < |a_2|$, repeating transformation at Equation (8.63) we either find that an integer solution does not exist, or obtain a unit coefficient of some unknown. In the latter case we multiply the transformed equation by a suitable integer and subtract it from another equation following the Gauss reduction technique.

The backward transformation from Equation (8.63) to the explicitly given unknowns $x_1, x_2, ..., x_n$ is unambiguously executable, as we can see from

$$
\begin{aligned}
\xi_1 &= x_1 \\
\xi_2 &= \left\lfloor \frac{a_1}{a_2} \right\rfloor x_1 + x_2
\end{aligned}
\qquad (8.67)
$$

It is reasonable to start the solution of Equation (8.59) with the equation containing the least coefficient. Let this coefficient in system at Equation (8.59) be a_{ij}. For such a general case we have the following transformation:

$$
\begin{pmatrix} x_1 \\ x_2 \\ \cdot \\ \cdot \\ \cdot \\ x_j \\ \cdot \\ \cdot \\ \cdot \\ x_m \end{pmatrix} = \begin{pmatrix} 1 & 0 & 0 & \cdot & \cdot & \cdot & \cdot\cdot\cdot\cdot & & & 0 \\ 0 & 1 & 0 & \cdot & \cdot & \cdot & \cdot\cdot\cdot\cdot & & & 0 \\ 0 & 0 & 1 & 0 & \cdot & \cdot & \cdot\cdot\cdot\cdot & & & 0 \\ \cdot & \cdot & \cdot & \cdot & \cdot & \cdot & \cdot & \cdot & & \cdot \\ 0 & \cdot & 0 & 1 & 0 & & \cdot\cdot\cdot\cdot & & & 0 \\ -\left\lfloor\dfrac{a_{i1}}{a_{ij}}\right\rfloor & \cdot & \cdot & -\left\lfloor\dfrac{a_{i(j-1)}}{a_{ij}}\right\rfloor & 1 & -\left\lfloor\dfrac{a_{i(j+1)}}{a_{ij}}\right\rfloor & \cdot\cdot\cdot\cdot & & & -\left\lfloor\dfrac{a_{in}}{a_{ij}}\right\rfloor \\ 0 & 0 & \cdot & \cdot & 0 & 1 & 0 & \cdot\cdot\cdot & & 0 \\ \cdot & \cdot & & & \cdot & \cdot & \cdot & \cdot & & \cdot \\ \cdot & & & & & & \cdot & \cdot\cdot & 0 & 1 & 0 \\ 0 & \cdot & \cdot & & \cdot & & & \cdot\cdot\cdot & 0 & 1 \end{pmatrix} \begin{pmatrix} \xi_1 \\ \xi_2 \\ \cdot \\ \cdot \\ \cdot \\ \xi_j \\ \cdot \\ \cdot \\ \cdot \\ \xi_m \end{pmatrix}
$$

(8.68)

The coefficients of the i-th equation of the transformed system are

$$
a'_{ip} = a_{ip} - a_{ij}\left\lfloor\frac{a_{ip}}{a_{ij}}\right\rfloor = \mathbf{Mod}\left(a_{ip}, a_{ij}\right) \quad \text{for} \quad p = 1,2,...,n \tag{8.69}
$$

and

$$
\left|a'_{ip}\right| < \left|a_{ij}\right| \tag{8.70}
$$

The reader can see that the divisibility conditions are similar to before and the value of coefficients in the equations can be lowered step-by-step to one, if of course an integer solution exists. As far as this is achieved, the equation is used for the reduction of the number of unknowns.

It can be shown that a reverse transformation of step-by-step (8.68) leads unambiguously back to the original unknowns of the original system of the equations. In terms of the algebraic structure theory the described method applies only operations over the ring of integers. Division of the integer numbers is not defined there. Such a division requires the rational numbers.

Interested reader can find more about the method in Abel (1990) containing further references. The described method is clearer with the help of a simple example.

Example 8.9. Find the integer solutions of

$$3x_1 - 2x_2 = -5 \tag{8.71}$$

There are two unknowns in just one equation. Evidently,

$$GDIV(a_{11}, a_{12}) = GDIV(3, -2) = 1 \tag{8.72}$$

The right-hand side of Equation (8.71) is divisible by 1, hence an integer solution exists. The transformed equation is

$$\mathbf{Mod}(3, -2)\,\xi_1 + (-2)\,\xi_2 = -5 \tag{8.73}$$
$$1\,\xi_1 - 2\,\xi_2 = -5 \quad \Rightarrow \quad \xi_1 = 2\,\xi_2 - 5 \tag{8.74}$$

Let ξ_2 be a free unknown. Its value can be a number k, then

$$\xi_2 = k \quad \text{and} \quad \xi_1 = 2k - 5 \tag{8.75}$$

Returning to the original unknowns we have

$$
\begin{aligned}
x_1 &= \xi_1 & x_1 &= 2k - 5 \\
x_2 &= \xi_2 - \left\lfloor \frac{a_{11}}{a_{12}} \right\rfloor \xi_1 & x_2 &= k + 1(2k - 5) = 3k - 5
\end{aligned}
\tag{8.76}
$$

The integer solution is illustrated in Table 8.6.

Table 8.6. Integer solutions of equation: $3x_1 - 2x_2 = -5$

x_i \ k	-1	0	1	2	3
x_1	-7	-5	-3	-1	+1
x_2	-8	-5	-2	+1	+4

Example 8.10. Two equations ($m=2$) in three unknowns ($n=3$) are given:

$$
\begin{aligned}
3x_1 + 2x_2 - 4x_3 &= 0 \\
-2x_1 + 5x_2 + 3x_3 &= 0
\end{aligned}
\tag{8.77}
$$

We are searching integer solutions of Equation (8.77). The rank of the coefficient matrix $r = 2$, $r = m < n$, i.e., the integer value of one unknown, can be chosen arbitrarily and the values of the remaining two are determined. Choose coefficient $a_{ij} = a_{12}, i = 1, j = 2$. According to Equation (8.68) the transformation equation is

$$\begin{pmatrix} x_1 \\ x_2 \\ x_3 \end{pmatrix} = \begin{pmatrix} 1 & 0 & 0 \\ -1 & 1 & 2 \\ 0 & 0 & 1 \end{pmatrix} \begin{pmatrix} \xi_1 \\ \xi_2 \\ \xi_3 \end{pmatrix} \tag{8.78}$$

The second row of the transformation matrix is

$$-\left\lfloor \frac{a_{11}}{a_{12}} \right\rfloor = -\left\lfloor \frac{3}{2} \right\rfloor = -1; \qquad 1; \qquad -\left\lfloor \frac{a_{13}}{a_{12}} \right\rfloor = -\left\lfloor \frac{-4}{2} \right\rfloor = -(-2) = 2 \tag{8.79}$$

and

$$\begin{aligned} x_1 &= \xi_1 \\ x_2 &= -\xi_1 + \xi_2 + 2\xi_3 \\ x_3 &= \xi_3 \end{aligned} \tag{8.80}$$

And by substituon into Equation (8.77)

$$\begin{aligned} \xi_1 + 2\xi_2 \qquad &= 0 \\ -7\xi_1 + 5\xi_2 + 13\xi_3 &= 0 \end{aligned} \tag{8.81}$$

Multiplying the first equation by 7 and adding it to the second one yields

$$19\xi_2 + 13\xi_3 = 0 \tag{8.82}$$

or

$$-13\xi_3 = 19\xi_2 \quad \Rightarrow \quad \xi_3 = -\frac{19}{13}\xi_2 \tag{8.83}$$

When choosing

$$\xi_2 = -13k, \quad k \in I \tag{8.84}$$

the solution is kept in the integer domain, as it is evident from

$$\xi_3 = -\frac{19}{13}(-13k) = 19k \tag{8.85}$$

The last unknown will be

$$\xi_1 = -2\xi_2 = -2(-13k) = 26k \qquad (8.86)$$

Returning to the original unknowns, all integer solutions of the equation system at Equation (8.77) are obtained as follows:

$$
\begin{aligned}
x_1 &= \xi_1 = 26k \\
x_2 &= -\xi_1 + \xi_2 + 2\xi_3 = -39k + 38k = -k \\
x_3 &= 19k
\end{aligned}
\qquad (8.87)
$$

where $k \in I$.

8.11 Concurrency and Conflict

The notions "concurrency" and "parallelism" are frequently used in the DEDS. They are related to the time evolution of the system events. In such considerations a certain time scale is necessary. There can be troubles with time relations if systems are distributed in space because of the physical relativity phenomena. One can imagine the problems when time synchronization signals are transmitted over long distance among individual components of a system.

Petri nets as a tool for the DEDS representation can reflect the considered notions in some way. Let us introduce the following definition coping with the problem.

Definition 8.16. Consider Petri net $PN = (P,T,F,W,M_0)$. Let $S \subseteq T$ be a subset of its transitions with cardinality greater than 1 and $\mathbf{m}$ be a reachable marking in PN. S is called the concurrent subset of the transitions at $\mathbf{m}$ if

$$\mathbf{s}^- = \sum_{t \in S} \mathbf{t}^- \leq \mathbf{m} \qquad (8.88)$$

Vectors $\mathbf{t}^-$ has been defined in Section 7.2. The meaning of the definition will be illustrated in the following example.

Example 8.11. Consider a Petri net depicted in Figure 8.25 with the initial marking specified. Let us analyze whether $S_1 = \{t_1, t_2\}$ is a concurrent subset of the transitions by $\mathbf{m}_0$. We have

$$\mathbf{t_1^-} = \begin{pmatrix} 1 \\ 1 \\ 0 \\ 0 \\ 0 \end{pmatrix}, \quad \mathbf{t_2^-} = \begin{pmatrix} 0 \\ 0 \\ 1 \\ 0 \\ 0 \end{pmatrix} \tag{8.89}$$

$$\mathbf{s_1^-} = \sum_{t \in S_1} \mathbf{t^-} = \begin{pmatrix} 1 \\ 1 \\ 0 \\ 0 \\ 0 \end{pmatrix} + \begin{pmatrix} 0 \\ 0 \\ 1 \\ 0 \\ 0 \end{pmatrix} = \begin{pmatrix} 1 \\ 1 \\ 1 \\ 0 \\ 0 \end{pmatrix} \le \begin{pmatrix} 1 \\ 1 \\ 1 \\ 0 \\ 0 \end{pmatrix} = \mathbf{m_0} \tag{8.90}$$

It can be concluded that S_1 is a concurrent subset of the transitions at $\mathbf{m_0}$. On the other hand,

$$S_2 = \{t_1, t_2, t_3\} \tag{8.91}$$

is not a concurrent subset of the transitions at $\mathbf{m_0}$ because

$$\mathbf{s_2^-} = \begin{pmatrix} 1 \\ 1 \\ 0 \\ 0 \\ 0 \end{pmatrix} + \begin{pmatrix} 0 \\ 0 \\ 1 \\ 0 \\ 0 \end{pmatrix} + \begin{pmatrix} 0 \\ 0 \\ 1 \\ 0 \\ 0 \end{pmatrix} = \begin{pmatrix} 1 \\ 1 \\ 2 \\ 0 \\ 0 \end{pmatrix} > \begin{pmatrix} 1 \\ 1 \\ 1 \\ 0 \\ 0 \end{pmatrix} \tag{8.92}$$

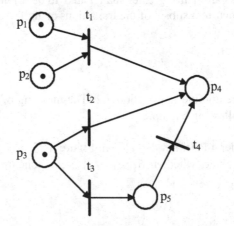

Figure 8.25. Petri net for concurrency analysis

The main idea of Definition 8.16 is that a marking, as $\mathbf{m}_0$ in the example, should have so many tokens in the pre-places of transitions belonging to set S_1 that all transitions in S_1 can fire simultaneously. In other words, firing a transition from S_1 does not influence the possibility to fire any of the remaining transitions in it. We remember that a standard way of the Petri net modeling technique is based on the assumption that at a time point just one transition fires. When considering concurrency, the possibility of simultaneous firing is investigated. Obviously, the modeling can be based on the assumption of simultaneous transition firing but the risk of conflicts and behavior complexity grow considerably.

The concurrency is closely related to the possibility of arbitrary order of the firing, which is dealt with in the following theorems.

Theorem 8.5. Consider a Petri net $PN = (P,T,F,W,M_0)$. Let S be a concurrent subset of transitions at a reachable marking $\mathbf{m}$. Let a firing sequence $\tilde{\sigma}$ start in $\mathbf{m}$ and contain any transition of S just once. Then the order of transition firing is arbitrary.

Proof. Concurrency of transitions in S enables simultaneous firing of all transitions at $\mathbf{m}$, *i.e.*, for a single firing of transitions from S there are enough tokens in the pre-places of transitions belonging to S, regardless of a firing order.

A slightly more complicated situation is with the reverse theorem.

Theorem 8.6. Consider a pure Petri net $PN = (P,T,F,W,M_0)$. If each transition of a subset S is fireable once in an arbitrary ordered firing sequence beginning in $\mathbf{m}$ and containing just transitions of S (the order of the transition firing is arbitrary), then S is a concurrent subset of transitions at $\mathbf{m}$.

The proof of the theorem can be found in Starke (1990). The role of the Petri net purity can be illustrated by a simple Petri net in Figure 8.26. Transitions t_1 and t_2 can fire in an arbitrary order by $\mathbf{m}_0$. However, despite this, the subset $S = \{t_1, t_2\}$ does not meet the requirement of Definition 8.16, as easily checked:

$$\mathbf{s}^- = \mathbf{t}_1^- + \mathbf{t}_2^- = (1) + (1) = (2) > \mathbf{m}_0 = (1) \tag{8.93}$$

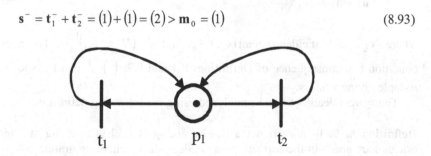

Figure 8.26. Non-pure Petri net and concurrency

It is also evident that both t_1 and t_2 cannot fire simultaneously. The next definition deals with the concurrency and conflict.

Definition 8.17. Consider a Petri net $PN = (P,T,F,W,M_0)$. A conflict is said to be in a subset S of the PN transitions at a reachable marking m if all transitions in S are fireable at m but S is not a concurrent subset of the transitions at m. In particular, two transitions t_i, $t_j \in T$ of PN are said to be in the conflict at the reachable marking m if there is a conflict in the subset $S=\{t_i, t_j\}$.

Definition 8.18. A Petri net $PN = (P,T,F,W,M_0)$ is said to be conflictless iff there is no such a reachable marking at which two transitions of PN are in conflict.

8.12 Analysis of Petri Net Properties

Several properties of Petri nets can be analyzed using the reachability graph or the coverability graph. Using them may be different. The difference will be treated. Next, we consider a particular Petri net $PN = (P,T,F,W,M_0)$ to be analyzed.

Marking Reachability

Let a marking represented by a vector **m** be given. Its dimension is $|P|=n$. Consider a bounded Petri net PN so that the reachability graph can be constructed. The vector **m** is not reachable if there is no node corresponding to **m** in the reachability graph. If such a node exists, **m** is reachable.

If the considered PN is unbounded, the coverability graph can be constructed instead the reachability graph. In the coverability graph a node covering a given **m** is searched. Its existence is a necessary but not a sufficient condition for the reachability of **m**. The strings leading from $\mathbf{m}_0$ to the nodes covering **m** should be analyzed and checked whether **m** is contained in some of them.

A necessary condition for **m** to be reachable is the existence of an integer solution of the linear algebraic system of equations

$$\mathbf{m} - \mathbf{m}_0 = \mathbf{N}_\Delta \mathbf{x} \tag{8.94}$$

where $\mathbf{N}_\Delta$ is a Δ-incidence matrix of PN and $\mathbf{x} \in (N^+ \cup \{0\})^{|T|T}$. The necessary condition is a consequence of Definitions 8.1 and 8.7. $(\ )^{|T|T}$ is a transposition of m-tuple giving vectors.

The system deadlock can be analyzed using the following definition.

Definition 8.19. In a Petri net a dead marking is such a marking at which no oriented arc goes out the corresponding node in the reachability graph.

Boundedness

Very often, the inspection of a given Petri net and a simulation of transition firing by means of a Petri net graphical editor reveals the Petri net boundedness.

Another possibility is to use the reachability graph. A Petri net is bounded if the reachability graph exists, *i.e.,* if the number of nodes does not grow to infinity on constructing the graph. The coverability graph construction algorithm can be directly used. In such a case if the Petri net is bounded, the result will be the reachability graph. Otherwise the coverability graph is obtained and the Petri net is unbounded.

Liveness

A Petri net *PN* is *L0*-live or dead if there exists a transition $t \in T$ that does not occur as a label of any arc of *PN*'s reachability or coverability graph. Transition *t* is *L1*-live if it appears as a label of at least one arc in the reachability or coverability graphs. It is *L2*-live if in the reachability or coverability graphs there exists an oriented path $a_{i_1} a_{i_2} \dots a_{i_k}$, containing at least two arcs labeled with *t*. *t* is *L3*-live if *t* is a label of an arc in the reachability or coverability graphs and the arc is an element of a cycle.

The *L4*-livenes is more complicated. It can be resolved by means of the strong connectivity of the reachability or coverability graphs. First, a sufficient condition for the *L4*-livenes can be formulated as a theorem.

Theorem 8.7. Consider a Petri net *PN*. If either its reachability or coverability graph is strongly connected and *PN* is *L1*-live then *PN* is *L4*-live.

Proof. In the strongly connected reachability or coverability graph each pair of nodes is connected by an oriented path in both directions. Each transition occurs at least once in the graphs because *PN* is *L1*-live. There is always at least one oriented paths going out of each reachable marking and reaching the initial marking and from there continuing with paths, in which according to *L1*-liveness there are all transitions as labels at least once, as it is required by the *L4*-liveness.

It is not necessary that the whole reachability or coverability graph be strongly connected as shown in the following theorem. The notion of the graph strong component was treated in Chapter 2. The extended notion of the sink strong component used in the theorem denotes such a strong component for which no arc of its component nodes goes out to a node not belonging to it.

Theorem 8.8. A Petri net *PN* is live (*i.e., L4*-live) iff there is at least one sink strong component in the reachability or coverability graph of *PN* ,and all sink strong components are *L1*-live.

Proof. The *L1*-liveness of the strong component is defined analogously as that of the whole Petri net. A slight difference consists in that in case of the *L1*-liveness of the strong component, for each transition *t* there is a firing sequence going out of a node of the component and containing transition t as a label. Then, considerations of the proof of Theorem 8.7 apply.

Obviously, all results in this section hold for Petri nets with capacities, too. In order to diversify provided examples, the following one uses this kind of Petri net.

Example 8.12. An illustrative example from Abel (1990) is used to show the point of the last two theorems. The analyzed Petri net is shown in Figure 8.27. Its reachability graph is in Figure 8.28.

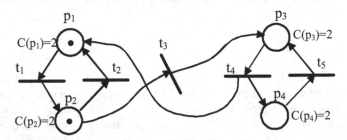

Figure 8.27. Petri net with capacities illustrating meaning of strong components for the liveness

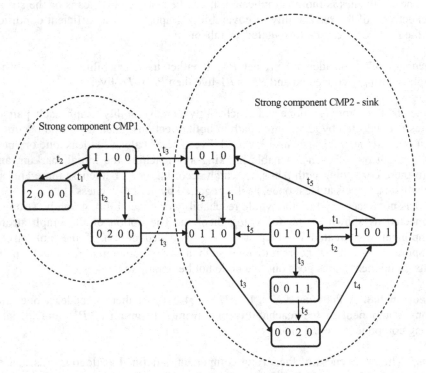

Figure 8.28. One $L1$-live sink strong component in the reachability graph for Petri net in Figure 8.27

Reversibility

In terms of reachability/coverability graph, a Petri net is reversible if from each node of the graph a directed path exists ending in the initial node. The reversibility and connectivity are dealt with in the following theorem.

Theorem 8.9. A Petri net is reversible if its reachability or coverability graph is strongly connected.

Proof. The theorem is a direct result of the graph strong connectivity property. On the other hand, an unbounded Petri net can be reversible even if its coverability graph is not strongly connected. We refer to Example 8.6 with Figure 8.15 in Section 8.7 illustrating the case. In such cases reversibility can be analyzed analogously as liveness using the properties of the sink strong components of the coverability graph.

A necessary and sufficient condition for reversibility is formulated in the following theorem.

Theorem 8.10. A bounded Petri net is reversible iff its reachability graph is strongly connected.

Proof. If a bounded Petri net is reversible, then from each node of its reachability graph a firing sequence leads to the initial marking $\mathbf{m}_0$ and from it to each reachable marking. A consequence of this is the strong connectivity of the reachability graph. The inverse implication is dealt with in the Theorem 8.9.

Finally, an important relation of the main Petri net properties is given next.

Theorem 8.11. All three Petri net properties, *i.e.*, liveness, boundedness and reversibility, are mutually independent.

Proof. The proof through counterexamples is applied. Some examples showing independence have already been introduced. See live, unbounded, non-reversible *PN* in Figure 8.5; non-live, unbounded, non-reversible *PN* in Figure 8.10; non-live, bounded, non-reversible *PN* in Figure 8.12; live, unbounded, reversible *PN* in Figure 8.14; *etc.*

8.13 Structural Properties

The main properties of Petri nets were analyzed in the previous section. However, it is necessary to mention briefly basic structural properties of Petri nets. A structural property does not depend on the initial marking. Let us define the Petri net structure for this purpose.

Definition 8.20. Consider a Petri net $PN = (P,T,F,W,M_0)$. The structure of the Petri net PN is given by

$$PS = (P,T,F,W) \tag{8.95}$$

Some chosen structural properties based on PS only are listed below.

Structural Liveness

PN is said to be structurally live if there exists an initial marking M_0 at which PN is live.

Structural Boundedness

PN is said to be structurally bounded if it is bounded given any initial marking M_0.

Structural Conservativeness

PN is structurally conservative if for any initial marking, PN is conservative with respect to a vector $\mathbf{v}_i$.

Siphons and Traps

Siphons and traps are two important structural objects in a PN and closely related to the Petri net properties, especially deadlock and liveness. Before their definition, the following notation is introduced. The pre-set of a place p, denoted as $^\bullet p$, is the set of p's input transitions, $i.e.$, $^\bullet p = \{t \in T, O(p, t) \neq 0\}$ formally. Its post-set is $p^\bullet = \{t \in T, I(p, t) \neq 0\}$. Consider a set of non-empty places $S \subseteq P$. Its pre-set is $^\bullet S = \bigcup_{p \in S} {}^\bullet p$ and and post-set $S^\bullet = \bigcup_{p \in S} p^\bullet$.

Definition 8.21. A set of place $S \subseteq P$ is called a siphon if $^\bullet S \subseteq S^\bullet$. It is a trap if $S^\bullet \subseteq {}^\bullet S$.

Their physical meanings are explained as follows. A siphon can keep or lose its tokens during any transition firing. Once it loses all tokens, it remains empty and thus disables all of its output transitions. An empty siphon is, therefore, the cause of partial or complete deadlock. A trap can keep or gain tokens during any transition firings. Once it receives tokens or is marked, it remains marked regardless which transition fires.

Example 8.13. Consider the Petri net in Figure 7.18 and $S = \{p_{M1}, p_{M2}, p_{A2}, p_{B2}\}$. It is easy to find that:

$$^\bullet p_{M1} = \{t_{A2}, t_{B3}\}, \quad {}^\bullet p_{M2} = \{t_{A3}, t_{B2}\}, \quad {}^\bullet p_{A2} = \{t_{A2}\}, \quad {}^\bullet p_{B2} = \{t_{B2}\}$$

$$p_{M1}{}^\bullet=\{\,t_{A1},\,t_{B2}\},\ p_{M2}{}^\bullet=\{\,t_{A2},\,t_{B1}\},p_{A2}{}^\bullet=\{\,t_{A3}\,\},p_{B2}{}^\bullet=\{\,t_{B3}\}$$

Thus

$${}^\bullet S=\{\,t_{A2},\,t_{A3},\,t_{B2},\,t_{B3}\}\ \text{and}\ S^\bullet=\{\,t_{A1},\,t_{A2},\,t_{A3},\,t_{B1},\,t_{B2},\,t_{B3}\}.$$

Clearly, ${}^\bullet S{\subset}S^\bullet$. Hence, S is a siphon. Initially it is marked with two tokens. Starting at the initial marking in Figure 7.18, after firing transitions t_{A1} and t_{B1} respectively, S is empty and the net enters a deadlock marking.

Now consider $S=\{p_{M1},\,p_{A1},\,p_{B2}\}$. We can easily find that ${}^\bullet S=S^\bullet=\{\,t_{A1},\,t_{A2},\,t_{B2},\,t_{B3}\}$. Hence it is a siphon and trap as well. It is initially marked with a token and remains so regardless of marking evolution.

A siphon is minimal iff it contains no other siphons as its proper subset. A minimal siphon is strict if it contains no marked trap. A strict minimal siphon may become empty during the marking evolution. Hence, to make such net live is control such siphons so that they are never empty. A P-invariant-based control method can be developed to achieve this purpose. By adding a control place (called monitor), these siphons can be well controlled (Ezpeleta *et al*. 1995). Unfortunately, the number of such siphons grows exponentially with the size of a Petri net and thus leads to very complex control structure for a sizable system. To reduce the control complexity, Li and Zhou (2004, 2006) invented the concept of elementary siphons whose control can prevent all other siphons from being emptied. They number is bounded by the smaller of $|P|$ and $|T|$.

There are other structural properties related to non-structural ones studied by many researchers. Often, they are studied in relation to a particular Petri net class, *e.g.,* marked graphs, free-choice nets, assembly Petri nets, disassembly Petri nets, augmented marked graphs, and production Petri nets. A very good systematic treatment of them can be found in an excellent tutorial paper written by Murata (1989) and books such as Zhou and Venkatesh (1998).

8.14 Problems and Exercises

8.1. A pure Petri net is given by the vectors

$$\mathbf{t}_1=\begin{pmatrix}-1\\1\\1\\1\\0\\0\end{pmatrix},\ \mathbf{t}_2=\begin{pmatrix}0\\-1\\-1\\0\\1\\0\end{pmatrix},\ \mathbf{t}_3=\begin{pmatrix}0\\0\\-1\\-1\\0\\1\end{pmatrix},\ \mathbf{t}_4=\begin{pmatrix}1\\0\\0\\-1\\-1\\0\end{pmatrix},\ \mathbf{t}_5=\begin{pmatrix}1\\-1\\0\\0\\0\\-1\end{pmatrix}$$

corresponding to transitions t_1–t_5. The initial marking is $\mathbf{m}_0=(1\,0\,0\,0\,0\,0)^{\mathrm{T}}$.

Represent the Petri net in the graphic form. Using vector representation determine if t_1 is fireable at $\mathbf{m}_0$. Construct the reachability graph. Analyze the Petri net properties: boundedness, liveness and reversibility.

8.2. A Petri net is depicted in Figure 8.29. Draw the coverability graph for it.

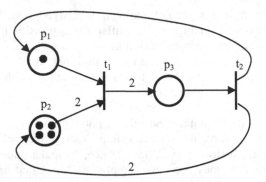

Figure 8.29. A Petri net the coverability graph to be drawn for Exercise 8.2

8.3. Analyze basic properties of the following Petri net in Figure 8.30.

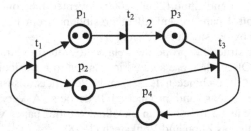

Figure 8.30. A Petri net for Exercise 8.3

What graph is it possible to construct: the reachability or coverability one? Use Theorem 8.2 to show whether the Petri net is unbounded.

8.4. A Petri net is given in Figure 8.31. Draw the reachability graph for it and using the graph determine its following properties: boundedness, liveness and reversibility. Find a P–invariant for the given Petri net.

8.5. Consider a computer processor with an input buffer having capacity 1 for waiting task to be processed by the processor. If a task requires processing, either the buffer is free and the task is put into it or if the buffer is occupied the task is refused. If a task is in the buffer and the processor is free, the task is moved from the buffer to the processor. Only one task can be processed in the processor. After processing, the task is removed from the processor and the processor is free.

- Describe the behavior of the specified buffer-processor system using a deterministic finite automaton.

- Describe the same with a Petri net and compare both representations.
- Modify the Petri net for the case when the buffer capacity is two.
- Analyze the properties of both Petri nets.

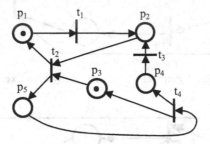

Figure 8.31. A Petri net for Exercise 8.4

8.6. The Petri net in Figure 8.32 has the initial marking $\mathbf{m}_0 = (0 \ 0 \ 1 \ 1)^T$. Find a concurrent subset of transitions at $\mathbf{m} = (2 \ 2 \ 1 \ 0)^T$ where the number of elements in the subset is greater than 1. Prove the concurrency using vectors.
    Is there a conflict in the Petri net at some reachable marking?

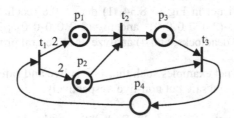

Figure 8.32. A Petri net for concurrency analysis

8.7. Is it possible that a Petri net live at level 4 is not reversible? Find a counter-example.

8.8. Prove that the existence of T-invariant is a necessary condition for a Petri net to be reversible.

8.9. Derive all the structural properties of the Petri net in Figure 8.31.

8.10. Derive all the structural properties of the Petri net in Figure 8.32.

8.11. Given the Petri net in Figure 8.33, 1) derive the reachability graphs when initial marking is $m_0=(2 \ 0 \ 0 \ 0)^T$ and $m_0=(2 \ 1 \ 0 \ 0)^T$; 2) derive P and T-invariants; 3) find all deadlocks; and 4) analyze its structural properties.

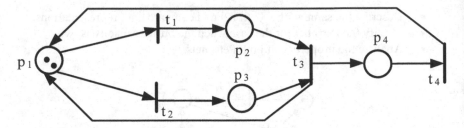

Figure 8.33. A Petri net for property analysis in Exercise 8.11

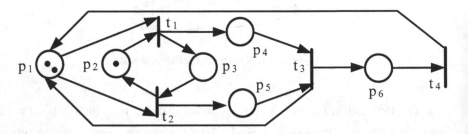

Figure 8.34. A Petri net for property analysis in Exercise 8.12

8.12. Given the Petri net in Figure 8.34, 1) derive the reachability graphs when initial marking is $m_0 = (2\ 1\ 0\ 0\ 0\ 0)^T$ and $m_0 = (2\ 2\ 0\ 0\ 0\ 0)^T$; 2) derive P and T-invariants; 3) find all deadlocks; and 4) analyze its structural properties.

8.13. Construct Petri net examples such that 1) it is live and safe but non-reversible; and 2) it is reversible and safe but non-live, respectively.

8.14. Given the Petri net in Figure 8.33, find all the minimal siphons.

8.15. Given the Petri net in Figure 8.34, find all the minimal siphons. Define a minimum trap as one that contains no trap as its proper set. Find all the minmim traps for the net in Figure 8.34.

8.16. Given the Petri net in Figure 8.33, when the net evovles to a deadlock, *e.g.*, $(0, 2, 0, 0)^T$, prove that $\{p_1, p_3, p_4\}$ is a siphon.

8.17. Prove that given any Petri net with an initial marking, at any deadlock marking, all the places with no token form a siphon.

9

Grafcet

9.1 Basic Grafcet Components

Grafcet is designed as a specification tool for logic control to be implemented preferably on programmable logic controllers (PLC). It is a tool related closely to the binary safe Petri nets interpreted for control (Section 7.5). The marking of such Petri nets is formally expressed by

$$M_k(p_i) = v, \quad v \in \{0,1\} \ \forall M_k \in R_{PN}(\mathbf{m}_0), \ \forall p_i \in P \tag{9.1}$$

Equation (9.1) results in the weights of the Petri net given by $W : F \to \{1\}$, *i.e.*, the weights are units.

The syntax of Grafcet components and elements has been precisely elaborated in order to support effective and correct implementation of the control policy into final control programs. In that context the position of the Grafcet is similar to that of the finite automata, Petri nets, state charts, *etc.*, being a tool standing between the system control requirements and the instruction codes realizing the control programs in the used hardware environment.

Like the Petri nets interpreted for control, Grafcet may be viewed as an extension of standard Petri nets defined in Definition 7.2. Quoting David and Alla (1994), the extension makes it possible to describe not only what "happens" but also "when it happens".

A series of international standards like IEC 848, ISO 7185 establishes concepts and guidelines for PLC recommended properties and programming technology. The standardization efforts in this field resulted in a complex standard IEC 61131, which supports design of industrial automation systems using programmable logic controllers both in hardware and software aspects. Basic features of hardware and software automation means are specified in the standard IEC 61131. Grafcet belongs to the tools keeping in line with concepts and ideas of that standard.

IEC 61131, part 3, provides three textual PLC programming languages and three graphical ones (John and Tiegelkamp 2001). The Sequential Function Chart graphical programming language is in essence close to Grafcet, while it has several additional language constructs (*e.g.,* sequence, loop or divergent path with user-defined priority) and it is framed into broader structural context with other PLC programming languages. We refer readers for more details to an excellent book (John and Tiegelkamp 2001).

The aim of this chapter is to follow up concepts of the Petri nets interpreted for control, which are melted into Grafcet, and not all PLC programming languages used in practice.

It is possible to define a class of Petri nets, which corresponds to a set of Grafcet models. The class is characterized as Petri nets with capacities and weights equal to one, and the binary initial marking. Firing rules of the Petri net class corresponding to Grafcet models should have been slightly adapted in order to ensure the correspondence.

We emphasize that Grafcet is intended for the DEDS control specification within the structure of Figure 4.1, enabling one to consider the system inputs and outputs, synchronization of events by external inputs and generation of output control commands. Particular graphic models, which are put together by means of Grafcet are called grafcets and written with the small letter. A grafcet is a simple labeled oriented mathematical graph with two disjunctive sets of nodes: steps and transitions, *i.e.,* it is a bipartite oriented graph.

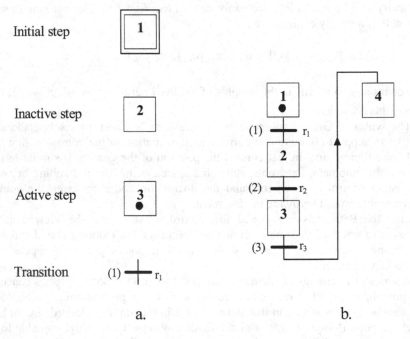

Figure 9.1. Grafcet steps, transitions, and oriented arcs

Steps in a grafcet are connected with the transitions and *vice versa* transitions with steps *via* oriented arcs. Steps correspond to Petri net places and Grafcet transitions to Petri net transitions. A step can be active or inactive. An active step indicates some partial situation or state in a system and a Grafcet transition, when fired, implies occurrence of an event. A step activity is marked by a token located in the step. The step is represented by a square, the initial step by a double square, and the transition by a short bar (Figure 9.1a). The initial step is automatically set active at the beginning of the system control based on Grafcet.

In Grafcet the arcs are oriented always from the top down; in such case arrows are not used. An arrow is added only to the bottom-up running part of an arc (Figure 9.1b). A layout of a simple sequence of steps and transitions is shown in Figure 9.1b.

A natural question arises about arcs joining or branching. Their syntax is ruled in the following way:

a.   Two or more arcs coming in a transition can be joined only by a double bar as shown in Figure 9.2a. The joining is called junction AND.
b.   An arc going out of a transition can be branched into arcs and go to steps only through a double bar as shown in Figure 9.2b. The branching is called distribution AND.

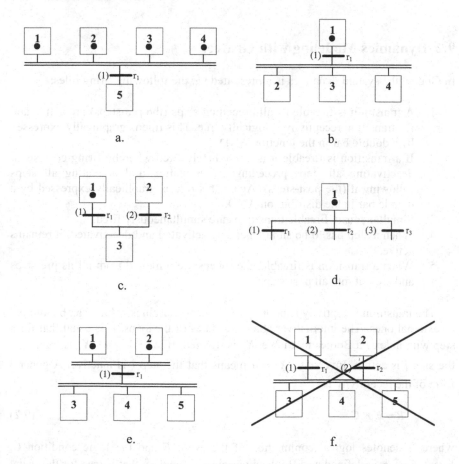

Figure 9.2a–f. Syntax of the Grafcet graphical components

    c.   Two or more arcs each coming out from a transition can join and go in a step as shown in Figure 9.2c. Such joining is called junction OR.

    d.   An arc going out of a step can branch into arcs going each in its transition as shown in Figure 9.2d. The branching is called distribution OR.

The combination of cases a and b from Figure 9.2a, b is possible, resulting in Figure 9.2e. It is to be underlined that component connections other than those presented in Figure 9.2a–d are not allowed. For example, the structure depicted in Figure 9.2f is not allowed.

Step and transition indexing is clear from Figure 9.2. Steps and transitions without input (output) arcs are allowed and are called source (sink) steps or source (sink) transitions.

A logic expression has to be associated with each transition. The expression is a Boolean variable or function. It is called the receptivity. If being true, it expresses that the transition firing condition is met. In Figure 9.2 the receptivity is denoted as r_i, i=1, 2 and 3.

9.2 Dynamics Modeling with Grafcet

In Grafcet the system dynamics is represented *via* the following firing rules:

1.   A transition is fireable iff all preceding steps (the pre-steps) are active and the transition receptivity is logically true. This rule is graphically expressed by a double bar in the junction AND.

2.   If a transition is fireable it is immediately fired whereby firing consists in deactivating all steps preceding the transition and activating all steps following it (the post-steps). Again this rule is graphically expressed by a double bar in the distribution AND.

3.   Simultaneously fireable transitions are simultaneously fired.

4.   When a step has to be simultaneously activated and deactivated, it remains active.

5.   When a transition is fireable, the tokens are removed from all its pre-steps and are put into all post-steps.

The transition receptivity is built up of Boolean variables, which can be internal or external ones. The internal variables are states of the steps. It is usual that for a step with index i a Boolean variable X_i is defined. If $X_i = 1$, *i.e.*, X_i is true, then the step i is active. Inversely, $X_i = 0$ means that the step i is inactive. A general form of the receptivity is

$$R = E \wedge C \tag{9.2}$$

where $\wedge$ denotes logical conjunction of the event E and the logic condition C. Event $E = \uparrow a$ is defined as a Boolean variable or function that is true for the rising edge of another Boolean variable a. The variable is true in a discrete time point, *i.e.*, for an infinitely short time interval. In other words, $\uparrow a = true$ iff a changes from 0 to 1. Similarly, $\downarrow a$ is related to the falling edge of an external logic variable or function. C in Equation (9.2) is a logic variable or function, which can be external or internal one with respect to a given Grafcet. Condition C may not contain events.

The distribution OR elementary structure (see Figure 9.2d) may bring about an indeterminism in the case when receptivities r_1, r_2, and r_3 are not mutually logic exclusive. PLC hardware implementations of the receptivities may not be simultaneous in spite of the designer's assumption they are when step 1 is active (Figure 9.2d) and, *e.g.*, receptivities r_1 and r_2 are true in "the same time". In such a case, despite of the assumption, hazardous dynamics causes the post-step of transition (1) can be activated, step 1 deactivated and post-step of transition (2) may not be active. Designers have to analyze hardware features in order to obtain the required behavior. A safer way is to use mutual logic receptivity exclusion in divergence OR.

Active influence on the controlled and control systems is modeled in the Grafcet by the so called actions. They are graphically represented by rectangles positioned to the right from the steps (Figure 9.3). There are two kinds of actions: level and impulse actions. An action is set if its associated step is active.

A level action is realized by means of a Boolean variable, *e.g.*, *switch_on_motor_M* where *switch_on_motor_M*=1 means that motor M is switched on; *switch_on_motor_M*=0 means that there is no signal for keeping motor M on; similarly *switch_off_motor_M*=1 represents a signal for motor switching off while *switch_off_motor_M*=0 means that there is no signal for motor switching off. The Boolean variable *switch_on_motor_M* is set to logic 1 if the corresponding step is active. The action for deactivation of switching has the variable $\overline{switch\_on\_motor\_M}$ (logic inversion).

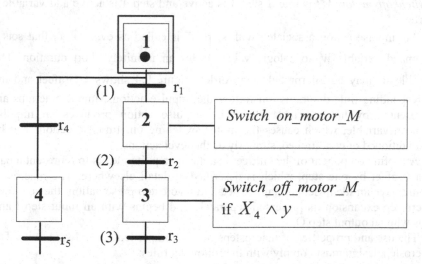

Figure 9.3. Level actions: unconditioned for step 2, conditioned for step 3

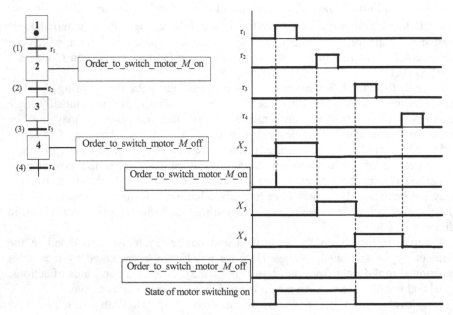

Figure 9.4. Impulse actions

Boolean variables associated with level actions are actually control commands that force either the external system to be controlled to some action by the Grafcet, or internally influence the control system itself *via* internal variables. The level actions may be unconditioned or conditioned by a Boolean variable or Boolean function. The use of the level actions is illustrated in Figure 9.3. The variable "*switch_off_motor_M*" is true if step 3 is active and step 4 is active and variable y is true.

An impulse action associated with step X_i is called as event $\uparrow X_i$ that sets a command variable w_i to a logic value w_i for an infinitely short duration. The variable w_i may be interpreted as an order. Figure 9.4 shows a Grafcet and the corresponding time diagrams illustrating the impulse actions. Impulse actions are represented by hatched rectangles. The impulse action produces an impulse Boolean variable, which causes the motor switching on. Impulse actions can be unconditioned or conditioned, similarly as the level actions.

A useful component of the Grafcet is a macrostep. Its idea is to represent a part of a Grafcet by one step, which is described in detail elsewhere. The macrostep should have an identification tag. A detailed macrostep presentation, the so-called macrostep expansion, is presented separately and begins with an input step I and ends with an output step O.

The use and properties of macrosteps depend on particular hardware units. The macrostep design must comply with the following rules:

1. The macrostep expansion contains just one input and one output step.
2. Each transition firing before a macrostep activates the input step of the macrostep expansion.
3. The output step of the macrostep expansion contributes in enabling the downstream transitions according to the Grafcet structure.
4. There are no arc connections between the macrostep expansion (of course with the exception of the input and output steps) and the rest of the grafcet

An example of a macrostep is given in Figure 9.5.
Grafcet considers time *via* time logic variable denoted as

$$v = t / i / \Delta \tag{9.3}$$

where t indicates a time variable, i refers to the logic variable X_i and Δ is a time interval. If $v = 0$ before X_i becomes 1, then after event $\uparrow X_i$ the interval Δ elapses and v changes from 0 to 1. If $v = 1$ and X_i changes from 0 to 1, v becomes 0 and, similarly as before, v will be 1 after the interval Δ. If after $\uparrow X_i$, event $\uparrow X_i$ repeats in time period shorter than Δ, time counting starts from the last event $\uparrow X_i$. Figure 9.6 illustrates the use of a time variable in a Grafcet. A Grafcet with a time variable $t / 4 / 6$ is given. The time unit is 1 s. Time diagrams for the grafcet in Figure 9.6 is given in Figure 9.7.

A situation when several transitions can be fired immediately one after another is called unstable. If a step is activated and after its activation the related transition is fireable, the situation is unstable. Such an iterated firing ends in a step whose deactivation depends on the next transition's receptivity that has not yet been true, or on activities of other pre-steps of this transition. Such a situation is stable. A level action can be realized only in the stable situation while an impulse action can be realized both in a stable as well as in an unstable situation.

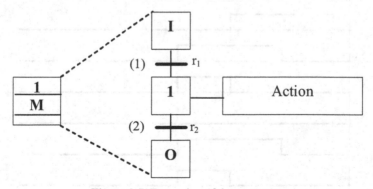

Figure 9.5. Expansion of the macrostep

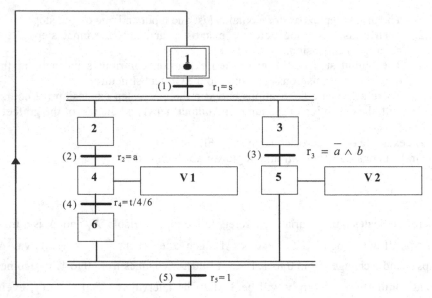

Figure 9.6. Work with time in Grafcet

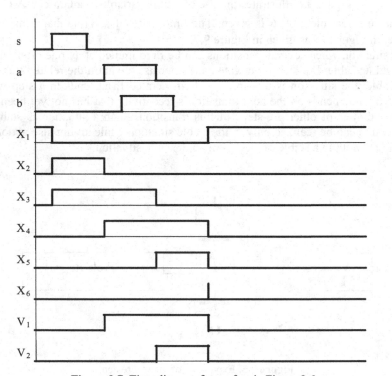

Figure 9.7. Time diagram for grafcet in Figure 9.6.

9.3 Comparison of Petri Nets and Grafcet

Petri nets interpreted for control (Section 7.4) and Grafcet have many common features. In fact, Grafcet may be viewed as being derived from the Petri nets. Both tools produce models that are bipartite oriented labeled mathematical graphs. Petri nets have places and transitions, while Grafcet has steps and transitions as graph nodes. Distributed and parallel activities are specified through markings and the system dynamics through transition firings subdued to firing rules.

Grafcet has a few specific properties differentiating it from Petri nets interpreted for control (*PNC*). The differences are as follows.

Marking of grafcets is a binary one whereas marking of *PNC* can be numerical. Firing rules in Grafcet are consequently subjected to the binary marking case. A step can only be active or inactive. Figure 9.8 illustrates how it works. Transition (1) is fireable and its firing only confirms activation of step 2 as shown in Figure 9.8b.

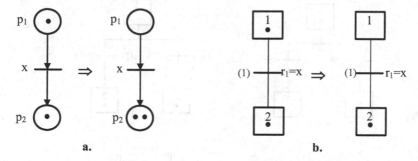

a.   b.

Figure 9.8a, b. The marking in Grafcet is strictly binary

Fireable transitions in Grafcet fire simultaneously while in the Petri nets fireable transitions can fire only one at a time. Figure 9.9 shows the marking development in *PNC* and in grafcets; x is a logic expression constituting a receptivity in grafcets or a logic firing condition in *PNC*. The marking result in Figure 9.9c is reached after firing t_1 and then t_2. Another possibility is firing t_2 first and then t_1.

Mutual relation of Petri nets and Grafcet is the following. If a Petri net is safe, then an equivalent Grafcet exists. On the other hand, it is not possible to represent every grafcet by an equivalent PNC according to Definiton 7.2. This is true if in a grafcet there is not any of the structures, which differentiate Grafcet from *PNC* (depicted in Figures 9.8 and 9.9), as then a PNC exists being equivalent to it. In such a case, all tools for Petri net analysis are applicable for Grafcet as well. Otherwise, the analysis tools are to be adapted for Grafcet. More comparisons between Petri nets and Grafcet can be found in Giua and DiCesare (1993) and Zhou and Twiss (1996).

Example 9.1. Figure 9.10 shows the crossing of cars and pedestrians. They all need to pass through the narrow part of the street. Either one car at a time can pass through the narrow part in one of the two directions, or pedestrians may cross the street there. The crossing control ensures a cyclic alteration of the car directions. Pedestrians are allowed to cross the street only after pushing the button on any side of the street when the time interval for car passing has expired.

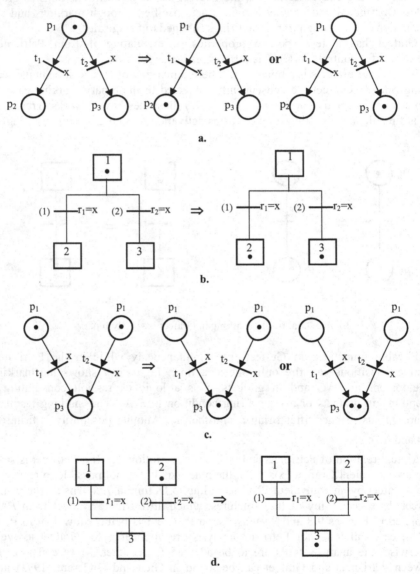

Figure 9.9. Comparison of the firing rules

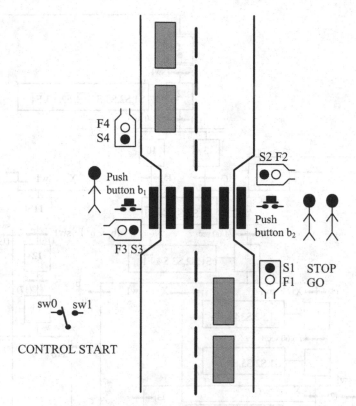

Figure 9.10. Situation on a crossing

Grafcet specifying the control of the crossing is shown in Figure 9.11. If a push button is pressed, the pedestrian is allowed to cross after the time interval for cars expires. Then a car can pass through. If sw1 is activated and the running time interval is finished, the system returns to the initial state. Other details are evident from the Grafcet.

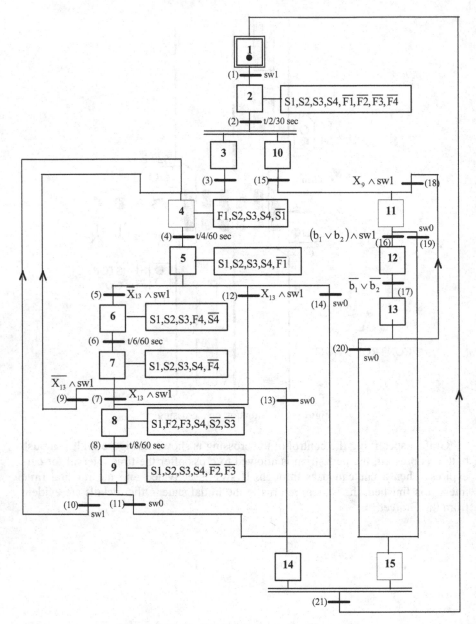

Figure 9.11. Grafcet for the example with crossing

Example 9.2. Control of the manufacturing cell in Figure 5.1 is written by a Grafcet in Figure 9.12. The reader can compare it with the Petri net in Figure 7.16. As an exercise it is possible to complete receptivities in the Grafcet (as is done for transitions (1) and (2) in Figure 9.12).

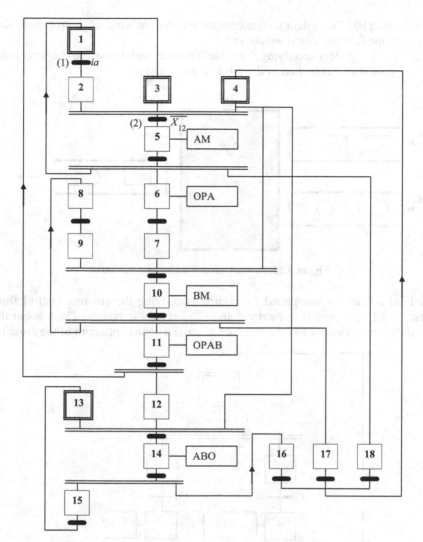

Figure 9.12. Grafcet for the manufacturing cell in Figure 5.1

9.4 Problems and Exercises

9.1. Complete all receptivities in Figure 9.12. Compare the grafcet with the corresponding Petri net.

9.2. Machine M serves for the production of products C from input workpieces A and B, respectively as Figure 9.13 shows. One workpiece A and one B must be available for the start of production. Solve the following design problems:

a. Complete the system with necessary sensors enabling its control according to the function described above.
b. Draw a grafcet specifying the system control with respect to the workpiece transfer and execution of the production.

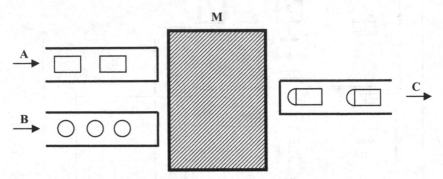

Figure 9.13. Production cell with one machine

9.3. Find a Petri net interpreted for control specifying the crossing control from Example 9.1. Compare both Petri net and Grafcet of this syetem. Think about the use of the found Grafcet and the Petri net to write a control program of the crossing.

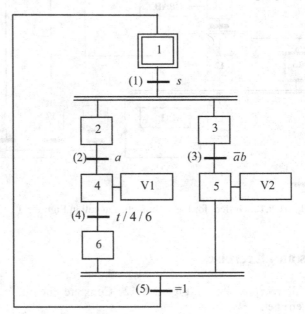

Figure 9.14. A grafcet for Exercise 9.4

9.4. A Grafcet is given in Figure 9.14. Analyze the control specified by the Grafcet. Figure 9.15 is a diagram of the logic variables s, a, and b. Complete the diagram with the time courses of the logic variables $X_1,..., X_6$ corresponding to the grafcet steps and the time course of the logic time variable $t/4/6$. Time is given in seconds.

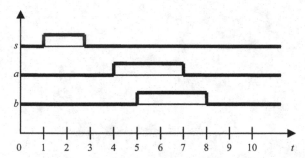

Figure 9.15. Timing diagram of logical variables s, a and b

10

Timed and High-level Petri Nets

10.1 From Standard to Higher-level Petri Nets

Petri nets in the standard form as considered until now are an effective tool for DEDS modeling and control design. They enable one to specify powerfully the system function. Analysis methods are used for testing Petri net model properties and hence to check the correct system function (Desel 2000). Very often quantitative properties of the system behavior are another point of interest. In other words, a kind of system function performance or system efficiency is dealt with. In order to make the performance analysis feasible, additional values, parameters, and variables are used within the Petri nets (Čapkovič 1993, 1994, 1998). Another reason for additional values to be built in the Petri nets is to make the Petri net models more transparent and understandable even for large and complex DEDS. Such extensions are often denoted as high level Petri nets (Struhar 2000) or generalized Petri nets (Juhás 2000).

Standard Petri nets are not suitable for performance analysis. Undoubtedly, for performance analysis, an important system variable is time. Time enriches information by telling in what time or time interval a particular event occurs or should occur (Čapek and Hanzálek 2000). There are three ways to embed time into Petri nets. The first is to map the Petri net places into time intervals given as real or integer numbers; the second is to map them analogously into the Petri net transitions; and the last is to map into the arcs (Zhou and Venkatesh 1998). The options can be used separately or together. The given time intervals cause delays in firing the respective transitions. Time intervals can be considered in deterministic or stochastic ways. The deterministic case of the timed Petri nets will be studied in Section 10.2 and the stochastic case in Section 10.3.

Section 10.4 deals with a class of high level Petri nets called colored Petri nets. The main idea is that each token in a colored Petri net has its individuality represented by a specific data value called color. Places, transitions and arcs of a Petri net can be equipped with logic conditions respecting the particular color of each token. Section 10.5 deals with a class of the high level Petri nets including the fuzziness property. Adaptive Petri nets are studied in Section 10.6 and Petri net-based design tools are presented in Section 10.7.

10.2 Deterministic Timed Petri Nets

Time may be associated either with the Petri net places or transitions, or with both. We will follow a general approach in (Zhou and Venkatesh 1998) covering three associations either together or separately in a deterministic way. The deterministic time association is a Petri net model extension enabling performance analysis using time relations. Deterministic approach is not applicable for all Petri nets defined by Definition 7.2. The problem originates mainly from the Petri net conflicts not excluded in the definition. A typical Petri net class with conflicts is the class of free-choice nets described in Section 7.4. Two or more arcs outgoing from a place bring about conflicts. The uncertainty about the continuation of the transition firing being in a conflict needs to model time behavior in a stochastic way. The deterministic way is very difficult or rather impossible to apply in praxis. Therefore, the deterministic time association is mostly restricted to the class of the marked graphs (see Section 7.4) – also named event graphs. The timed marked graphs are delimited by the following definition.

Definition 10.1. A timed marked graph is given by

$$TMG = (P, T, F, W.M_0, \pi, \tau) \tag{10.1}$$

where the meaning of P, T, F, W, M_0 is the same as in Definition 7.2, $MG = (P, T, F, W, M_0)$ is a marked graph $(i.e., \forall p \in P : |\bullet p| = |p \bullet| = 1)$, and π is the place delay function $\pi : P \to R^+$ (the set of non-negative real numbers), τ is the transition firing time function $\tau : T \to R^+$ and

1. A token, which arrives in a place, is not available for the connected transition with the place during the time associated with the place.
2. A transition is fireable and fires if all its pre-places contain the available tokens (*i.e.,* tokens not time blocked) required by the arc weights.
3. If a transition is fireable, its firing starts by removing the respective number of tokens from pre-places and firing completes after the time associated with the transition expires and the tokens are deposited in the respective post-places.

In what follows, we will show a paradigm for the system performance analysis *via* deterministic timed Petri nets. For this purpose the timed marked graphs will be considered having all arc weights equal to one. In other words, we consider Petri nets belonging to the class of timed binary marked graphs. Moreover, we consider strongly connected timed binary marked graphs where for the graph connectivity property both places and transitions are considered as graph nodes.

In the above delimited Petri net sub-class, the performance analysis is based on directed simple cycles contained in the particular Petri net, which is taken as a mathematical graph with places and transitions given as a set of nodes. In a directed simple cycle, the total time delay is a sum of times associated with all places and transitions comprised in the cycle. In a simple cycle the total number of tokens is the number of tokens present in all the places in the cycle. Note that a directed simple cycle is one that contains no repeated nodes except the beginning and ending ones. The minimum cycle time of the analyzed marked graph as a whole is

$$\mu = \max_{i} \frac{D_i}{N_i} \qquad\qquad (10.2)$$

where D_i is the total time delay of the i-th directed simple cycle and N_i is the total number of tokens in this cycle, D_i/N_i is the cycle time.

The bottleneck cycle is the j-th one where $D_j/N_j = \mu$ holds. A system may have multiple such cycles. When additional resources are available to improve the system productivity, one should certainly invest into the facility causing the bottleneck. The acquisition of a same machine can be reflected through the increase of a token in a loop. The improvement in the speed of a process can be reflected through the reduction of the delay in a place or transition. The delay can also be associated with the arcs in a marked graph, simulating the time for a token to flow through the arc. This extension is useful in modeling transportation of goods over conveyors, or fluid flowing through a pipe in process industry.

The use of the above approach *via* the enumeration of cycles is of exponential computational complexity. In other words, it is not applicable to large-size marked graphs. Fortunately, the minimum cycle times can be obtained, *e.g.*, by linear programming (Morioka and Yamada 1991; Campos *et al.* 1992; Zhou and Venkatesh 1998).

The described analysis method *via* the cycle enumeration is illustrated in the following example.

Example 10.1. The Petri net in Figure 10.1 is a model of a manufacturing system where p_1 stands for the processed part availability (if marked with a token), p_2, p_3, p_4 stand for manufacturing process on the machines A, B, C, respectively, and p_5 for availability of machine A. Maximum two parts can be prepared for processing at the input. Time delays associated with places and transitions are introduced in the Petri net. A part from input is deposited with delay $\tau_1 = 3$ in the working range of machines A and C. When A completes its job, the processing continues in machine B, which starts its required operation. When machines B and C complete their operations, the product is transferred to the output and the next part is deposited to the input. The delays connected with the places mean the lengths of operations and are denoted as δ_i. Simple cycles and delays are in Table 10.1. The resulting minimum cycle time is 13 time units. The manufacturing process can start again not earlier than after 13 time units. The bottleneck takes place at cycle $p_1 t_1 \, p_2 \, t_2 \, p_3 \, t_3 \, p_1$.

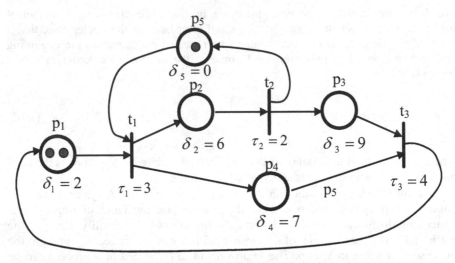

Figure 10.1. Timed marked graph with time delays

Table 10.1. Time delays of simple cycles in the Petri net of Figure 10.1

Simple cycle	Total time delay	Token sum	Cycle time
$p_1 t_1 p_2 t_2 p_3 t_3 p_1$	26	2	13
$p_1 t_1 p_4 t_3 p_1$	16	2	8
$p_5 t_1 p_2 t_2 p_5$	11	1	11

The timed marked graphs can be developed as a powerful tool for bottleneck analysis and thus help identify where one should invest and where one should not. For the above example, adding another machine of type A, *i.e.*, p_5 receiving one more token, contributes none to the cycle time reduction. On the other hand, doubling Machine A's processing speed, *i.e.*, reducing p_2's delay to 3, can reduce the system cycle time from 13 to 11.5 time units.

10.3 Stochastic Timed Petri Nets

In stochastic timed Petri nets, firing rates and time delays associated with Petri net transitions are assumed to be random variables. In this section, we are restricted to the cases when the stochastic time variables are associated with transitions only and exponentially distributed. Such models are termed stochastic Petri nets, SPN for short. Primarily, firing rates associated with transitions are considered. They determine firing repetitions when firing conditions are permanently fulfilled. The reciprocals of average firing rates are average time delays and *vice versa*. Stochastic timed Petri net models are related to the models based on the Markov chains (Zhou and Zurawski 1995; Bause and Kritzinger 1996). A thorough treatment of this topic can also be found in (Ajmone Marsan *et al*. 1995) and

(Wang 1998). Basic properties and application of the stochastic timed Petri nets are illustrated through an example.

There are a number of extensions to the above discussed stochastic Petri nets. If some transitions can fire much faster than others, their firing rate can be viewed as an infinite value. In other words, firing them takes nearly zero time. Such transitions are called immediate transitions. They always fire before any timed transitions if enabled at the same time. The resulting model is called Generalized Stochastic Petri Nets, GSPN for short (Ajmone Marsan *et al.* 1995). It is proved that both SPN and GSPN can be converted into their equivalent Markov chain models. Hence, the technique used to solve Markov chain models can be utilized to solve both models. Under certain conditions, some transitions are allowed to have deterministic time delay, resulting in Deterministic Stochastic Petri Nets (DSPN). They can assume to have arbitrary distributed time delay and lead to Extended Stochastic Petri Nets (ESPN). Both DSPN and ESPN can be converted into their equivalent semi-Markov chains for their solutions. When a transition is associated with a delay of arbitrary distribution, the resulting timed Petri nets cannot be analytically analyzed in general. The in-depth treatment of the topic can be found in (Wang 1998). The following is an exhibit of solving a stochastic Petri net *via* an example.

Example 10.2. Consider a manufacturing layout, which is modeled with a Petri net (Figure 10.2). Marked place p_1 represents a work-piece available at the input; p_2, and p_4 represent operations executed during processing a work-piece with machines A and B, respectively. When both operations are finished (a token is both in p_3 and p_5) the processed workpiece is unloaded and a new part is deposited in the input. Places $p_6 - p_7$ represent states when A or B is in repair.

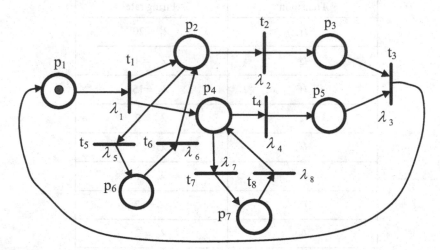

Figure 10.2. Stochastic timed Petri net with firing rates

Average firing rates λ_i are associated with transitions. Reciprocals of the rates are the average times of the respective operations. For example, when a token arrives in place p_2, transition t_2 starts firing, which takes a delay comprising operation at machine A and unloading the workpiece from the machine when the operation is finished. The time is a random variable with an average equal to $1/\lambda_2$. For other transitions, the situation is similar. Table 10.2 describes the transition meanings.

Table 10.2. Meaning of transitions in the Petri net of Figure 10.2

Transition	Meaning
t_1	Loading work-piece from input into the processing range of machines A and B
t_2	Processing a work-piece by machine A and unloading
t_3	When both operations are finished, removing the processed work-piece and loading a new work-piece at the input
t_4	Processing of a work-piece by machine B and unloading
t_5	Machine A breaks down
t_6	Machine A is being repaired
t_7	Machine B breaks down
t_8	Machine B is being repaired

Table 10.3. Firing rates in the example

Transition	Firing rate
t_1	$\lambda_1 = 20$
t_2	$\lambda_2 = 4$
t_3	$\lambda_3 = 15$
t_4	$\lambda_4 = 2$
t_5	$\lambda_5 = 2$
t_6	$\lambda_6 = 1$
t_7	$\lambda_7 = 2$
t_8	$\lambda_8 = 2$

By comparing the net in Figure 10.2 with the one in Figure 10.1 with respect to machine A it is evident that p_6 has been added. It represents a state when A is in repair after a breakdown. The place is connected to new transitions t_5 and t_6. Each transition in the net represents a whole process that takes some randomly distributed time, and during it the tokens are blocked in the transition. Table 10.2 shows that more actions can be covered by one transition.

The i-th average firing rate associated to t_i is denoted as λ_i, the i-th time delay is z_i. Firing rates for our example are given in Table 10.3. When a transition starts its firing, tokens from pre-places are taken and when firing ends the tokens are deposited in post-places. All weights are equal to one. The firing rules are usual.

The reachability graph for the Petri net is shown in Figure 10.3. Each arc of the graph is labeled as usual with t_i leading to the passage from one marking to its successor. Each arc is additionally labeled with the average firing rate λ_i associated with the corresponding transition. The markings are the states of the system. A Markov chain can be generated for the states. Its topology is the same as of the reachability graph. In the Markov chain transit arcs between states are equally labeled with the firing rates λ_i. The transition rate matrix for the Markov chain is

$$
\mathbf{A} =
\begin{pmatrix}
-a_1 & a_1 & 0 & 0 & 0 & 0 & 0 & 0 & 0 \\
0 & -a_2-a_4-a_5-a_7 & a_2 & 0 & 0 & a_4 & 0 & a_5 & a_7 \\
0 & 0 & -a_4-a_7 & a_4 & a_7 & 0 & 0 & 0 & 0 \\
a_3 & 0 & 0 & -a_3 & 0 & 0 & 0 & 0 & 0 \\
0 & 0 & a_8 & 0 & -a_8 & 0 & 0 & 0 & 0 \\
0 & 0 & 0 & a_2 & 0 & -a_2-a_5 & a_5 & 0 & 0 \\
0 & 0 & 0 & 0 & 0 & a_6 & -a_6 & 0 & 0 \\
0 & a_6 & 0 & 0 & 0 & 0 & 0 & -a_6 & 0 \\
0 & a_8 & 0 & 0 & 0 & 0 & 0 & 0 & -a_8
\end{pmatrix}
$$

The first row and first column correspond to marking (state) $\mathbf{m}_0$, the second ones to $\mathbf{m}_1$ etc.; λ_k is assigned to the matrix entry $(i, j), i \neq j$, if there is a transit from state $\mathbf{m}_i$ to $\mathbf{m}_j$ via transition t_k. For $i = j$, the negative sum of firing rates of the rest of entries in the i-th row is assigned to the entry (i,i). The assignment is evident from the reachability graph and matrix $\mathbf{A}$. The following matrix equation is well known from the theory of Markov chains

$$\left(\pi_0 \ \pi_1 \ \pi_2 \ \ldots \ldots \ \pi_8\right)\mathbf{A} = 0 \tag{10.3}$$

and, of course,

$$\pi_0 + \pi_1 + \pi_2 + \ldots + \pi_8 = 1 \tag{10.4}$$

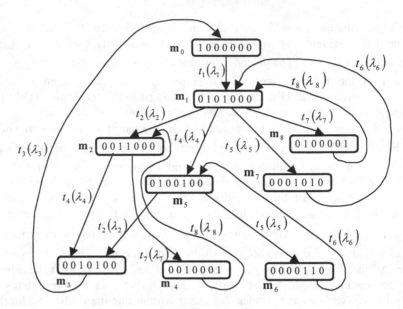

Figure 10.3. Reachability graph for Petri net in Figure 10.2

where π_i is a probability that the system is in a state represented by a marking $\mathbf{m}_i$. To explain Equation (10.3), consider, *e.g.*, the first column of matrix $\mathbf{A}$: entries for $\mathbf{m}_1 - \mathbf{m}_8$ correspond to transits from them to $\mathbf{m}_0$; the entry for $\mathbf{m}_0$ in this row according to the construction of $\mathbf{A}$ corresponds to all transits from $\mathbf{m}_0$ to other markings. The sum of products of probabilities and firing rates for passes into the state $\mathbf{m}_0$ should be balanced with the same sum for the passes out of $\mathbf{m}_0$.

The solution to Equations (10.3) and (10.4) with the firing rates in Table 10.3 provides probabilities given in Table 10.4.

Table 10.4. Calculated probabilities for Example 10.2

π_0	π_1	π_2	π_3	π_4	π_5	π_6	π_7	π_8
0.029	0.098	0.196	0.039	0.196	0.050	0.098	0.196	0.098

Various performance characteristics can be calculated from the model, *e.g.*, exploitation of machine A is 24.6% as follows from

$$\pi_1 + \pi_5 + \pi_8 = 0.246 \tag{10.5}$$

The system throughput given as a rate can be calculated as follows:

$$\pi_3 \, \lambda_3 = 0.585 \tag{10.6}$$

which corresponds to

$$\frac{1}{\pi_3 \lambda_3} = 1.71 \text{ time units} \tag{10.7}$$

A breakdown of the machine B is characterized by probability

$$\pi_4 + \pi_8 = 0.294 \tag{10.8}$$

or 29.4% of the production execution time.

10.4 Colored Petri Nets

If the relations between the system states given by markings are complex and/or the system consists of many identical subsystems, then the Petri nets in the standard form become very complicated and difficult to read. Each subsystem requires its own Petri net subset. The colored Petri nets (CP-nets) were introduced by Jensen (1981) in order to solve the problem of the Petri net invariants for the so-called high-level Petri nets presented by Genrich and Lautenbach (1981). Jensen's improved version of the high level Petri nets was later developed into a nice tool-CPN (Jensen 1997).

The basic idea is that in colored Petri net tokens have their own individuality or identity represented by data values of some prescribed type called colors. Logic expressions and functions can be built up using the token colors and can be associated with places, transitions and arcs of a CP-net.

An exact description of colors has to be attached to each colored Petri net. Nowadays, colored Petri net designers often use a language called CPN ML for the CP-net design. CPN ML is closely related to the constructions and declarations used in ordinary high level programming languages. In what follows we give a basic introduction to CP-nets using the following example. In its development we rely on reader's intuition bearing in mind that the notation of CPN ML is familiar.

Example 10.3. Consider a system with two processes sharing two different kinds of resources. There are one resource of type R and three resources of kind S available. These may be for instance a robot and machines in a manufacturing line, concurrently manufacturing two kinds of products. The process p (manufacturing of a product of type p) needs two machines of type S to be assigned in a certain time. The process q needs, besides two machines S, also robot R for its finishing. Both processes are running cyclically. A model of the system represented by a P/T Petri net is shown in Figure 10.4.

Each process is represented by one "subnet", the subnets are mutually interconnected through places R and S representing the shared resources. A standard Petri net representation of this kind of system would be very complicated due to a higher number of processes and resources. In this situation modeling by a colored Petri net is very helpful (Figure 10.5).

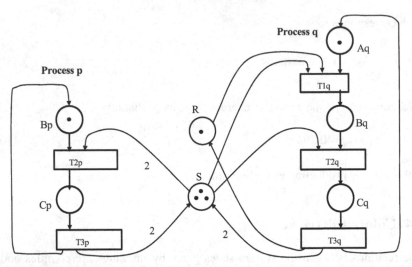

Figure 10.4. A system with two processes modeled by a P/T Petri net

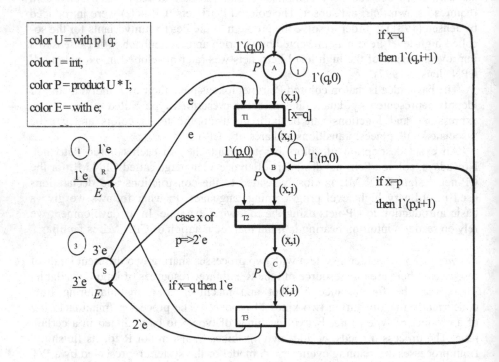

Figure 10.5. A system with two processes modeled by a colored Petri net

By inspecting the two figures, it is clear that colored Petri net notation is slightly different compared with the standard P/T Petri net conventions. A colored net consists of three parts: a *net structure* (places, transitions and arcs), *declarations* (listed in a frame in the left upper corner) and *net inscriptions* connected with the net elements. A fundamental difference with respect to P/T Petri nets lies in the token color. Each token is assigned its own *color* as some value of certain data type. This data type can also be complex, *e.g.,* a structure or a record, where for instance the first item is a real number, the second is a text string and the third could be a record of integer pairs. In this example we have used tokens of two color types (Figure 10.5, in a frame): *P* and *E*, where color *P* is a Cartesian product of colors *U* and *I*. Color *U* contains a binary value of two options: p and q, corresponding to the type of the process. Color *I* is introduced in addition to the standard P/T net representation and contains an integer value, which counts the total number of finished cycles for each process. By introducing a color *I*, it is demonstrated how it is possible to extend simply modeling convenience of the standard P/T nets using the token colors. Now, color *P* contains information about the process and also about the number of finished products − outcomes of the process. Color *E* contains just information about the type of the shared resource. Hence, there are two different kinds of tokens in the net, but in each place only tokens of one certain type is possible in this example.

A possible color, called *color set*, is expressed by inscription in italic associated with each place. Thus places R and S can contain tokens of color *E* and places A, B, and C tokens of color *P*. A careful reader has surely noted different marking inscriptions. The names of places are written inside the places instead of tokens. Because of the need to know both the number of tokens and their color, the marking is written near the places in such a manner that the total number of tokens in the respective place is written as a number in a small ring followed by a multiset inscription representing color and number of tokens. For example, next to place A there is a marking $\left(1\right)$ 1'(q,0) , which means that in the place, there is one token in total, namely one token with color (q,0) (q is the process type and 0 means the number of finished products of type q). By convention, we omit the marking of places with no token. The initial marking is represented by underlined expressions placed next to the respective places. The system in Figure 10.5 is in the initial state. Hence, the actual marking is equal to the initial one.

Assigning a color to each token and a color set to each place allows one to use a smaller number of places than in standard P/T nets. In this example, places Bp, Bq and Cp, Cq have been joined. Using colors we do not lose the possibility to distinguish the process types. This possibility brings about an important benefit in more complicated systems. However, by introducing colors the Petri net dynamics becomes more complex. It is necessary to introduce more complex expressions associated with arcs to describe fine possibilities of the marking evolution. Therefore arcs contain expressions whose results are elements of multi-sets, namely colored tokens. Sometimes such an arc can "transmit" a token without any change, *e.g.,* the arc connecting A and T1, where arc (x,i) moves the token from place A *via* transition T1 into place B without change. We can use an abbreviation. Instead of 1'e it is sufficient to write the symbol e as, *e.g.,* for the arc connecting R and T1. A transition is enabled/fireable iff all its pre-places contain tokens with proper colors as specified by its input arcs. There is a more complex arc inscription: "if x=p then 1'(p,i+1) else empty" at the arc leading from T3 to B. The token passes this arc only if it is the token concerning process p. This arc also increases the number of finished products of type p. There is a similar inscription at arc connecting S and T2: "case x of p=>2`e |q=>1`e". In case when the value of color x of the token passing transition T2 is p, two tokens of type e are withdrawn from place S (if available). If it is a token of color q, only one token is withdrawn (according to the structure in Figure 10.4). A condition [x=q] at transition T1 means that there is a part of the net concerning only process q and that no token of the type p may pass this transition (such a kind of token is not possible here because of the net structure, initial marking and arc expressions). When a transition fires, tokens with colors specified by its output arcs are deposited to its post-places.

It is further possible to develop and adapt the described CP-net, *e.g.,* by joining places concerning the shared resources into one place and adding the necessary arc inscriptions, *etc.* However, such subsequent adaptations could reduce the transparency or readability of the net.

Example 10.4. Consider an automatic guided vehicle system depicted in Figure 10.6. The system consists of fixed tracks divided into sections. This is a long practice to ensure transportation safety. A collision-free function is achieved by the condition that only one vehicle can be in a section. The vehicles move through the sections in both directions. The switches are routing vehicles according to the chosen vehicle path. A normal stop is not allowed in the switch area except between switches SW2 and SW1, and between SW7 and SW8. There a vehicle can stop and change its direction, if necessary. Sensors detect the presence of a vehicle in a section. There are no sensors in the switch sections, with the exceptions mentioned above. Processing centers are situated along some sections.

A control system provides a collision-free guidance of the vehicles to fulfil their transportation tasks. The route optimization problem has not been considered here. We have adopted an acceptable solution to determine for each vehicle's section position, possible continuations to the goal section avoiding collisions (see more in Hrúz *et al.* 2002).

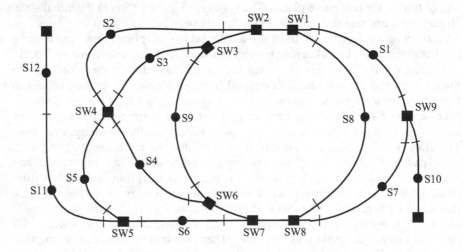

Figure 10.6. AGV system with fixed tracks

An ordinary Petri net modeling the transportation system is given in Figure 10.7. There are three subnets in it corresponding to three vehicles moving in the system. An identical subnet has been added for each further vehicle. Places p14 and p13 correspond to the special switch sections SW1–SW2 and SW7–SW8. Other switch sections are not represented by places. A vehicle presence is modeled by a deposit of a token in the place of a corresponding subnet. Analyze the movement control of the 1st vehicle. According to its task we have the following: for a transit from sections S_i to S_j, a command is released for its motion to the next chosen section that has no other vehicle. It is clear that many vehicles will make the ordinary Petri net very complicated and cumbersome to model and solve the transportation control problem.

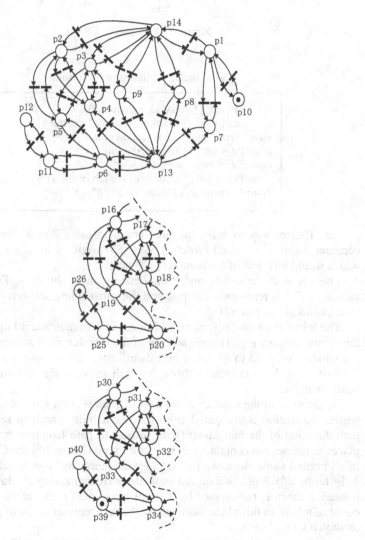

Figure 10.7. Petri net for the transportation system

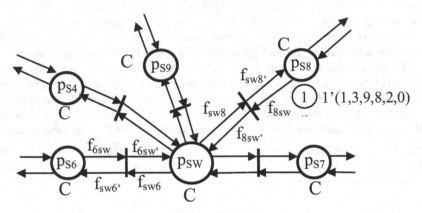

Figure 10.8. A part of CP-net

Table 10.5. Definitions of token colors

color Id = int; (* AGV id – number *)
color Start = int; (* starting section of transfer *)
color Dest = int; (* final section of transfer *)
color Pd = int; (* partial destination *)
color Job = int; (* assigned job *)
color Prio = int; (* priority of the job (vehicle)*)
color C = product Id*Start*Dest*Pd*Job*Prio;

An efficient way to solve the problem is to use CP-nets. For the illustration consider a situation around switches SW6–SW8 with help of a colored Petri net, which would be a part of the complete CP-net.

The net with the colors and their description is shown in Figure 10.8. The switch section is represented by p_{SW} and the surrounding sections by p_{Si}. The color inscriptions are in Table 10.5.

The token color contains more additional information according to the needs of the control system, *e.g.,* starting section of the transfer, final section of the transfer, job number assigned to a vehicle and its priority. We also add a value for "partial destination" – the next section, through which an AGV should move to execute the actual transfer.

At the start of the transfer as well as in each section crossed by the AGV, the partial destination is computed from the reachability graph to select the optimal path direction of the movement (with respect to path length or transfer time). All places in the net can contain a token of color type C, which is the Cartesian product of all needed value elements. For safety we assume the capacity of all places to be 1. In terms of CP-nets we do not indicate current marking by dots put in places; instead, a token is represented by a small circle next to the place showing overall count of tokens in this place followed by the text representation of particular values as shown with place p_{S8}.

Table 10.6. Arc expressions

```
f8sw :
(* call next section planning algorithm and set pd *)
f8sw' :
if (pd in {4,6,7,9} then 1'(Id,Start,Dest,Pd,Job,Prio);
fsw8 :
1'(Id,Start,Dest,Pd,Job,Prio);
fsw8' :
if (pd=8) then 1'(Id,Start,Dest,8,Job,Prio);
```

The main functionality of this CP-net is realized by arc expressions represented by functions f_{ij}. By means of f_{ij} the next section is computed, to which the vehicle will be directed and the movement is realized by setting appropriate external signals.

A simplified example of arc expressions for Section 8 is shown in Table 10.6, and expressions for other arcs are generated similarly (not shown in Figure 10.8.).

When the net is externally synchronized with the process, we obtain a control algorithm represented by the CP-net. It is easy to see that the movements of all vehicles can be represented in one net. It is not necessary to have a separate subnet for each vehicle. The inscriptions of colors are easy to to understand: „int" means values of the type integer; "Id" is a color identifying a vehicle, with three vehicles having the colors 1, 2, and 3. Similarly, it is with starting and goal section of a transfer job; "Pd" is a color for the partial destination calculated by an optimization program. The calculation considers possible partial destinations and tries to find the best way respecting the occupation of the continuation places. Further, there are colors "Job" and "Prio". The tokens in the CP-net (three in the example) are given colors according to the product Id*Start*Dest*Pd*Job*Prio. The arc expression constructs are self-explanatory. An idea is that the expressions set the condition of the arc to the token color if the achievable partial destination is in the vicinity.

To model a smart card associated with a batch of parts/materials to be processed and meet the need to analyze and control deadlocks in automated production, a token's colors are introduced into Petri nets, which are different from the above introduced in (Wu 1999 and Wu and Zhou 2001, 2004, 2005, 2007). They represent the output transitions of a place, which a token in the place intends to enable following the determined part routes. The resulting models are called colored resource-oriented Petri nets. They has been applied to DEDS in flexible manufacturing and assembly, Automated Guided Vehicle (AGV) systems, track systems and cluster tools in semincoductor fabrication, and oil-refinary scheduling problems. The work has great significance in simplifying the deadlock modeling, analysis and control complexity. It can also facilitate the scheduling and help derive optimal schedules.

10.5 Fuzzy Petri Nets

The fuzziness concept can be incorporated in Petri nets. Some additional aspects should be supplemented for that purpose. Fuzzy Petri nets are useful as models for expert rule-based decisions, temporal reasoning and many others. A rich collection of contributions to various aspects of the fuzziness used with Petri nets can be found in Cardoso and Camargo (1999).

In this section we have chosen from many possibilities a kind of fuzzy Petri nets adapted for the temporal problem solutions. First of all, let us introduce a few notions from the fuzzy set theory.

Definition 10.2. A fuzzy set A in a universe of discourse U, written A in U, is defined by a set of pairs

$$A = \{(\mu_A(x), x)\} \tag{10.9}$$

where $\mu_A : U \to [0,1]$ (a real number interval) is a membership function, which represents the element's $x \in U$ degree of membership (by mapping U into interval $[0,1]$) of the fuzzy set A.

The notion of a fuzzy number is further necessary for data operations in fuzzy Petri nets. If a fuzzy set in the domain U consisting of real numbers is

a) normal, *i.e.,* $\max\limits_{x \in U} \mu_A(x) = 1$ and

b) convex, *i.e.,* $\forall x, x_1, x_2 \in U, x_1 \lessgtr x < x_2, \quad \mu_A(x) \geq \min(\mu_A(x_1), \mu_A(x_2))$

then it is a fuzzy number. Binary fuzzy operations are defined for fuzzy numbers (recall that a binary operation on a set M is a mapping $M \times M \to M$); they are similar to ordinary binary operations on the real number domain. Fuzzy number operations $\oplus, (-), \otimes, (:), \max, \min$ can be defined similarly as ordinary arithmetic operations $+, -, \times, :, \max, \min$. For example, fuzzy operation $\oplus$ for fuzzy numbers A, and B is defined by

$$\mu_{A \oplus B}(z) = \sup_{z = x + y} \left[\min(\mu_A(x), \mu_B(y)) \right], \ x, y, z \in U \tag{10.10}$$

The resulting fuzzy number $Z = A \oplus B$ is given by the membership function $\mu_{A \oplus B}(z)$. Consider its value for one particular z. It is calculated as a supremum of all pairs of ordinary numbers x and y, which give the value z by taking a minimum of the membership functions of x and y; and the supremum of those minimums defines the resulting membership value for one value of z. In this way, all points of the membership function of Z can be calculated.

Consider the timed fuzzy Petri nets from (Ribarič and Bašič 1998). The time value is given on a time scale T, which is a linearly ordered set like R^+ and N^+. Following the fuzzy concept there is an uncertainty in determining a time point. This uncertain knowledge about the time a (when some event occurs) can be expressed by a possibility distribution function $\pi_a : T \to [0,1]$, *i.e.,* $\pi_a(t) \in [0,1]$ for

$\forall t \in T$. π_a is a numerical estimate of a possibility that the time point a is precisely t, whereby the time as a physical variable varies independently. π_a is equivalent to $\mu_A(t)$ under the assumptions that π_a is normal and convex, and $\mu_A(t)$ is normal and convex, too. Then the fuzzy set A is associated with the considered fuzzy time point a and A is a fuzzy number determining the time point a in a fuzzy way. Denote by $D(T)$ the set of all normal and convex possibility distributions π defined on T.

Now we are ready for the following example taken from Ribarič and Bašič (1998): Fred, John and Mark have a meeting as soon as all arrive at work. Fred leaves home about 7:00 in the morning. He goes by car and arrives to work about 20 minutes later. John comes to work a few min earlier than Fred. Mark leaves the house approximately at the same time as Fred. He takes a bus. The bus takes about 20 min to reach the bus stop nearest to the office. Then it takes him a few minutes more to get to the office. The question is: What are possible starting times of the meeting?

We can well present the fuzzy temporal relations by means of a fuzzy Petri net in Figure 10.9. Marked places can represent partial states as described in Figure 10.9. Transitions represent actions described by fuzzy temporal linguistic expressions. The places can be marked with fuzzy tokens. Each token has its identity given by data values as in colored Petri nets. A possible marking of place p_i is one of the following ordered pairs: $(\pi_{b(i)}, \varnothing), (\varnothing, \pi_{e(i)}), (\pi_{b(i)}, \pi_{e(i)})$, and $(\varnothing, \varnothing)$. The last case means that no token is in place p_i. $\pi_{b(i)}$ and $\pi_{e(i)}$ are possibility distribution functions. $\pi_{b(i)}$ stands for the time point of the beginning of a token presence in place p_i, $\pi_{e(i)}$ is for the end time point. A token is assigned the distributions as values on its arrival in a particular place.

In the treated fuzzy Petri nets, three functions have been used:

1. Function $\tau : P \rightarrow D(T) \cup \varnothing$. Let τ_i be associated with place p_i. It determines in a fuzzy way the detainment of a token in p_i.

2. Function Θ related to function τ. When a token arrives in a place p_i it is assigned value $(\pi_{b(i)}, \varnothing)$. Function Θ changes according to τ $(\pi_{b(i)}, \varnothing)$ on $(\pi_{b(i)}, \pi_{e(i)})$ where $\pi_{e(i)} = \pi_{b(i)} \oplus \tau_i$.

3. Function λ , which maps the set T of transitions to a set of fuzzy operations $\oplus \kappa, (-)\kappa, \underline{\min}, \underline{\max}$, etc., where κ is a fuzzy number. The operations are applied to a pair (π_b, π_e). Practically, the beginning given by the fuzzy number in the post-place $\pi_{b(j)}$ is obtained by a fuzzy operation, e.g., $\pi_{b(j)} = \pi_{b(i)} \oplus \kappa_k$ where p_i is a pre-place of a transition t_k .

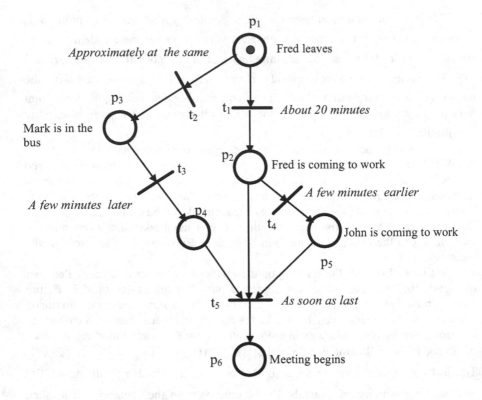

Figure 10.9. Fuzzy Petri net model of temporal relations

Table 10.7 lists the results of functions τ and λ. Fuzzy numbers in Table 10.7 are represented in the so-called triangular form when the membership function is given as a triangle (see Figure 10.10 for κ_1). Using the triangle membership function, the statement "*about 20 min later*" is interpreted according to Figure 10.10 as "within $\pm$ 5 min around 20 min".

Starting from the initial marking $\mathbf{m}_0 = \left[\left(\pi_{b(1)}, \varnothing \right), \varnothing, \varnothing, \varnothing, \varnothing, \varnothing \right]^T$ and performing fuzzy operations by stepwise applying the functions τ and λ we obtain the final marking

$$\left(\varnothing, \varnothing, \varnothing, \varnothing, \varnothing, m_6 \right)$$

where $m_6 = \left(\underline{\max} \left(\pi_{b(2)}, \pi_{b(4)}, \pi_{b(5)} \right), \varnothing \right)$. $\underline{\text{Max}}$ is a fuzzy operation of maximum as illustrated in Figure 10.11. It is the fuzzy expression of the time of the meeting beginning by using the membership function or the corresponding fuzzy number. It is easy to imagine that similar fuzzy temporal relations can occur in various kinds of DEDS.

Table 10.7. Description of fuzzy relations

State, action	Infliction of τ and λ
State: Fred leaves home	$\tau(p_1) = 0$ because it is the start of the process
Action: about 20 min later	$\lambda(t_1) = \oplus \kappa_1$, $\kappa_1 = (15,20,25)$
State: Fred is coming to work	$\tau(p_2) = 0$ because he is ready for the meeting immediately after the arrival of all others
Action: approximately at the same time	$\lambda(t_2) = \oplus \kappa_2$, $\kappa_2 = (-5,0,5)$
State: Mark is in the bus	$\tau(p_3) = (15,20,25)$, Mark is about 20 min in bus, therefore $\pi_{e(3)} = (15,20,25)$
Action: a few minutes later	$\lambda(t_3) = \oplus \kappa_3$, $\kappa_3 = (0,5,10)$
State: Mark is coming to work	$\tau(p_4) = 0$
Action: a few minutes earlier	$\lambda(t_4) = \oplus \kappa_4$, $\kappa_4 = (-10,-5,0)$
State: John is coming to work	$\tau(p_5) = 0$
Action: as soon as last	$\lambda(t_5) = (\underline{max}, \varnothing)$

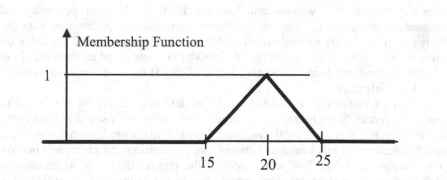

Figure 10.10. Membership function of a fuzzy number

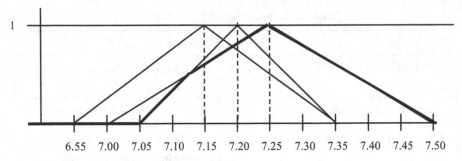

Figure 10.11. Resulting fuzzy expressed time of the meeting beginning expressed by the membership function

10.6 Adaptive Petri Nets

Incorporating learning capability into a Petri net framework leads to adaptive Petri nets. Broadly speaking, intelligent techniques such as artificial neural network, fuzzy logic and knowledge based systems together can bring adaptable feature to Petri nets. Consequently, adaptive Petri nets can become a framework for dynamic knowledge inference under changing environments (Asar *et al.* 2005). The basic conditions need to be defined under which a Petri net can be modeled to qualify for adaptive task similar to biological neural network. The related approaches have borrowed the concepts from the work based purely on biological brain model. Thus the developed models can mimic a biological brain in terms of its distributed function feature. Some work involves synergy of Petri nets and intelligent techniques where ideas are motivated from the concepts of fuzzy logic and neural networks through the weights and learning features. A small percentage of researchers are active in applying intelligent techniques in conjunction with the Petri net methodology on real world problems. This section intends to focus on presenting the concept and examples of adaptive Petri nets based on the work (Li *et al.* 2000; Yeung and Tsang 1998; and Gao *et al.* 2003) for the purpose of dynamic knowledge inference.

In many situations, it is difficult to capture data in a precise form. In order to represent certain knowledge, fuzzy production rules are used for knowledge representation (Chen *et al.* 1990, Gao *et al.* 2003). A fuzzy production rule is a rule which describes the fuzzy relation between two propositions. Its antecedent portion may contain "AND" or "OR" connectors. If the relative degree of importance of each proposition in the antecedent contributing to the consequent is considered, a Weighted Fuzzy Production Rule is needed (Yeung and Tsang 1998). For example,

R_1: IF it is dark (p_1) and Vision Processing System (VPS) works well (p_2) THEN the data from VPS is not dependable (p_3) with certainty factor μ_1=0.9, threshold λ_1=0.5, and weights w_1=0.6 and w_2=0.4.

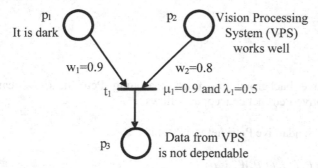

Figure 10.12. The Petri net representation of a weighted fuzzy production rule

This rule means that

1. p_1 and p_2 are two antecedent propositions and p_3 a consequent one (may become an antecedent proposition of other rules).
2. This rule's certainty factor is 0.9.
3. If the sum of p_1 and p_2's truth degrees (not given) weighted by their weights w_1=0.6 and w_2=0.4 exceeds the firing threshold value λ_1=0.5, this rule is executable.
4. p_1 and p_2's weights are 0.6 and 0.4-implying that p_1 is more important than p_2. Their sum should be one given a conjunctive rule.

The importance weight should be one by default for the cases of a single antecedent proposition, or multiple propositions connected with OR. Please note that in (Yeung and Tsang 1998), the threshold value is defined for each proposition and then an exactly same or close statement is as an input to the rule's antecedent propositions. The similarity value between an input proposition and that in a rule has to be computed. The rule can be executed only if it exceeds the threshold for each proposition.

This example rule can be easily converted to a Petri net as shown in Figure 10.12 where each place represents a proposition and transition t_1 represents rule R_1. Suppose that p_1 and p_2's truth degrees are given as θ_1=0.4 and θ_2=0.5. Since $y=w_1\theta_1+w_2\theta_2$=0.24+0.2=0.44<$\lambda_1$=0.5, this rule cannot be executed. However, if θ_1 increases to 0.6, y=0.36+0.2=0.56>λ_1=0.5. Hence, this rule can generate the results. If p_3 has no other input transitions, its truth degree is: $\theta_3=y\mu_1$=0.56×0.9=0.504.

In addition, different from the other "non-reasoning" Petri nets, θ_1=0.4 and θ_2=0.5 remain unchanged at p_1 and p_2 unless new updates arrive. Note that some previous work removes them as they are treated as tokens (Li *et al.* 2000). When p has multiple input transitions fired, e.g., t_{1-k} with y_j as the weighted truth degrees and μ_j as the certainty factor of t_j, j=1, 2, .., k, p's truth degree is derived as the center of gravity of these fired transitions, *i.e.*,

$$\theta(p) = \frac{\sum_j y_j \mu_j}{\sum_j \mu_j}$$

With the above background of fuzzy reasoning Petri nets, we can now introduce an adaptive Petri net concept as follows.

Definition 10.3. An adaptive Petri net is a 9-tuple

$$APN = (P,T,I,O,\theta,W,\lambda,\mu)$$

where *P*, *T*, *I*, and *O* defines APN structure, or more specifically

1. $P = \{p_1, p_2, \cdots, p_n\}$ is a finite set of propositions or called places.
2. $T = \{t_1, t_2, \cdots, t_m\}$ is a finite set of rules or called transitions.
3. *I*: $P \times T \rightarrow \{0,1\}$, is an $n \times m$ input matrix defining the directed arcs from propositions to rules. $I(p_i, t_j) = 1$, if there is a directed arc from p_i to t_j; and $I(p_i, t_j) = 0$, if there is no directed arcs from p_i to t_j, for $i = 1,2,...,n$, and $j = 1,2,...,m$.
4. *O*: $P \times T \rightarrow \{0,1\}$, is an $n \times m$ output matrix defining the directed arcs from rules to propositions. $O(p_i, t_j) = 1$, if there is a directed arc from t_j to p_i; $O(p_i, t_j) = 0$, if there is no directed arcs from t_j to p_i for $i = 1,2,...,n$, and $j = 1,2,...,m$.
5. $\theta: T \rightarrow [0,1]$ is a truth degree vector. $\theta = (\theta_1, \theta_2 \cdots, \theta_n)^T$, where $\theta_i \in [0,1]$ means the truth degree of p_i, $i = 1, 2, ..., n$. The initial truth degree vector is denoted by θ^0. It is treated as a marking (no longer integer but any real number between 0 and 1).
6. $W: P \times T \rightarrow [0,1]$ is the weight function that associates a weight with an input arc from a place to a transition. If a transition *t* has multiple input places, the sum of all the weights from these places to *t* must equal one. If *t* has a single input place *p*, then $W(p,t) = 1$.
7. $\lambda: T \rightarrow [0,1]$ is the function that assigns a threshold value to t_i.
8. $\mu: T \rightarrow [0,1]$ is the function that assigns a certainty factor value to t_i.

Assume that the weights need to be learned given the input and output date of place truth degrees. To do so, we need to define the execution rule first.

Definition 10.4. Given APN, $\forall t \in T$, *t* is enabled if $\forall p \in {}^\bullet t$, $\theta(p) > 0$. Firing an enabled *t* produces the new truth degree for its output place(s), denoted by $y(t)$:

$$y(t) = \begin{cases} \sum_j \theta(p_j) \cdot W(p_j,t) \ , & \sum_j \theta(p_j) \cdot W(p_j,t) > \lambda(t) \\ 0, & \sum_j \theta(p_j) \cdot W(p_j,t) < \lambda(t) \end{cases} \qquad (10.11)$$

1. If p has no input transition, its truth degree must be given initially.
2. If p has only one input transition t, $\theta(p) = y(t)\mu(t)$.
3. If p has multiple fired input transitions, t_{1-k} with y_j as the weighted truth degrees and μ_j as the certainty factor of t_j, $j=1, 2, .., k$, p's truth degree is derived as the center of gravity of the fired transitions, $i.e.,$

$$\theta(p) = \frac{\sum_j y_j \mu_j}{\sum_j \mu_j}$$

According to the above definitions, a transition t is enabled if all its input places have their truth degrees positive. If the sum of their weighted truth degrees is greater than its threshold $\lambda(t)$, t fires. Thus, through firing transitions, truth degrees can be reasoned from a set of known antecedent propositions to a set of consequent propositions step by step. We may use a continuous function $y(t,x)$ to approximate $y(t)$ in Equation (10.11). Let

$$y(t,x) = x \cdot F(x)$$

where

$$x = \sum_j \theta(p_j) \cdot W(p_j,t)$$

$F(x)$ is a sigmoid function that approximates the threshold of t,

$$F(x) = 1/\left(1 + e^{-b(x-\lambda(t))}\right)$$

where b is a large constant called steepness. If b is large enough, when $x > \lambda(t)$, $e^{-b(x-\lambda(t))} \approx 0$, then $F(x) \approx 1$, and when $x < \lambda(t)$, $e^{-b(x-\lambda(t))} \rightarrow \infty$, then $F(x) \approx 0$. This approximation is essential to equip the net with learning capability to be shown later.

Algorithm 10.1. (Fuzzy Reasoning)
INPUT: APN with initial truth degrees of a set of antecedent propositions, $i.e.,$ θ^0.
OUTPUT: The truth degrees of consequence propositions
        Initialization: $k = 0$.

Step 1. Let $k = k+1$. Find and fire all enabled transitions and update truth degree of places according to Definition 10.4 to obtain θ^k.

Step 2. If $\theta^k \neq \theta^{k-1}$, go to Step 1.

Theorem 10.1. If an APN is acyclic, *i.e.*, it contains no cycles, then Algorithm 10.1 terminates in a finite number of steps.

Proof. It is clear that when input places to a transition have the same truth degrees, the result from firing it changes no truth degrees. As the net is acyclic, only limited number of steps will be needed such that all places will end up with their constant truth degrees.

Example 10.5. Suppose that an expert system has the following weighted fuzzy production rules:

R_1: IF p_1 THEN p_4 with certainty factor μ_1 and threshold λ_1.

R_2: IF p_2 *AND* p_4 THEN p_5 with certainty factor μ_2, threshold λ_2, and weights w_1 (input arc from p_2 to t_2) and w_2 (input from p_4 to t_2).

R_3: IF p_3 *OR* p_5 THEN p_6 with certainty factors μ_3 for t_3 between p_3 and p_6, μ_4 for t_4 between p_5 and p_6, and thresholds λ_3 and λ_4.

The system is converted into Figure 10.13 where R_1 and R_2 are clear while R_3 is converted into a subnet with two transitions t_3 and t_4.

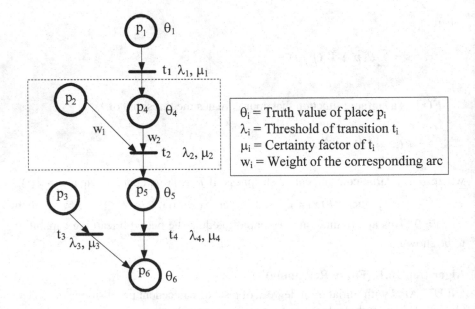

θ_i = Truth value of place p_i
λ_i = Threshold of transition t_i
μ_i = Certainty factor of t_i
w_i = Weight of the corresponding arc

Figure 10.13. APN of a given expert system

Suppose that the data are given as follows:

$$\mu_1 = 0.80, \quad \mu_2 = 0.85, \quad \mu_3 = 0.75, \quad \mu_4 = 0.82$$
$$\lambda_1 = 0.50, \quad \lambda_2 = 0.55, \quad \lambda_3 = 0.60, \quad \lambda_4 = 0.40$$
$$w_1 = 0.63, \quad w_2 = 0.37$$

We use the following sigmoid functions as

$$F_i(x) := \frac{\mu_i}{1 + e^{-b_i(x - \lambda_i)}}, \quad i = 1, 2, 3, 4$$

to approximate the four thresholds λ_{1-4} where steepness b_i is selected as $b = 200$. These functions vs x are drawn in Figure 10.14. For transition t_2, $y(t_2, x) = xF_2(x)$ where $x = \theta(p_2)w_1 + \theta(p_4)w_2$.

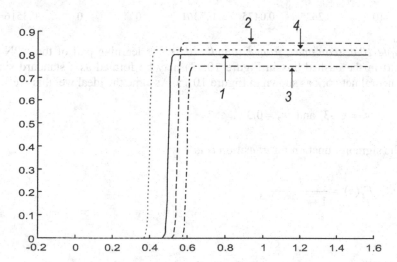

Figure 10.14. Four sigmoid functions in Example 10.5

Table 10.8 gives the reasoning results of APN given the above data, and truth degrees of places p_{1-3}. Each process involves only two steps. If multiple transition fires, they can fire together following Definition 10.4. This is one major difference between standard Petri nets and "Petri nets for reasoning".

One can see that some truth degrees of places are zero. This means that the corresponding thresholds are not passed. For example, in Group 1, since $\theta(p_1) = 0.219 < \lambda_1 = 0.5$, t_1 cannot fire, leading to $\theta(p_4) = 0$.

In Example 10.5, we assume that weights are known. Suppose that weights are unknown but we have sufficient data obtained from the expert system. We can then introduce the neural network concept and related back-propagation algorithm to learn these weights. These weights can be updated when new data are introduced.

Table 10.8. APN Reasoning results given the weights, threshold, certainty factor, steepness, and truth degrees $\Theta(p_1)-\Theta(p_3)$

Group No.	$\Theta(p_1)$	$\Theta(p_2)$	$\Theta(p_3)$	$\Theta(p_4)$	$\Theta(p_5)$	$\Theta(p_6)$
1	0.2190	0.0470	0.6789	0	0	0.3243
2	0.6793	0.9347	0.3835	0.5434	0.5850	0.3055
3	0.5194	0.8310	0.0346	0.4072	0.4517	0.2359
4	0.0535	0.5297	0.6711	0	0	0.3206
5	0.0077	0.3834	0.0668	0	0	0
6	0.4175	0.6868	0.5890	0	0	0.0279
7	0.9304	0.8462	0.5269	0.7443	0.6647	0.3472
8	0.0920	0.6539	0.4160	0	0	0
9	0.7012	0.9103	0.7622	0.5610	0.5867	0.6705
10	0.2625	0.0475	0.7361	0	0	0.3516

Example 10.6. (continued from Example 10.5) The learning part of the APN (see the part in the dashed box in Figure 10.13) may be formed as a standard single-layer neural network as shown in Figure 10.15. Assume the ideal weights are

$$w_1 = 0.63, \text{ and } w_2 = 0.37 .$$

The sigmoid function for transition t_2 is

$$F_2(x) := \frac{0.85}{1 + e^{-200(x-0.55)}}$$

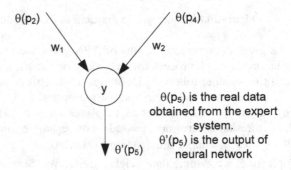

Figure 10.15. The neural network translation of the learning part in Example 10.6

Suppose that inputs $\theta(p_1)$ and $\theta(p_2)$ are given random data from 0 to 1, and the real output $\theta(p_5)$ is obtained according to the expert system. Given any initial condition for w_1 and w_2, put the same inputs to the neural network. The error between the output of neural network $\theta'(p_5)$ and that of the expert system $\theta(p_5)$ can be used to modify the weights with the following learning law:

$$W(k+1) = W(k) + y\delta e(k)\Theta(k)$$
$$e(k) := \theta(p_5)(k) - \theta'(p_5)(k)$$

where

$W(k+1) = [w_1(k+1), w_2(k+1)]$ is the weight at iteration $k+1$;

$W(k) = [w_1(k), w_2(k)]$ is the weight at iteration k;

$$y = \frac{0.85x}{1 + e^{-200(x-0.55)}}$$

δ is the learning rate whose small value assures that the learning process converges. Select $\delta = 0.07$;

$\Theta(k) = [\theta(p_2)(k), \theta(p_4)(k)]$, and $\theta(p_2)(k)$ is a given truth degree of p_2, $\theta(p_4)(k)$ is the reasoned truth degree of p_4;

$\theta'(p_5)(k)$ the reasoned truth degree of p_5 (the output of neural network); and $\theta(p_5)(k)$ is the given truth degree of p_5 from the expert system, all at iteration k.

After a training process ($k > 400$), the weights converge to real values. Figure 10.16 shows simulation results.

In this example there is only one learning layer. A more complicated case with two learning layers can be found in (Li *et al.* 2000).

The development of theory and applications of Adaptive Petri Net (APN) still represents a hot research direction. APN promises to solve the knowledge learning problem in expert systems and other application problems.

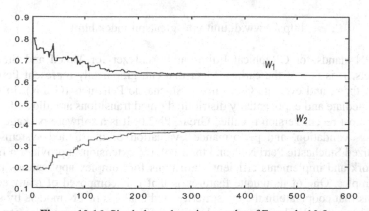

Figure 10.16. Single layer learning results of Example 10.5

10.7 Petri Net-based Design Tools

This section presents a brief overview of many significant tools that have been constructed and applied in academia and industry. The Petri net community has well maintained an active website, *i.e.,* the Petri Nets World, and e-mail list. The website address for about seventy registered Petri net-based design tools is

http://www.informatik.uni-hamburg.de/TGI/PetriNets/tools/quick.html

Most of these tools are free of charge for academic research. There are also many tools in use, which may be found in a variety of publications. The authors have directly or indirectly used the following tools.

SPNP

http://www.ee.duke.edu/~kst/

SPNP supports the analysis of stochastic Petri nets. It was developed by Ciardo *et al.* (Hirel *et al.* 2000). The model type used for input is a stochastic reward net (**SRN**). SRNs incorporate several structural extensions to **GSPNs** such as marking dependencies (marking dependent arc cardinalities, guards, *etc.*) and allow reward rates to be associated with each marking. The reward function can be marking dependent as well. They are specified using **CSPL** (C based SRN Language) which is an extension of the C programming language with additional constructs for describing the SRN models. SRN specifications are automatically converted into a Markov reward model which is then solved to compute a variety of transient, steady-state, cumulative, and sensitivity measures. For SRNs with absorbing markings or deadlocks, mean time to absorption and expected accumulated reward until absorption can be computed. This tool has been widely used for performance evaluation of various discrete event systems (Zhou and Venkatesh 1998).

GreatSPN

http://www.di.unito.it/~greatspn/index.html

GreatSPN stands for GRaphical Editor and Analyzer for Timed and Stochastic Petri Nets. It is one of the earliest tools that can graphically represent Petri nets, simulate them, and evaluate Generalized Stochastic Petri nets (GSPN). In GSPN, both immediate and exponentially distributed timed transitions are allowed.

The most recent version is called GreatSPN2.0. It is a software package for the modeling, validation, and performance evaluation of distributed systems using Generalized Stochastic Petri Nets and their colored extension. It provides a friendly framework and implements efficient algorithms for complex applications not just toy examples. One of its unique features is that it is composed of many separate programs that cooperate in the construction and analysis of PN models by sharing files. Using network file system capabilities, different analysis modules can be run on different machines in a distributed computing environment. Its modular structure allows easy addition of new analysis modules. All modules are written in C programming language to guarantee portability and efficiency on different Unix machines. All solution modules use special storage techniques to save memory both for intermediate result files and for program data structures. Its applications to various DEDS are well documented in a book (Ajmone Marsan *et al.* 1995).

INA

http://www2.informatik.hu-berlin.de/~starke/ina.html

INA represents Integrated Net Analyzer. INA is a tool package supporting the analysis of Petri nets and colored Petri nets. It consists of:

- A textual editor for nets
- A by-hand simulation part
- A reduction part for Petri nets
- An analysis part to compute

    - Structural information
    - Place and transition invariants
    - Reachability and coverability graphs

The by-hand simulation part allows starting at a given marking to forward fire either single transitions or maximal steps; the user can thus traverse parts of the reachability graph.

The reduction part can be used to reduce the size of a net (and of its reachability graph) whilst preserving liveness and boundedness.

The analysis can be carried out under different transition rules (normal, safe, under capacities), with or without priorities or time restrictions (three types), and under firing of single transitions or maximal sets of concurrently enabled transitions. INA can compute the following structural information:

- Conflicts (static, dynamic) and their structure (*e.g.,* free choice property)
- Deadlocks and traps (deadlock-trap-property)
- State machine decomposition and covering

Invariant analysis can be done by computing generator sets of all place/transition invariants and of all non-negative invariants. Vectors can be tested for invariance properties.

For bounded nets, the reachability graph can be computed and analysed for liveness, reversability, realizable transition invariants, and livelocks. The symmetries of a given net can be computed and used to reduce the reachability graph size. It is also possible to apply stubborn reduction. Furthermore, minimal paths can be computed, and the non-reachability of a marking can be decided.

Some external graphical editors and tools can export nets to INA. This tool has been proven robust and very useful. For example it is used for deadlock control design in (Li and Zhou 2006) and (Uzam and Zhou 2006).

Another influential tool is **CPN Tools** at the following site:

http://www.informatik.uni-hamburg.de/TGI/PetriNets/tools/db/cpntools.html

It is a widespread tool for editing, simulating and analyzing colored Petri nets. It features incremental syntax checking and code generation which take place while a net is being constructed. A fast simulator efficiently handles both untimed and timed nets. Full and partial state spaces can be generated and analyzed for boundedness and liveness properties. It is possible to specify and check system-specific properties. The tool also provides support for simulation-based performance analysis.

10.8 Problems and Exercises

10.1. Figure 10.17 models a discrete event system. Initial marking is shown in the figure, times delay j time units is associated with place p_j and i units with t_i. Please show all the cycles and find the system cycle time delay. Given two extra tokens, which place(s) should receive them to minimize the cycle time?

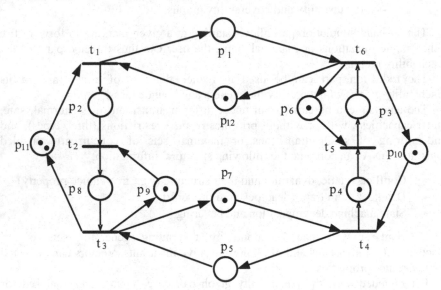

Figure 10.17. A marked graph for Exercise 10.1

10.2. Assume initial marking of the stochastic Petri net is shown in Figure 10.18 and transition t_i has its firing rate λ_i. Please derive the productivity if firing t_4 signifies the completion of a product. Do so for $m_0 = (4\ 1\ 0\ 0\ 0\ 0)^T$ when $\lambda_i = i$.

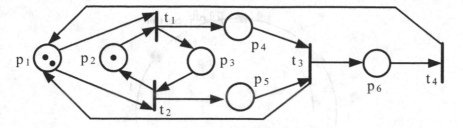

Figure 10.18. A stochastic Petri net for Exercise 10.2

10.3. A serial manufacturing system is schematically given in Figure 10.19. It consists of three cells. Products are coming one by one to the system *via* input. After processing in cell 3 they are moved onto the output. Draw a color Petri net specifying the function of the manufacturing system.

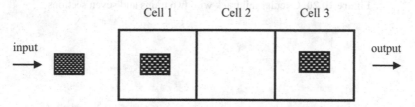

Figure 10.19. A serial manufacturing line

10.4. For Example 10.3 create a Petri net modeling the behavior of the given system. Consider a serial manufacturing system with ten cells and compare the Petri net and the color Petri net models.

10.5. A circle rail track is shown in Figure 10.20. Two trains are moving on the track clockwise. A train can pass in the next section only if next two sections are free.

   a.   Find a Petri net describing the prescribed moves of the trains.
   b.   Analyze the basic properties of the derived Petri net.
   c.   Find a color Petri net describing the prescribed train behavior.

10.6. Suppose that in Example 10.6, w_1 and w_2 are fixed but the certainty factors μ_2 and μ_4 are unknown. Assume that the random input and accurate output data from the expert system can be obtained with their ideal values at

$$\mu_2 = 0.8 \text{ and } \mu_4 = 0.5.$$

Derive the back-propagation algorithm to learn these two parameters, and investigate the convergence with different initial values of μ_2 and μ_4 between 0 and 1, different steepness, and learning rates.

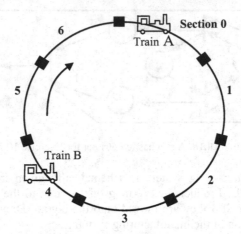

Figure 10.20. Circular rail track with two trains and seven sections

11

Statecharts

11.1 Introduction

So far it could be observed how some shortcomings of finite automata used for DEDS modeling were overcome by Petri nets and Grafcet. This is mainly due to the latter's ability to specify the parallel activities of subsystem states and concurrency of events. However, a certain imperfection of both persists. Difficulties in visualizing large and complex DEDS require to apply a sort of hierarchical decomposition. Another tool for coping with this problem is statecharts (Havel 1987 and Fogel 1997, 1998).

Statecharts use the same notions as the tools mentioned above, namely state and event, and are based on the basic transition system as well. They represent structure and dynamic behavior in a drawn graphical form. Their set-theoretic and functional description is possible, too. The next section introduces their concepts and Section 11.3 presents their applications to DEDS.

11.2 Basic Statechart Components

In a statechart, states are represented by rounded rectangles marked with a symbol, usually a letter as illustrated in Figure 11.1.

The state Q in Figure 11.1a, which can be decomposed into three sub-states A, B, and C as shown in Figure 11.1b, can be viewed as a superstate. Transition from Figure 11.1a to Figure 11.1b is an example of state refinement as opposed to state clustering (corresponding to the transition from Figure 11.1b to Figure 11.1a.

According to the principle of state encapsulation, the state Q encapsulates states A, B, and C. Figure 11.1c depicts a three-level state hierarchy. The semantics of the decomposition depicted in Figure 11.1 is *exclusive-or* (XOR) whereby in Figure 11.1b the situation "state Q is active" means that one and only one of states A, B, and C is active at some time. If saying "the system is in state Q" we mean that it is, *e.g.*, in state B. If the system in Fig, 11.1c is in state Q it can be in state A

being in state V. In such a case Q can be neither in state B nor C, nor A nor U at the same time.

Transition from one state into another one is represented by an arrow labeled with abbreviation of an event. The event label "e" may be completed with a condition in parentheses e(P) indicating that through event "e" the system transits from one state to the other if the additional condition P holds on the event occurrence (Figure 11.2). An option is that an arrow is labeled only with condition (P).

Figures 11.2a and b are equivalent as for the representation of the state transfer $Q \rightarrow R$, which is carried out if condition P holds on event e_1. Arrows can be directed either into a superstate, *e.g.*, the arrow labeled with e_2 in Figure 11.2b or out of a superstate, *e.g.*, the arrows labeled with e_1.

Thus far, time was not explicitly considered in the described statechart components. There are several possibilities to include time conditions in the statecharts, for example, similarly to those in Petri nets. The order of state transitions is specified like that in untimed Petri nets.

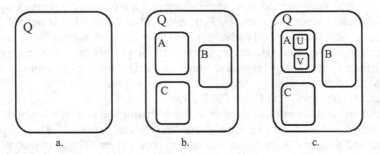

Figure 11.1. Representation of states for exclusive-or activities with hierarchical involvement

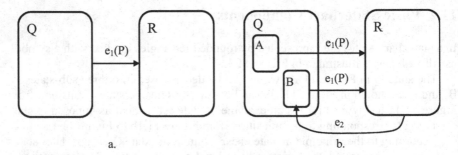

Figure 11.2. State transit representation using arrows labeled with events

In statecharts a default state activation can be specified with an arrow and a dot as shown in Figure 11.3. The state transfer is performed according to the arrows pointing at default states. In Figure 11.3a, the default transit to state Q by event "e" is accomplished through the transit in A given by the transit to the default state V. As a whole, event "e" transfers the system from state R to V. Other needed arrows are left out.

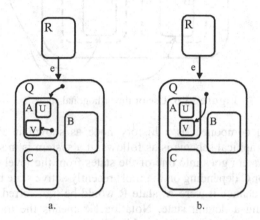

a.   b.

Figure 11.3. Default state activation

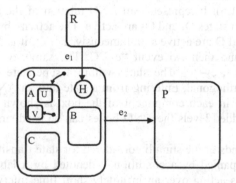

Figure 11.4. Use of the history node

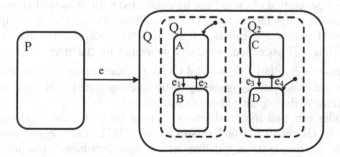

Figure 11.5. Orthogonal states Q_1 and Q_2

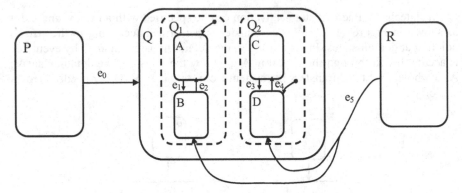

Figure 11.6. Use of the orthogonal states

Another useful component is a history node as shown in Figure 11.4. The semantics of the graphical scheme is as follows: if a system is in state R and event "e_1" occurs, the system goes into one of the states from the level of the superstate Q, which is A, B or C depending on the most recently active state before Q was left by event e_2. If by chance it were A, state R would be transferred in V because of the arrow indicating a default state. Notation H* means the transit in the most recently active state on the lowest hierarchical level.

Concurrency and independency are represented by dashed rectangles within a superstate (Figure 11.5). It represents an AND relation of the state activities. On entering state Q, both states Q_1 and Q_2 are active. The activity by refinement means that both states A and D are active simultaneously. For Q_1 it is A and then *via* "e_1" B, for Q_2 it is D and then *via* event "e_4" C. The same event can effect both components of Q if, , $e_2 = e_4$. The states within a superstate being in the AND relation are called orthogonal. Entering from a state into an AND box ensures the activation of a state in each component of the box as shown in Figure 11.6. In hierarchically embedded levels the AND boxes can be used similarly as the XOR boxes.

There are two kinds of the stimuli considered for state transitions: the event "e" alone or "e" accompanied by a condition P denoted by a label e(P). The event represents a stimulus acting over an infinitely short time interval or in a discrete time point. State transfer by an event is realized if a durable (level kind) condition P is met. The generation of events can be connected with transits of states (notation e_1/s) or with the states themselves. The possibilities are explained in Figure 11.7. The generated events are called actions and the changes of values are called activities. Thus s is an action, which is generated by the transfer $P \rightarrow Q$ and is firing transit $A \rightarrow B$. Start (X) and end (X) are special events setting and resetting X. An example of an activity is setting Y by entering state C, setting Z by the exit from C, setting W during the activity of state C.

The reader can find more information about the statecharts and their formal definitions in (Harel 1987) and (Harel *et al.* 1987). Their applications to the modeling of reactive systems together with a statechart-based structured analysis method called STATEMATE are in detail presented in (Harel and Politi 1998). A thorough treatment of converting Petri nets into statecharts is given in (Eshuis

2006). An algorithm of polynomial complexity is proposed to translate a class of Petri nets into an equivalent statechart. Meanwhile, the Petri net structure is preserved. Some recent applications to industrial automation can be found in (Lee *et al.* 2005).

11.3 Statechart Application

A printed circuit board (PCB) assembly cell in Figure 7.10 is used as an illustrating example. The Petri net interpreted for control and a statechart specification are compared. Figure 11.8 shows the Petri net interpreted for the cell control. The alteration of the robots by both the component picking and inserting has been chosen. The cell and the Petri net specification are described in Example 7.3, Section 7.3. The meaning of additional transition conditions and control commands associated with places is given in Table 11.1. If a transition bears a condition, CR1_a, it is a shortage for a test if variable CR1_a $==$ true (logic one).

The same control as that from the Petri net has been specified using the statechart depicted in Figure 11.9. Both representations offer the similar complexity graphically.

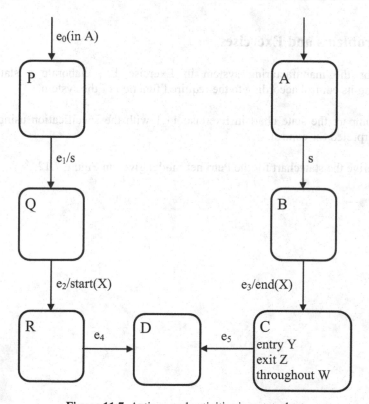

Figure 11.7. Actions and activities in a statechart

Table 11.1. Petri net logic conditions and control commands

Logical conditions attached to transitions	
CR1_a (CR2_a)	Electronic component for R1 (R2) is available
R1_ep (R2_ep)	End of component picking from feeder by robot R1 (R2)
AR1_b (AR2_b)	Arm of robot R1 (R2) pulling back done
R1_pcb (R2_pcb)	Robot R1 (R2) is near to PCB area prepared for inserting
R1_cins (R2_pcb)	Inserting of a component by R1 (R2) done
R1_fa (R2_fa)	Robot R1 (R2) is near to feeder area prepared for picking
Control commands	
R1_P (R2_P)	Command for R1 (R2) to pick an electronic component
AR1_B (AR2_B)	Command pull back the arm of robot R1 (R2)
R1_MPC (R2_MPC)	Command for R1 (R2) to move to PCB area
R1_IC (R2_IC)	Command to insert a component
R1_MFA (R2_MFA)	Command to move to feeder area

11.4 Problems and Exercises

11.1. For the manufacturing system in Exercise 1.5, elaborate a statechart specifying its control according to the required function of the system.

11.2. Compare the state chart in Exercise 11.1 with the specification using Petri nets interpreted for control.

11.3. Derive the statechart for the Petri net model given in Figure 7.12.

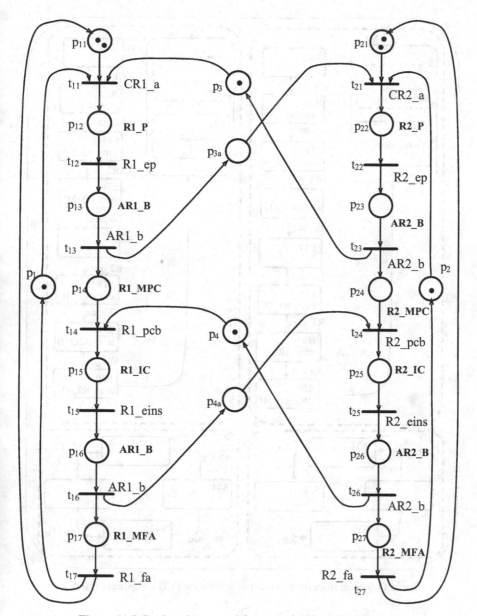

Figure 11.8. Petri net interpreted for control of the PCB assembly

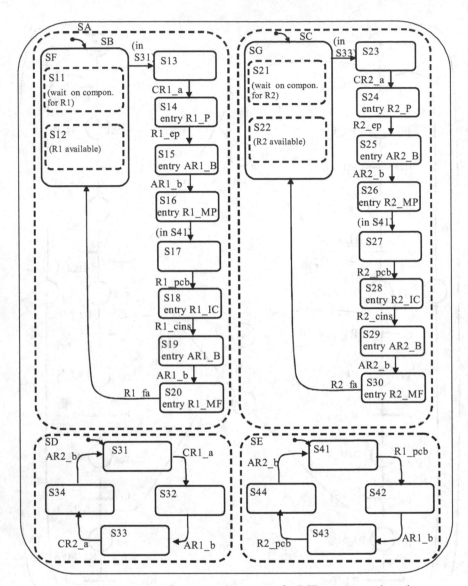

Figure 11.9. Statechart for a two-robot system for PCB component insertion

12

DEDS Modeling, Control and Programming

12.1 Modeling Methodology

Finite automata, Petri nets, Grafcet, statecharts, *etc.,* are tools for DEDS specification and analysis. They represent a system as a whole or its chosen parts. The function of the control system to be designed requires specific extensions of the expression means in order to enable a reactive performance of the control in a feedback system structure.

Generating a DEDS model is a highly creative process. There are many ways how to achieve this goal. However, precise and exhausting hints for it do not exist. Rather there are several supporting methodologies. One of useful practical ways is to first elaborate a model of the complete system as it appears to an observer with the control included. Second, a model of the system control is to be elaborated, which may serve for the control function analysis and writing control programs.

Operations in execution represent partial system states, which change through events. From the abstract system viewpoint the basic DEDS features are generally as follows:

a. Concurrency of operations
b. Synchronization of operations
c. Sharing common resources such as machines, robots, storages, data blocks, *etc.*
d. Limitation of resources
e. Cyclic behavior, *i.e.,* the repetition of some procedure schemes
f. Failure-safe processing without shutdowns, deadlocks, *etc.*
g. Achieving maximum efficiency with respect to some criteria, *e.g.,* right product mix, maximum throughput, minimum make-span and other, *via* optimal routing of objects in the system and scheduling of operations.

Considering Petri nets as a main DEDS modeling tool in this book, it is easy to verify that the individual features are ensured as follows:

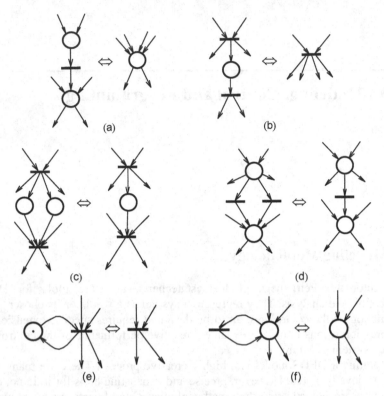

Figure 12.1. Basic Petri net transformation rules

Features a through c: they are contained in the very substance of Petri nets;
Feature d: by the boundedness;
Feature e: by the reversibility;
Feature f: by the liveness;
Feature g: by the Petri nets interpreted for control.

Then a basic goal is to create a bounded, reversible and live Petri net as a DEDS model having all required states or markings reachable. There are two ways to achieve it:

1. To generate a Petri net without any constraints and then to check the analyzed features and make necessary corrections.
2. To use elementary building blocks preserving the required features in bottom-up composition or in top-down refinement.

Frequently, the complexity of a real DEDS leads to a large and complex Petri net. Analyzing such nets is practically very difficult and time consuming. Therefore, availability of a systematic approach to the Petri net design is highly desirable. The substance of such approaches is model reduction and composition techniques. Such a transformation, either reduction or composition, helps to simplify or create Petri nets preserving the required properties, and makes an analysis of Petri nets easier.

A basic list of the most frequently used transformation rules is presented graphically in Figure 12.1. The rules (Murata 1989; Zhou and Venkatesh 1998) include:

a. Fusion of series places;
b. Fusion of series transitions;
c. Fusion of parallel places;
d. Fusion of parallel transitions;
e. Elimination of self-loop places;
f. Elimination of self-loop transitions.

Often it is advantageous to use a hierarchical decomposition of a Petri net (a higher level) into subnets (lower level). A standard way is to develop either a transition of a hierarchically higher Petri net into a subnet or place into a subnet (Abel 1990). In the former case, the developed transition splits into two transitions between which the subnet is located, whereby no arcs are permitted between the subnet and the nodes of the higher level Petri net. The latter case is treated similarly using place splitting. Decomposition of a transition is illustrated in Figure 12.2.

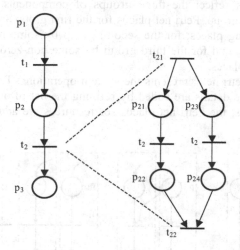

Figure 12.2. Decomposition of a transition

Consider a composition method (Ferrarini 1992, 1995 and Ferrarini *et al.* 1994) as a model of such efforts. The basic idea is to represent an elementary control task using a strongly connected binary Petri net state machine (SCSM, see Section 7.4) with one and only one marked place within the initial marking. A complex control function is designed using a composition of a group of elementary control tasks represented by a composition of the corresponding elementary SCSMs. The SCSM interactions have to satisfy various relations, *e.g.*, synchronization and blocking. Properties like liveness and boundedness are preserved by the composition.

Various kinds of interconnections are shown in Figure 12.3. There are: (a) self-loop, (b) inhibitor arc, and (c) synchronization. The functions of the interconnections are easy to understand from the graphical representation where the elementary SCSM are noted as E_1, E_2, ... , and E_k.

Zhou and DiCesare (1991) and Zhou *et al.* (1992) proposed one of the excellent reduction/composition methodologies. The core idea of their methodology is a top-down refinement of a relatively simple first-level Petri net, which models basic system operations, and a bottom-up completion of the net with places and transitions corresponding to the system resources. More discussions can be found in (Zhou and DiCesare 1993). Brief introduction to their results is presented below.

A common DEDS feature is the possibility to distinguish the following system components:

1. Operations performed in the system, *e.g.*, processing parts in a flexible manufacturing system (FMS), and processing pieces of data in a distributed computer system.
2. Subsystems representing fixed resources, *e.g.*, machines and robots in FMS, and computer processors.
3. Subsystems representing variable resources, *e.g.*, pallets or fixtures in FMS, and external memory devices in the computer systems.

Petri net models reflect the three groups of components. A set of places corresponds to each group. Petri net places for the first group are characterized by a zero initial marking places; for the second group by some non-zero binary or fixed marking place, and for the third group by some non-zero possibly variable numerical marking places.

The first-level Petri net represents the system operations. Then, using the top-down approach more details are added by refining the net of operations and their relations. Afterwards, the Petri net places for resources are added to the net in a bottom-up manner.

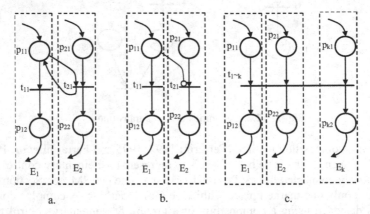

Figure 12.3. Three kinds of elementary structure connections by the composition

Figure 12.4. A sequence Petri net

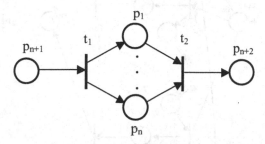

Figure 12.5. A parallel Petri net

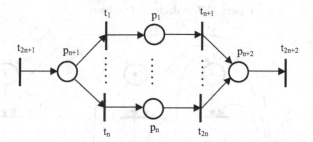

Figure 12.6. A choice Petri net

The first-level Petri net is composed of basic design modules. They are chosen to posses the three most important properties reflecting real-world systems, namely boundedness, reversibility and liveness. The basic design modules include:

a. Sequence Petri net (Figure 12.4),
b. Parallel Petri net (Figure 12.5),
c. Choice Petri net (Figure 12.6),
d. Decision-free choice Petri net (Figure 12.7).

It has been proved in the above-mentioned works of Zhou *et al.,* that any composition of basic design modules preserves the three mentioned basic properties: boundedness, reversibility and liveness. In the refinement process, modules a and b can replace a place, and modules c and d can replace a transition.

The bottom-up procedure resolves the problem of sharing the system resources. Two kinds of resource sharing can be distinguished: a parallel mutual exclusion of resources and a serial mutual exclusion of resources. Resource sharing situations in

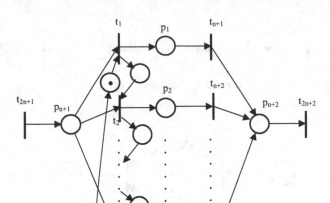

Figure 12.7. A decision-free Petri net

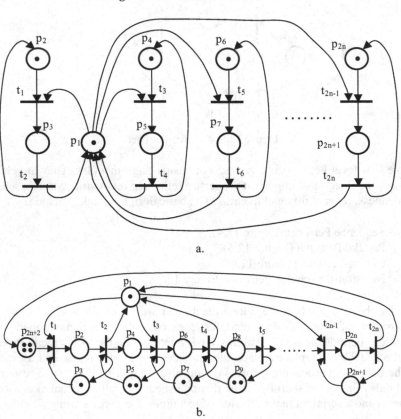

a.

b.

Figure 12.8a. Parallel mutual exclusion and **b**. Serial mutual exclusion, also called sequential mutual exclusion

DEDS expressed by means of the Petri net elementary structures are depicted in Figures 12.8. The initial marking in Figure 12.8 is one possible example. However, there are several possibilities depending on the actual requirements. It has been proved that structures in Figure 12.8 preserve the basic Petri net properties considered above both individually or in combination with the basic design modules under certain conditions.

The described methodology is applicable in the DEDS regardless of the system's substance. Next this methodology will be illustrated on a flexible manufacturing system depicted in Figure 12.9 (Niemi *et al.* 1992).

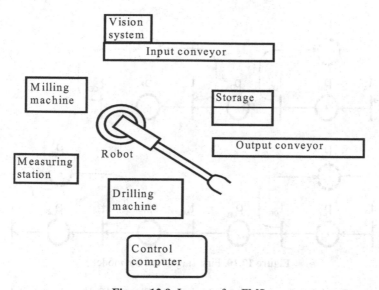

Figure 12.9. Layout of an FMS

This system comprises a semiconductor camera vision system for the recognition and location of parts coming into the system on an input conveyor. A central transferring server of the system is a robot. Further, there are two NC machining units: a drilling machine with two drills above an xy-table and a three-axis milling machine with a tool changer. A measuring station is located between the milling and drilling machines. There is an intermediate storage where the robot can temporarily put the parts, and an output conveyor. The intermediate storage can be used as a multiple storage for more types of the products. The system is equipped with control computers on both process and coordination levels.

A particular technological process to be cyclically realized in the system is as follows. There are two different types of raw parts at the input. The first is intended for the product of type A, the second for type B. The parts come irregularly, *i.e.,* with random time intervals between two parts and with a random number of consecutive parts of one type. The parts are recognized and located by the camera vision system. Product A is obtained by drilling followed by measuring. Product B is obtained by milling. The cell robot provides the necessary transfers of the parts or products. No intermediate storage is considered in this particular example.

Example 12.1. Figure 12.10 shows the first step of a Petri net system performance modeling. The places of Petri nets correspond to operations. For the sake of brevity the meaning of the system places is given in Figure 12.11, which shows the next stage-refined model. Transitions correspond to the starts and ends of operations. The first stage Petri net model is drawn in full lines (the Petri net is not a strongly connected one). The question of reachability, boundedness, liveness and reversibility is to be resolved. Undoubtedly a full-line Petri net is bounded, live and each of its marking is reachable. However, it is not reversible. The additional parts of the net drawn in dashed lines turn the Petri net into a reversible one.

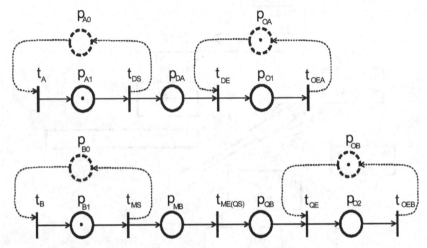

Figure 12.10. First stage Petri net model

The next refined Petri net model is shown in Figure 12.11. The availability of machines is given by marked places, *e.g.,* the drilling machine is free if place p_D contains a token. If the part of the net modeling the robot function is omitted, a marked graph (Section 7.4) is obtained which is evidently bounded, reversible and live for the given initial marking. The robot presents a shared resource with conflicts (there are more arcs outgoing of place p_R). The conflicts are of mutual exclusion type. Though the resulting net is no more a marked graph, nevertheless it is bounded, reversible and live as discussed earlier in this section.

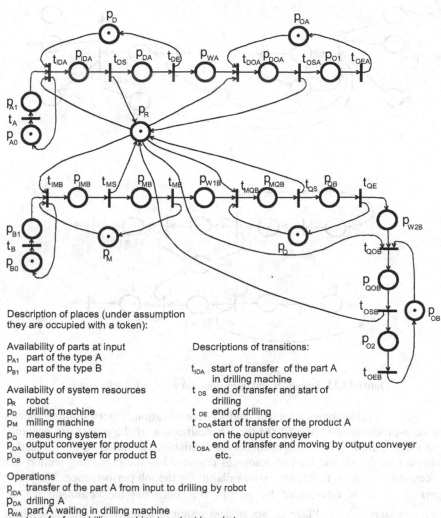

Description of places (under assumption they are occupied with a token):

Availability of parts at input
p_{A1} part of the type A
p_{B1} part of the type B

Availability of system resources
p_R robot
p_D drilling machine
p_M milling machine
p_Q measuring system
p_{OA} output conveyer for product A
p_{OB} output conveyer for product B

Operations
p_{IDA} transfer of the part A from input to drilling by robot
p_{DA} drilling A
p_{WA} part A waiting in drilling machine
p_{DOA} transfer from drilling machine to output by robot
p_{O1} transfer of the product A at output by conveyer
p_{IMB} transfer of the part B from input to milling by robot
p_{MB} milling B
p_{QB} measuring of the semiproduct B, *etc.*, analogously as for A

Descriptions of transitions:

t_{IDA} start of transfer of the part A in drilling machine
t_{DS} end of transfer and start of drilling
t_{DE} end of drilling
t_{DOA} start of transfer of the product A on the ouput conveyer
t_{OSA} end of transfer and moving by output conveyer
etc.

Figure 12.11. Petri net model for FMS

12.2 Resolution of Conflicts

System conflicts reflected as Petri net conflicts are to be removed by practical control solutions (Hrúz *et al.* 1996; Hrúz 1997). The following three figures show the basic Petri net interpreted for control structures corresponding to three basic conflict situations in which the resources can be involved.

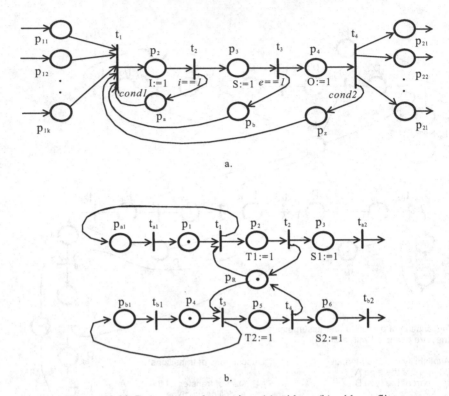

a.

b.

Figure 12.12. Resources and operations (a) without (b) with conflict

Figure 12.12(a) shows a typical conflict-free situation. Places $p_{11}-p_{1k}$ specify the semi-products s_1-s_k needed for the job realization. If the places are occupied with tokens, the required semi-products are available. Tokens in p_a, p_b, ..., and p_z indicate that the second kind of resources described in the preceding section like processing machines, robots *etc.,* are available for the job performance. The job or operation itself is represented by p_3. In this Petri net the command for the operation start is S:=1. Place p_2 specifies a required transfer of semi-products to the input of the processing machine (command I:=1), and p_4 (command O:=1) is connected with the transfer out of the manufacturing cell. Various events, *e.g.,* semi-product presence signals, start or end of the operations *etc.,* can be associated with the transitions. Output places $p_{21}-p_{2l}$ indicate that the operation brings about several different workpieces as the output. Let us follow a portion of the model dynamics. After firing t_1 (if *cond1* is met), a token goes to p_2 and the command I:=1 is generated forcing the transfer of semi-products to the processing machine, start of conveyors *etc.* When the process variable i gets the value 1, transition t_2 fires and a token comes to p_3 and p_a, respectively. The command S is set to one which triggers the operation. Analogously, after the end of operation (process variable $e==1$), the semi-products obtained in the operation are transferred to the output and the machine is free. The outgoing arcs from the transitions t_2,t_3,t_4

specify that, after the corresponding events, the resources are again available. The Petri net in Figure 12.12a belongs to the Petri net class called marked graphs with respect to the structure. There are no conflicts in the marked graphs.

The next typical situation in FMS is modeled by a Petri net interpreted for control as depicted in Figure 12.12b. One resource, *e.g.*, a robot (available if place p_R is occupied with a token) is used or shared by two operations, namely a workpiece transfer to the first machine (place p_2, control command $T1:=1$) and a workpiece transfer to the second one (p_5, control command $T2:=1$). The individual machine operations are initiated by means of commands $S1$ and $S2$. There can be more shared resources. For brevity, the elementary structure in Figure 12.12b is limited to one shared resource only. Obviously, it is a parallel mutual exclusion. Finally, Figure 12.13 shows a situation when a job decision or choice occurs in the system. Then, p indicates the availability of a semi-product for the next processing that can be performed either by operation $O1$ (place p_{OP1}) or by $O2$ (place p_{OP2}), *etc.*, up to the operation Oj (p_{OPj}). As is clear from Figure 12.13, machine $M1$ is required for the operation $O1$, machine $M2$ for $O2$, *etc.*

The composition of the above described elementary building blocks (Figures 12.12 and 12.13) can generate aggregated and more complex Petri net structures containing conflicts. Consider a Petri net interpreted for control (further denoted as a *PC*) is built up of the elementary structures shown in Figures 12.12b and 12.13. The conflicts which may occur in the *PC* are to be eliminated in order to achieve a required deterministic behavior *via* control.

Assume that a reversible, bounded and live (on level L4) *PC* has been designed using the methodology described in Section 12.1 and it specifies the desired control. Conflicts due to the shared resources with or without operation choice can be eliminated by means of inhibitors and incidentors. The elimination method consists of the following steps.

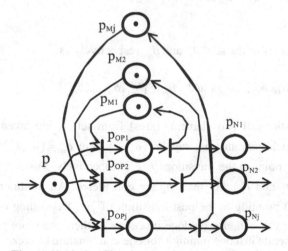

Figure 12.13. Conflict situation due to operation choice

Step 1
The given *PC* is inspected in order to find all places with two or more outgoing arcs. Let the set of such places be denoted as P_O. Then the set of places with more than one incoming arcs is found and denoted P_I. The two sets may not be disjunctive (they can intersect), *i.e.,* generally $P_O \cap P_I \neq \varnothing$. If the *PC* contains operation choice, it is assumed that according to the elementary structure in Figure 12.13, the resources for the operations are represented by the corresponding places in the net.

Step 2
For each place $p_k \in P_O$ the set of outgoing arcs is

$$A_k = \left\{ \left(p_k, t_{j_1} \right), \left(p_k, t_{j_2} \right), ..., \left(p_k, t_{j_k} \right) \right\}$$ (12.1)

where

$$\left(p_k, t_{j_1} \right) \in F, \left(p_k, t_{j_2} \right) \in F, ..., \left(p_k, t_{j_k} \right) \in F$$ (12.2)

Step 3
For each place $p_m \in P_I$ a set of incoming arcs is

$$B_m = \left\{ \left(t_{j_1}, p_m \right), \left(t_{j_2}, p_m \right), ..., \left(t_{j_m}, p_m \right) \right\}$$ (12.3)

where

$$\left(t_{j_1}, p_m \right) \in F, \left(t_{j_2}, p_m \right) \in F, ..., \left(t_{j_m}, p_m \right) \in F$$ (12.4)

Step 4
Denote the elements of the sets A_k and B_m respectively as

$$A_k = \left\{ a_{k1}, a_{k2}, ..., a_{kj_k} \right\} \text{ and } B_m = \left\{ b_{m1}, b_{m2}, ..., b_{mj_m} \right\}$$ (12.5)

All combinations of two elements (arcs) from set A_k are taken into account. Consider one such combination, say $\left\{ a_u, a_v \right\}$, where $a_u = \left(p_k, t_u \right), a_v = \left(p_k, t_v \right)$. The arcs a_u, a_v point to the transitions t_u, t_v, respectively. Further, the pre-places $\tilde{p}_u$ and $\tilde{p}_v$ of t_u and t_v other than p_k are considered. An inhibitor going out of $\tilde{p}_u$ is generated pointing to the post-transition of $\tilde{p}_v$. Depending on the required system behavior this can also be done inversely. If there are more pre-places, it is sufficient to generate just one inhibitor for one combination of arcs.

This step is repeated for all combinations of elements from set A_k ; the procedure is repeated for all sets A_k , $k = 1,2,....,$. If necessary, it is also repeated for sets B_m , $m = 1,2,.....,$. In case of shared resources it is sufficient to deal with sets A_k only.

The case when neither $\tilde{p}_u$ nor $\tilde{p}_v$ exists has to be specially treated. It should be noted that a formal application of the described method can generate redundant inhibitors because several inhibitors can eliminate the same conflict. After adding inhibitors, reversibility, boundedness and liveness of the obtained Petri net has to be analyzed.

Example 12.2. In this example a more complicated situation is shown using a technology layout of Figure 12.9. For the system in Figure 12.9 Let us consider two intermediate storages of parts and the following production task: one type of product is to be produced by a freely ordered milling and drilling of a raw part coming irregularly at the input. Input, output and storages are considered as operations in a broader sense. Possible prescribed sequences of operations are:

1. Input	1. Input	1. Input	1. Input
2. Milling	2. Milling	2. Drilling	2. Drilling
3. Drilling	3. Storage no. 1	3. Milling	3. Storage no. 2
4. Output no. 1	4. Drilling	4. Output no. 2	4. Milling
5. Output no. 1	5. Output no. 2		

An ordinary Petri net modeling the given system behavior is drawn in full line in Figure 12.14. Places and transitions are similar to those in Figure 12.12. The capacity of storages is 5, and the same are capacities of the corresponding places, namely p_{S1}, p_{WS1}, as well as of places p_{S2}, p_{WS2}. As in the previous case, sharing of resources appears with places p_R, p_M, and p_D . It brings about mutual exclusions, treated in the previous section.

The arcs (p_A, t_{11}) and (p_A, t_{21}) represent a conflict situation, namely an operation conflict. It occurs if the places p_A, p_R, p_M, p_D all contain a token. Analogously an operation conflict can appear with place p_{WM1} or p_{WD2}. The places associated with the operation conflict are crosshatched in Figure 12.14. This type of conflicts is called a branching operation conflict. On the other hand, p_{RMSD} and p_{RDSM} bring about the so-called merging operation conflict.

For control design it is necessary to eliminate the indeterminacy produced by any type of conflict. *PC* extends the modeling power of standard Petri nets, involving a larger class of flexible manufacturing systems. Control needs to deal with the conflicts of all three groups, *i.e.,* 1) branching operation conflicts, 2) merging operation conflicts, and 3) resource sharing conflicts that should be identified and eliminated using *PC*.

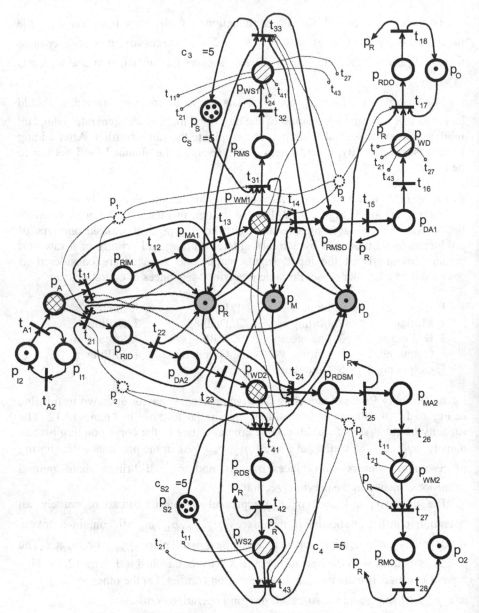

Figure 12.14. FMS Petri net model with conflicting operations

Let us consider the first group. In Figure 12.14 the involved places are crosshatched. Take p_A and extensions drawn in dashed lines. A token in p_1, the empty place p_2 and a token in all places p_A, p_R, p_M, p_D represent a situation when (under safe rules respecting capacities) an otherwise conflicting situation is removed. Let $M(p, r)$ denote the marking in p at the r-th time point.

A similar situation occurs if

$$M(p_A,r)=1, M(p_1,r)=0, M(p_2,r)=1,$$
$$M(p_R,r)=1, M(p_M,r)=1, M(p_D,r)=0$$

The incidentor (p_1, t_{21}) eliminates the conflict for the marking

$$M(p_A,r)=1, M(p_1,r)=0, M(p_2,r)=0,$$
$$M(p_R,r)=1, M(p_M,r)=1, M(p_D,r)=1$$

whereby the control policy giving priority milling over drilling in a symmetric situation has been used. The inhibitor (p_{WM1}, t_{21}) eliminates the conflict in case when drilling both as the first operation as well as the second one is possible. The branching operations conflicts associated with p_{WM1} and p_{WD2} are solved similarly as with p_A.

The merging operation conflict associated with p_{RMSD} is solved by means of the inhibitors (p_{WM1}, t_{33}) and (p_3, t_{33}) giving priority to the firing of t_{14}. The latter inhibitor "covers" the whole occupation of the drilling machine. Similarly it is with place p_{RDSM}.

Each resource sharing conflict has to be solved by analyzing the situation in the pre-places of the transitions involved in the conflict. The pre-places are hatched or crosshatched. A resource sharing conflict can be solved by an appropriate use of inhibitors, *e.g.*, if for the markings at the r-th time point

$$M(p_{WS1},r)=1, M(p_R,r)=1, M(p_{WM2},r)=1,$$
$$M(p_D,r)=1, M(p_3,r)=0, M(p_{O2},r)=1,$$

then the conflict is eliminated by means of the inhibitor (p_{WS1}, t_{27}) whereby a priority is given to the transfer to the drilling machine, according to the second operation sequence introduced above. Some resource sharing conflicts can be solved within the solution of operation conflicts. For example, a resource sharing conflict with (p_R, t_{11}) and (p_R, t_{21}) is solved within the solution of the branching operation conflict associated with p_A. The same happens if both p_R and p_{WM1} are occupied and a branching operation conflict for p_{WM1} is solved.

Inhibitors and incidentors eliminate conflicts and generate the final *PC*-Petri net with complete transition conditions and place control variables.

Example 12.3. A *PC* is to be generated (Figure 12.15) for the technological layout in Figure 12.9 and the following production:

two products are to be produced, each from its own input raw part

product X1:   1st operation–drilling
                   2nd operation–milling
product X2:   one operation–milling

After drilling, semi-product X1 can be stored in the intermediate storage if the milling machine is occupied; and product X2 can only go directly to the output. Each product has its own output. Drilling X1 is preferred to milling X2. Milling X2 is preferred, if possible, to storing. Other relations are obtainable, at least the authors hope, from the "talk" of the Petri net model.

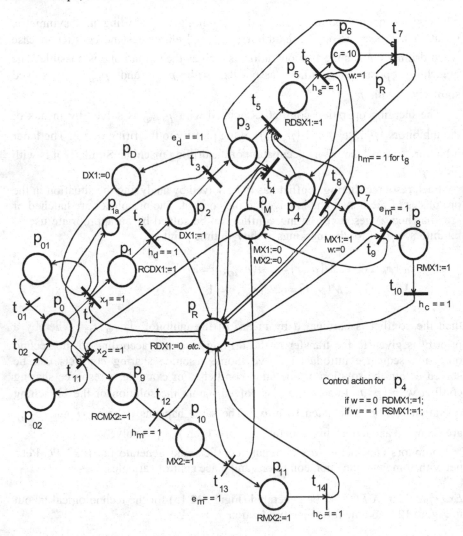

Figure 12.15. FMS job with operation conflicts due to the use of storage

Example 12.4. Figure 12.16 represents a robotic cell with one robot R, two processing machines M1 and M2, input (C1) and output (C2) belt conveyors. Workpieces gather at the stop S where a photo-sensor P1 detects a workpiece to be prepared for the processing in the cell. The machines perform different jobs with different operation times. A prepared workpiece is transferred to a free machine. If both machines are free, M1 has priority.

After finishing the job, the robot transfers the processed workpiece on the output conveyor C2 (assuming that there is free space for it). The robotic cell coordination control is represented by a *PC* in Figure 12.17.

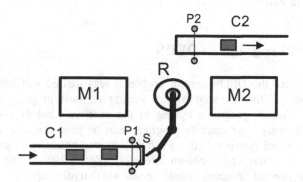

Figure 12.16. A robotic manufacturing cell

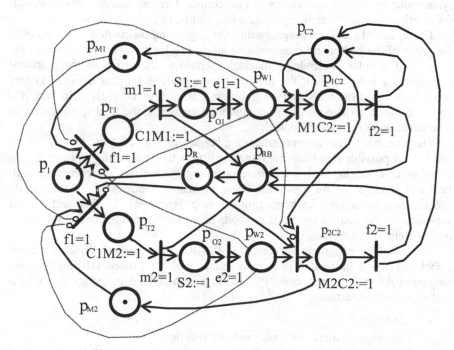

Figure 12.17. *PC*-Petri net for the control of robotic cell

In this example, the robot is a typical shared resource. The corresponding place in the PC in Figure 12.17 is p_R. Place p_{RB} if occupied with a token, specifies a control situation when the command is generated to return the robot to the home position. Places p_{M1} and p_{M2} stand for the availability of M1 and M2, analogously place p_{c2} for the conveyor C2. Condition $f1 == 1$ (or shortly $f1 = 1$) signals the presence of a workpiece at the input. The control commands are associated with places, *e.g.*, the command C1M1:=1 starts the transfer of the workpiece from a fixed position at conveyor C1 into machine M1, and the command C1M2 into M2.

12.3 Control Programs in DEDS

The framework for the DEDS control problems was treated in Chapter 4. After control specification, the next step towards writing a control program is choice of a suitable tool. Control programs belong to the reactive kind of programs. The programming language for reactive programs can be procedural or graphical. In both cases the control system reactivity can be achieved either by a cyclic sampling of the system variables or by system interrupts (Zöbel 1987). The former case means that the control program should repeat sufficiently rapidly the following control scheme: evaluation of the actual system data and successive production of the required actions. The latter requires a possibility to program responses to system interrupts. Petri nets interpreted for control, Grafcet, and statecharts support the creation of control reactive programs responding to external events.

The so-called real-time programming languages and instruction lists pertain to the group of the procedural programming languages for reactive programs. On the other hand popular ladder logic diagrams pertain to the group of the graphical programming languages for the reactive programs. The borders between system control specification tools and programming languages are fuzzy. In fact, the control program written using a programming language is the final specification or determination of the system control.

There are many real-time programming languages, *e.g.*, PEARL, Occam, Forth, and Ada. A possible real-time programming way is to use a standard programming language co-operating with a real-time operating system that completes the real-time and reactive control functions. There are many combinations of the control specification tools and real-time programming languages. In this section, the substance of the control synthesis methods using such combinations is illustrated on a few chosen cases.

The real-time programming language PEARL (Werum and Windauer 1989; Zöbel 1987) will be treated as a representative of the procedural programming languages for reactive programs. PEARL is a multitasking language with all the real-time specific features:

- Modularity
- Language elements and constructs for real time
- Processing of external events

- Programming of input/output operations
- Parallel processes
- Process synchronization
- Exceptions

A program module is delimited with instruction words MODULE <*name*>; and MODEND;. Modules are compiled independently. A module consists of a system division and/or problem division introduced by words SYSTEM; and PROBLEM; respectively. Data are allowed to be used only when defined. PEARL distinguishes local (on the module level) and global (for all modules) data and variables. The system division contains description of the system hardware components and their mutual connections in the particular configuration. The description has to meet the given syntax rules but its actual contents depends on the computer system used and the language compiler for the system.

The problem division contains specification of the input and output data stations and interrupt specification referring to the description in the system division. Further it contains the declaration of all variables used in the program. The specifications are introduced by the keyword SPECIFY or by short SPC. The data submitted by a data station can be specified in details with keywords IN, OUT, INOUT, ALPHIC, BASIC and others. The following statements provide data from the data stations: READ with its counterpart WRITE (for data transfer without transformation), GET – PUT (for data transfer with transformation from the external alphabetic form into the computer internal binary form), and TAKE – SEND (for transfering data with attribute BASIC without transformation). One can declare (DECLARE or DCL) all usual types of variables and types necessary for the reactive programs with the keywords FLOAT, FIXED, BIT, DURATION, CLOCK, CHARACTER, SEMA, STRUCT, *etc.*

The subroutines and function are available through the keyword pair PROCEDURE – RETURN. They are called using CALL <*identifier*> [list_of_actual_parameters]. The brackets [ ] denote an option, *i.e.,* a non-compulsory fraction of the statement.

PEARL has rich expression facilities like IF – THEN – ELSE – FIN block, CASE – ALT – OUT – FIN block, FOR – FROM – BY – TO – REPEAT – END block, and many others. PEARL enables one to write highly structured programs where transferring data via data stations and data manipulating in the statement expressions are possible only if the conform variables are correspondingly specified or declared. The reader can find a detailed description *e.g.,* in Werum and Windauer (1989), or in Reißenweber (1988).

The possibility of dealing with interrupts is an important property of the reactive programs. The following self-explanatory program example shows how it is possible to deal with interrupts in PEARL (keywords are written bold).

MODULE MOD1;
    SYSTEM;
        ALARM: **INT*3;** /*INT*3 is the name recognized by the compiler
                in the actual computer systems*/

```
          PROBLEM;
            SPECIFY ALARM INTERRUPT
            START:       TASK PRIORITY 10;
                                WHEN ALARM ACTIVATE EMERG1;
                         END;

       MODEND;
```

The **TASK** definition is an important property for the program reactivity. Tasks specified and started in PEARL represent autonomous processes competing mutually for the computer processor. The following instructions are available to define the time or event condition for activating a task, which after its activation competes for getting the processor:

Task definition

<name>:   **TASK [PRIORITY** *<priority level>***]**;
                          .
                       task statements
                          .
                       END;

Task activation or suspension

[*<schedule>*]   **ACTIVATE**   *<name>*;
                       SUSPEND   [*<name>*];
[*<simple schedule>*] **CONTINUE**   [*<name>*];
<simple schedule>   **RESUME;**

                       PREVENT   [*<name>*]; /*task schedule is suspended*/
                       TERMINATE[*<name>*]; /*task is terminated, schedule
                                                remained*/

Task control

                       AT   *<time>*
                       AFTER   *<duration>*
                       WHEN   *<interrupt>*
                       UNTIL   *<time>*
                       DURING   *<duration>*

The structure of the above-introduced options is graphically illustrated in Figure 12.18.

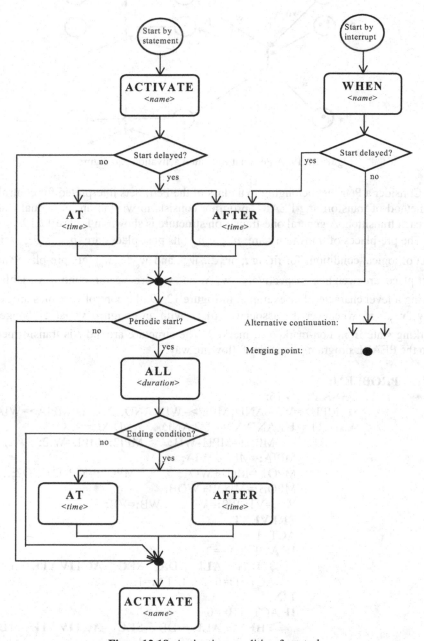

Figure12.18. Activation condition for a task

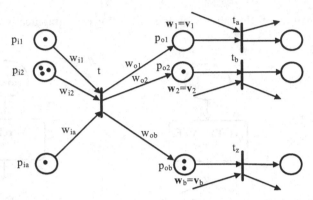

Figure 12.19. A Petri net transition activation paradigm

Consider a Petri net belonging to a class of the Petri nets interpreted for control. A method of transforming PEARL program consists in writing an individual task for each transition. A general one-transition structure is shown in Figure 12.19.

The pre-places of transition t are $p_{i1},...,p_{ia}$, its post-places are $p_{o1},...,p_{ob}$; L is a set of logical conditions for firing t; $w_{i1},...,w_{ia}$, and $w_{o1},...,w_{ob}$ are pre-place and post-place arc weights, respectively; $\mathbf{w}_1,\mathbf{w}_2,...,\mathbf{w}_b$ are vector control variables having a level character. For example, in Figure 12.19, the control variables are set to $\mathbf{v}_1,\mathbf{v}_2,...,\mathbf{v}_b$ whenever the associated place with an assigned variable changes marking state from non-marked to marked. The structure around t is transformed into the PEARL program part in the following way.

```
PROBLEM;
   T:  TASK PRIO 16;
          IF MPI1>=WI1 AND MPI2>=WI2 AND... AND MPIA>=WIA
          AND X1==U1 AND X2==U2 AND... AND XC==UC
              THEN    MPI1:=MPI1-WI1;    MPI2:=MPI2-WI2;    ...;
                      MPIA:=MPIA-WIA;
                      MPO1:=MO1+WO1;        MPO2:=MPO2+WO2;...,
                      MPOB:=MPOB+WOB;
                      W1:=V1; W2:=V2; . . ., WB:=VB;
                      PREVENT;
                      ACT_T:=0;
                      IF ACT_TA==0
                          THEN  ALL  DA  SEC  ACTIVATE  TA;
                          ACT_T:=0; ACT_TA:=1;
                      FIN;
                      IF ACT_TB==0
                          THEN  ALL  DB  SEC  ACTIVATE  TB;
                          ACT_T:=0; ACT_TB:=1;
                      FIN;
                          .
                          .
```

IF ACT_TZ==0

THEN   ALL   DZ   SEC   ACTIVATE   TZ;   ACT_T:=0;
ACT_TZ:=1;
FIN;

END;

Note that the notation of variables complies with the PEARL requirements, and therefore, they are written in capitals. Value 1 of the variable ACT_T means that the task T corresponding to transition t is active; analogously for transitions $t_a, t_b, ..., t_z$. These variables help to protect an active transition from being activated again. Some PEARL versions do not support this. A task T is deactivated by instruction **PREVENT** <*name*>. If *name* is not specified, the instruction refers to the task where it is located; otherwise it is applied to the task *name*. A task T ends with **END;**. Note that it finishes immediately with **TERMINATE;**, if this instruction is used, all its scheduling is cancelled so that it can no longer start. The structure of activation of the next transitions can be modified according to the possibilities described above. Time intervals DA, DB, ..., and DZ can be equal, which is a simpler case. Their values must correspond to the dynamics of the system variables. The above program part can be further improved following the introduced programming paradigm.We will show the use of the PEARL language program based on the Petri net in Example 12.5.

Example 12.5. The technological layout is given in Figure 12.20. The workpieces arrive separately and irregularly into the manufacturing line. A row of workpieces stops at the gate Gt being down. The gate is operated down and up by a pneumatic cylinder Cl2. A workpiece is pressed down by cylinder Cl1 located on the plate between belt conveyors B1 and B2 when there is a sufficient stock of parts at the input. The stock of parts is considered sufficient if all four photo-sensors P1 through P4 are signaling simultaneously a workpiece presence. Thus, the minimum number of parts is four. The variables associated with the photo-sensors P1 through P4 are denoted as F1 through F4, respectively. In case of a sufficient stock, the gate Gt driven by the cylinder Cl2 lifts up and the pair of workpieces moves on conveyor B2. When it reaches the inductive sensor IS, a manipulator Sb shifts the pair aside onto conveyor B3, which starts moving. When the pair is passing photo-sensor P5, Gt goes down and the pairing cycle can start again. A collision can occur if the manipulator is not in the basic position back when a pair is at sensor P5. In such a case motor M2 stops and the pair has to wait until the manipulator comes back. Then B2 is started again. The paired workpieces move to the working space of the 3-axial Cartesian robot, which picks it up and puts into one of the two processing machines or into the buffer to wait.

Now the system variable table to be put together (Table 12.1) and a Petri net interpreted for the control (Figure 12.21) specify the control of the manufacturing line. In the Petri net it is strictly assumed that just one transition fires at a time point. The meaning of the places and transitions of the Petri net is given in Table 12.2. It is a binary Petri net model.

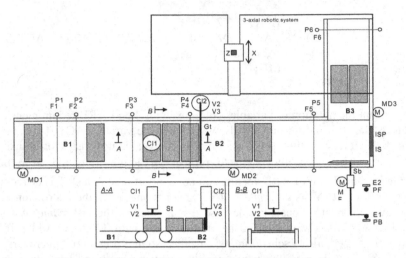

Figure 12.20. Manufacturing line for handling workpieces

Table 12.1. Meaning of the input and output variables for the control unit

Input variables (to the process computer)	Meaning
START=1	Start given by an operating personnel
STOP=1	Stop the input and empty the line
Fi=1 (0)	A workpiece is (is not) at sensor Pi, i=1, 2, ..., 6
ISP=1 (0)	A workpiece is (is not) at ind. sens. IS
PF=1 (0)	Manipulator Sb is (is not) shifted out
PB=1 (0)	Manipulator Sb is (is not) in the basic position

Output variables (commands from the control computer)	Meaning
MD1=1 (0)	The belt conveyor B1 on (off)
MD2=1 (0)	The belt conveyor B2 on (off)
MD3=1 (0)	The belt conveyor B3 on (off)
V1=1 and V2=0 and V3=0	Cylinder Cl1 pushes the break up and the gate Gt down
V1=0 and V2=1	Cylinder Cl1 down
V1=0 and V3=1	Gate Gt up
MF=1 (0)	Manipulator Sb moves forward
MB=1 (0)	Manipulator Sb moves back

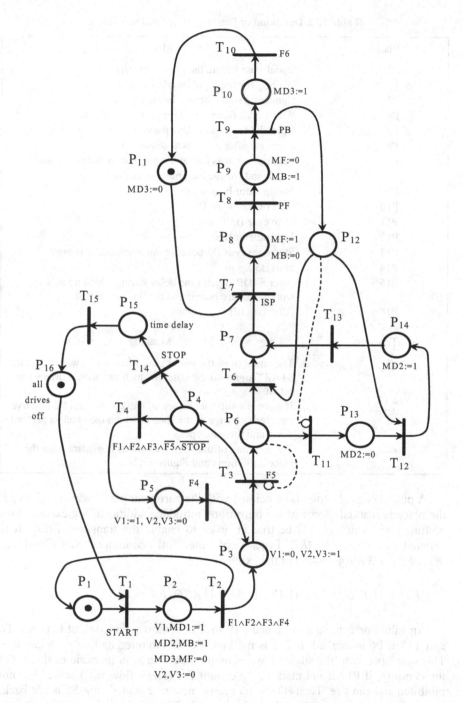

Figure 12.21. Petri net for the manufacturing line

Table 12.2. Definition of Petri net places and transitions

Place	Meaning
P1	Initial state before the operation starts
P2	Waiting for the input (workpieces)
P3	Command state: break down, gate up
P4	Wait for sufficient workpiece stock or STOP
P5	Open the break and close the gate
P6	Interstate after a pair is at sensor P5
P7	Manipulator was free and a pair moves toward sensor IS
P8	Command for the manipulator forward
P9	Manipulator backward
P10	Start of conveyor B3
P11	Conveyor B3 free
P12	Manipulator free
P13	Stop conveyor B3 because the manipulator is busy
P14	Start B3 again
P15	After STOP signal time delay during which no new workpieces are paired and the line is emptied
P16	All drives off
Transition	Meaning
T1	Transition from the initial state when it was waiting for the START signal to the state in which the needed drives are set on
T2	If there is a sufficient stock of workpieces and P2 is active (marked) passage to the state in which the break is pressed down and the gate opens
etc.	The meaning of the other transitions is evident from the place definitions and Figure 12.21.

A partial control action is associated with the particular place, which is active if the place is marked. Some of the transitions introduce additional logical variables or functions, which should be true in order to enable the transition firing. It is denoted as, *e.g., START* instead of the full condition *START*==1 or $F1 \wedge F2 \wedge F3 \wedge F4$ instead of the expression

$$(F1==1)\,AND\,(F2==1)\,AND\,(F3==1)\,AND\,(F4==1).$$

An inhibitor coming from place P12 to T11 removes the conflict between T6 and T11 if P6 is marked. If P12 is marked, T11 is inhibited and only T6 can fire. This corresponds to the situation when manipulator Sb is in its basic position. On the contrary, if P12 is not marked, T6 cannot fire (token flow rule) but T11 is not inhibited and can fire. Then B2 stops because new pair is at F5 but Sb is not back. A subtle problem is solved by inhibitor (P6,T3). If there is a gap between workpieces of a pair and P3 is marked (indicating that a new pair is on the way) the gap produces a false firing of T3 and the control collapses. A time delay introduced in the place P15 provides time for waiting until the manufacturing line is empty and all drives are switched off (place P16).

A part of the control program written in PEARL illustrates the method described in this section.

```
MODULE (MAIN);
    SYSTEM;
        TERM: CON;
        DIG_IN_0: DIGE(0)*1*0,15;
        DIG_OUT_0: DIGA(1)*1*0,15;
    PROBLEM;
        SPC DIG_IN_0 DATION IN BASIC;
        SPC DIG_OUT_0 DATION OUT BASIC;
        SPC DID_IO_INIT ENTRY GLOBAL;
        DCL SCREEN DATION INOUT ALPHIC DIM(2,24,80)
            DIRECT GLOBAL CREATED (TERM);
        DCL VARINP BIT(16) INIT ('0000'B4);
        DCL (PER1, PER2) DUR INIT (0,0);
        DCL (START, STOP) FIXED INIT (0, 0);
        DCL (F1, F2, F3, F4, F5, F6) FIXED INIT (0, 0, 0, 0, 0, 0);
        DCL (ISP, PF, PB) FIXED INIT (0, 0, 0);
        DCL (MD1, MD2, MD3) FIXED INIT (0, 0, 0);
        DCL (V1, V2, V3) FIXED INIT (0, 0, 0);
        DCL (MF, MB) FIXED INIT (0,0);
        DCL (P1, P2, P3, P4, P5, P6, P7, P8, P9, P10, P11, P12, P13, P14,
        P15, P16)
        FIXED INIT (1, 0, 0, 0, 0, 0, 0, 0, 0, 0, 1, 1, 0, 0, 0, 1);
        DCL (ACT_T1, ACT_T2, ACT_T3, ACT_T4, ACT_T5, ACT_T6,
        ACT_T7, ACT_T8, ACT_T9, ACT_T10, ACT_T11, ACT_T12,
        ACT_T13, ACT_T14, ACT_T15)
        FIXED INIT(0, 0, 0, 0, 0, 0, 0, 0, 0, 0, 0, 0, 0, 0, 0);
        DIN:   PROC RETURNS (BIT(16));
                DCL WD BIT(16);
                TAKE WD FROM DIG_IN_0;
                RETURN (WD);
            END;
        DOUT:PROC (B FIXED, H FIXED); /*INDEX OF THE BIT IN
        THE OUTPUT WORD FOR   MD1 IS 1,
                                MD2:2,MD3:3,V1:4,V2:5,
                                V3:6,MF:7,MB:8*/
                IF H==0
                THEN SEND 0 TO DIG_OUT_0 BY CONTROL((B-1),
                    (0));
                ELSE SEND 1 TO DIG_OUT BY CONTROL((B-1),(0));
                    FIN;
            END;
        DINALL:      TASK PRIO 6;
```

```
                                    VARINP:=DIN;
                                    F1:=VARINP.BIT(16);
                                    F2:=VARINP.BIT(15);
                                    F3:=VARINP.BIT(14);
                                    F4:=VARINP.BIT(13);
                                    F5:=VARINP.BIT(12);
                                    F6:=VARINP.BIT(11);
                                    START:=VARINP.BIT(10);
                                    STOP:= VARINP.BIT(9);
                                    ISP:= VARINP.BIT(8);
                                    PF:= VARINP.BIT(7);
                                    PB:= VARINP.BIT(6);
                        END;
STRT:           TASK PRIO 8;
                        CALL DID_IO_INIT;
                        ACTIVATE STARTTASK;
                        END;
STARTTASK: TASK PRIO 8;
                        OPEN SCREEN;
                        PUT  'START   WITH   START-PUSH-
                        BUTTON'   TO   SCREEN   BY
                        POS(1,2,2), A;
                        ALL PER1 ACTIVATE DINALL;
                        ALL PER2 ACTIVATE T1;
                        END;
        T1:     TASK PRIO 7;
                        IF P1==1 AND P16==1 AND START==1
                        THEN CLOSE SCREEN;
                        P1:=0; P16:=0; P2:=1;
                        DOUT (4,1); DOUT (5,0); DOUT (6,0);
DOUT (1,1); DOUT (2,1); DOUT (3,0); DOUT (7,0); DOUT (8,1);
                        PREVENT;
                        ACT_T1:=0;
                        IF ACT_T2==0
                        THEN ALL PER2 ACTIVATE T2;
                        FIN;
                        FIN;
                        END;
        T2:     TASK PRIO 7;
IF P2==1 AND F1==1 AND F2==1 AND F3==1 AND F4==1
THEN            P2:=0; P1:=1; P3:=1;
                        DOUT(4,0);
                        DOUT(5,1);
                        DOUT(6,1);
                        PREVENT;
                        ACT_T2:=0;
                        IF ACT_T3==0
```

```
                        THEN ACT_T3:=1; ALL PER2 ACTIVATE T3;
                        FIN;
                        IF ACT_T1==0
                        THEN ACT_T1:=1; ALL PER2 ACTIVATE T1;
            FIN;
      FIN;
   END;
T3: TASK PRIO 7;
      IF P3==1 AND P6==0 AND F5==1
      THEN        P3:=0; P4:=1; P6:=1;
                  PREVENT;
                  ACT_T3:=0;
                  IF ACT_T6==0
                     THEN ACT_T6:=1; ALL PER2  ACTIVATE
                     T6;
                  FIN;
                  IF ACT_T11==0
                     THEN ACT_T11:=1;
                  ALL PER2  ACTIVATE T11;
                  FIN;
                  IF ACT_T14==0
                     THEN ACT_T14:=1;
                  ALL  PER2 ACTIVATE T14;
                  FIN;
                  IF ACT_T4==0
                     THEN ACT_T4:=1;
                     ALL PER2 ACTIVATE T4;
                  FIN;

            FIN;
      END;

                        .
                        .

MODEND;
```

Let us make a few remarks about the program. The keywords and syntax signs are written in bold and the variables not in bold in order to make the program more transparent. DID_IO_INIT is an external procedure containing the driver of the input and output modules and their initialization. The task with the label STRT in the module MAIN is automatically activated after the program is loaded by the system. Other tasks should be activated internally in the program. Instruction w.**BIT**(k) picks out the k-th bit from the bit string word w.

There are many programming languages that enable to set programs similarly as shown above. The instruction lists can be more or less comprehensive but they have to enable to interpret the basic real-time or reactive features of a control system.

2.4 Ladder Logic Diagrams

The popular programming technique is based on the so-called ladder diagrams. Ladder diagrams are a graphical tool for presenting the instruction set. The reactivity of the control program is provided by cyclically repeating the ladder rungs from the beginning of the program to its end. Each rung is subordinated to the scheme

$$\text{CONDITIONS} \Rightarrow \text{ACTIONS}$$

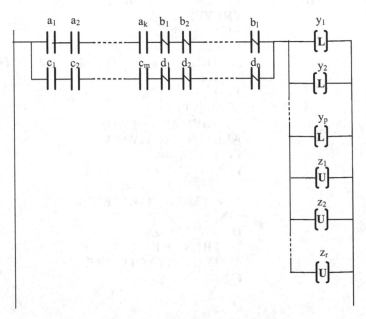

Figure 12.22. One rung of a ladder diagram

The conditions are given by states and actual events in the controlled system. A state is a consequence of its preceding events. The actions are programmed interventions of the control system to the controlled one. Figure 12.22 shows a generalized rung of a ladder diagram. The conditions are expressed using the Boolean logic. $a_1, a_2, ..., a_k$, $b_1, b_2, ..., b_l$, $c_1, c_2, ..., c_m$, $d_1, d_2, ..., d_n$, $y_1, y_2, ..., y_p$, $z_1, z_2, ..., z_r$ are Boolean variables. The sense of the logical condition at the left-hand part of the rung in Figure 12.22 expressed in the logic form is

$$C = \left(a_1 \wedge a_2 \wedge ... \wedge a_k \wedge \overline{b_1} \wedge \overline{b_2} \wedge ... \wedge \overline{b_l} \right) \vee \left(c_1 \wedge c_2 \wedge ... \wedge c_m \wedge \overline{d_1} \wedge \overline{d_2} \wedge ... \wedge \overline{d_n} \right) \vee ...$$

$$(12.6)$$

where the inversion of b_1 is denoted as $\overline{b_1}$, *etc.* A serial arrangement of the conditions expresses the logical conjunction of the Boolean variables, the parallel arrangement the logical disjunction. Obviously, the diagram in Figure 12.22 is based on the properties of the contact relay networks. If condition C or Equation (12.6) is true, then the values "logical true" are assigned to all variables $y_1, y_2, ..., y_p$ and the values "logical false" to all variables $z_1, z_2, ..., z_r$. In this way an IF .....THEN.....ELSE is incorporated in the ladder diagram. Letter **L** means that if the whole condition is true the rung circuit is latched, *i.e.*, the associated action or control variable is set. Letter **U** stands for unlatched.

The ladder logic diagram consists of rungs that are translated into the machine instructions of a particular PLC and executed (Hrúz *et al.* 2000; Mudrončík and Zolotová 2000). The possibilities and the form of the ladder diagrams depend on the properties of the used PLC and its programming facilities, *e.g.*, an action may set some variables and start the counters as well. The condition part of the rungs can contain the counter states, which can influence the generation of other actions.

The ladder diagrams have several drawbacks from the software engineering point of view. They are less transparent. Their readability is comparatively difficult by other programmers than by the program authors. This is very bulging when such a program is put into operation and debugged. Any change or adaptation of the ladder diagram program to other control computers is usually difficult. An inter-stage specification link like Petri nets is therefore useful. In what follows we will show the transformation of the Petri nets into the ladder diagrams.

Consider a binary and safe elementary Petri net given in Figure 12.23. L is the set of Boolean conditional expressions for firing transition t; $y_1, y_2, ..., y_{r+s}$ are control variables representing the control actions.

The ladder diagram presentation of the required control function ensuring the partial dynamic behavior related to t's firing is in Figure 12.24. In Figure 12.24 the block L is broken down by the corresponding serial and/or parallel structure of the contacts in a manner treated earlier. The Petri net marking is represented using the variables $v_{p1}, v_{p2}, ..., v_{pr}, v_{p(r+1)}, v_{p(r+2)}, ..., v_{p(r+s)}$ in the ladder diagram.

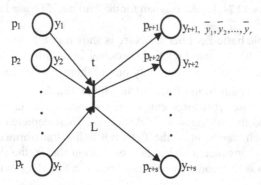

Figure 12.23. An elementary Petri net structure for transformation into a ladder program

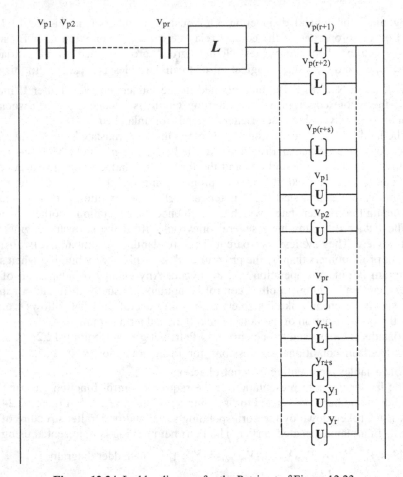

Figure 12.24. Ladder diagram for the Petri net of Figure 12.23

Another possible basic Petri net structure is shown in Figure 12.25. As before a binary and safe Petri net interpreted for control is assumed. The partial logical conditions $L_1, L_2, ..., L_k$ are such that only one of them is true. Then the diagram in Figure 12.26 corresponds to the Petri net in Figure 12.25. The reader can imagine how different Petri net structures can be transformed into ladder diagrams. An inverse situation to that of Figures 12.25 and 12.26 is depicted in Figures 12.27 and 12.28. For each transition of the Petri net being transformed into the ladder diagram, a rung is constructed under the assumption that in the Petri net there are no conflicts, and it is safe and only one transition can fire at a discrete time point.

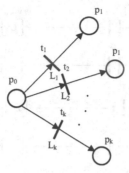

Figure 12.25. A Petri net elementary structure

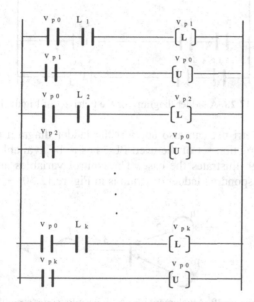

Figure 12.26. A ladder diagram for the Petri net of Figure 12.25

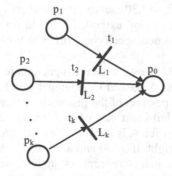

Figure 12.27. A Petri net elementary structure

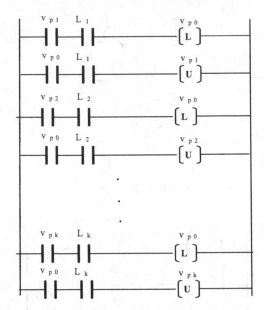

Figure 12.28. A ladder diagram for the Petri net of Figure 12.27

A non-binary Petri net can also support the ladder diagram programming. In such a case an instruction set of the used PLC has to be available. An elementary net in Figure 12.29 illustrates the case. The control variables are omitted in the example. The corresponding ladder diagram is in Figure 12.30.

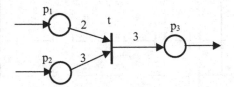

Figure 12.29. An elementary non-binary Petri net structure

The elementary structures and their transformation into ladder diagrams depicted in Figures 12.22–12.30 serve as paradigms for the transformation procedure. They can be used and extended to build up complex ladder diagram programs based on the Petri nets specifying DEDS control.

Example 12.6. Figure 12.31 shows an electro-pneumatic motion drive. It consists of three pneumatic pistons, A, B and C, operated by electro-pneumatic two-way solenoid valves. The basic position of the pistons is the terminal position on the left. Pressure air is fed to the left-hand part of each piston after the activation of the valve. The activation signal for A is $a^+=1$, while $a^-=0$, *etc.*, for others. On $a^+=1$, $a^-=0$ piston A moves to the right. If $a^+=0$ and $a^-=1$ piston A moves back to the basic position on the left. The inductive sensors indicate terminal positions of the

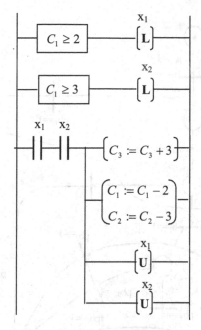

Figure 12.30. Ladder diagram for the Petri net of Figure 12.29

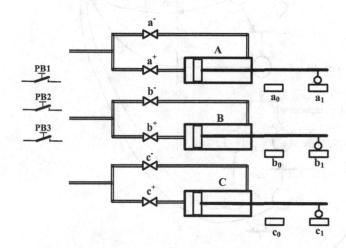

Figure 12.31. Three-piston electro-pneumatic motion system

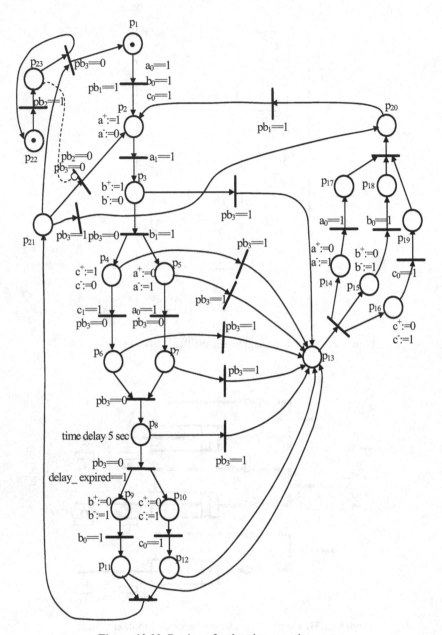

Figure 12.32. Petri net for the piston motion system

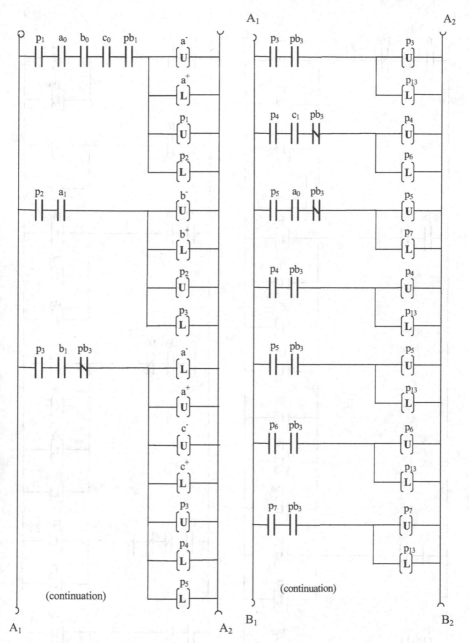

Figure 12.33. Ladder diagram program for the piston system: segments A and B

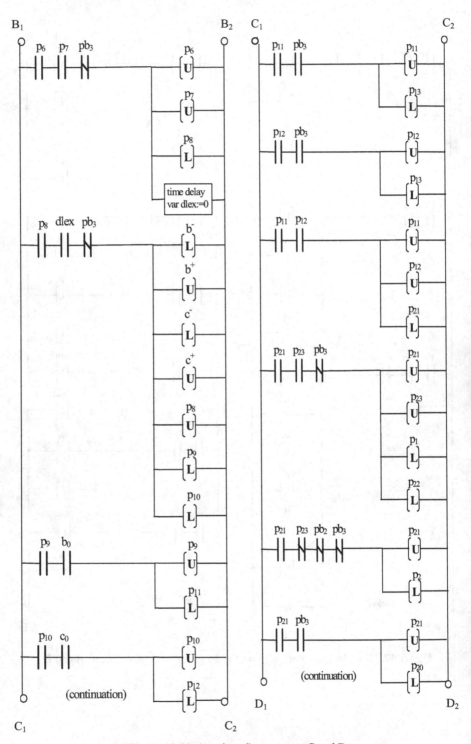

Figure 12.33. (continued): segments C and D

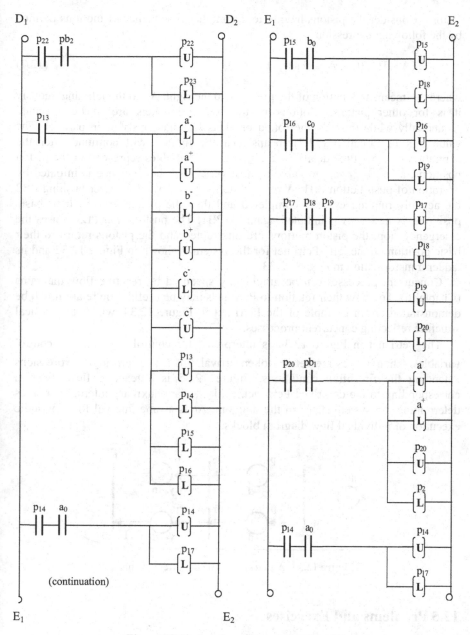

Figure 12.33. (continued): segments E and F

pistons. Consider the pistons having to accomplish a sequence of motions defined by the following expression:

$$[A+, B+, (C+, A-), 5 \text{ sec}, (B-, C-)] \qquad (12.7)$$

where A+ means the motion of the piston A to the right, A– to the left; analogously it is for other pistons. Motions in the square brackets are to be executed sequentially while those in the round brackets simultaneously or in parallel. The value 5 s in Equation (12.7) means that the motion will continue after the expiration of a 5-s time delay. a_0, a_1, b_1, *etc.*, are variables representing the piston positions, a^-, a^+, b^-, *etc.*, are valve control variables. The sequence is initiated by operation of push-button PB1. A regular stop is given by PB2. After pushing PB2 the actually running cycle is completed and then the pistons stop in their basic positions. A new start is initiated again by PB1. The push-button PB3 causes the emergency stop; the piston motions are interrupted and the pistons return to their basic position on the left. Petri net for the system is shown in Figure 12.32 and its ladder transformation in Figure 12.33.

Concurrent processes can be graphically specified by reactive flow diagrams (Chapter 6). To know their relation to Petri nets may be useful. The relation will be demonstrated on an example of the Petri net in Figure 12.34, which is a typical structure reflecting concurrent processes.

The Petri net in Figure 12.34 is interpreted for control; $y_1, y_2...$ are control variables. Their values are set by token arrival. L_1, L_2, … are logic expressions extending the fireability conditions. Figure 12.35 is a reactive flow diagram corresponding to the described Petri net. k_1, k_2, … are auxiliary internal variables determining the system states in the diagram, Δt is a time interval for a periodic execution of individual flow diagram blocks.

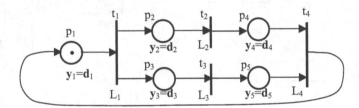

Figure 12.34. A Petri net with concurrent events

12.5 Problems and Exercises

12.1. Exercise 6.2 deals with a manufacturing cell using two robots. Derive a Petri net interpreted for control specifying the required function of the system. On the basis of the Petri net model, write a program in the form of the ladder logic diagram for the system control. For the comparison of the programming technique, write a part of the program in a real-time programming language, *e.g.*, in the language PEARL or in some other language you know.

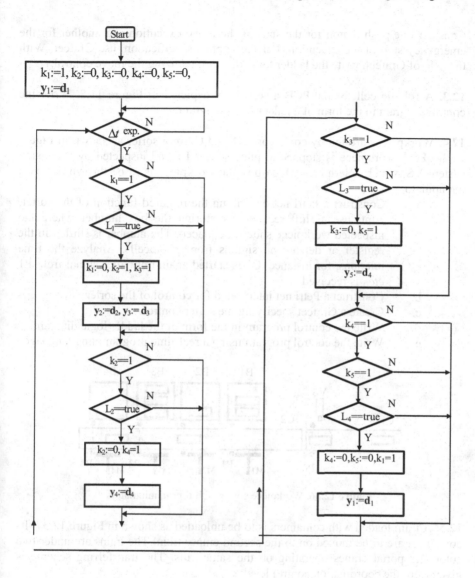

Figure 12.35. Reactive flow diagram corresponding to the Petri net in Figure 12.34

12.2. Consider a four-piston electro-pneumatic move system. Complete necessary sensors for the system control purpose. Let the pistons perform repeatedly the following cycle of movements:

[A+, B+, (C+, D+, A-), 5 sec. interval, (B-, C-, D-)]

Consider one push button for the start of the move execution and another for the emergency stop of the execution. For the control specification, use Grafcet. With the help of Grafcet, write the ladder logic diagram realizing the move control.

12.3. A robotic cell for the PCB assembly is depicted in Figure 11.8. Write the control program in the form of a ladder logic diagram.

12.4. Workpieces are fed by conveyors C1 and C2 in a sorter as shown in Figure 12.36. Each workpiece is stopped at photosensor P1 and inspected by the vision system VS, which determines which box the workpiece is to be thrown out by a manipulator.

    a.   Construct a Petri net describing the required function of the sorter. Hint: use a shift register for storing the box number where the inspected workpiece should be placed. The number is shifted in the register or deleted on signals from photocells. Analyze the time relations, for instance, C1 is started again only after signal from P1 *etc.,* is received.

    b.   Construct a Petri net interpreted for control of the sorter.

    c.   Create a Grafcet specifying the sorter control.

    d.   Write the control program in the form of the ladder logic diagram.

    e.   Write the control program using a real-time programming language.

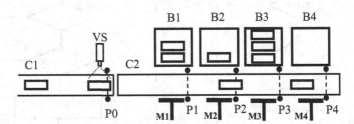

Figure 12.36. Workpiece sorter with four manipulators

12.5. A train loaded with containers is to be unloaded as shown in Figure 12.37. Its containers are to be moved on to the second empty train. The trains are under two automatic portal cranes operating on the same rails. The transferring scheme is given from the coordination control level.

    a.   Express the collision-free movements of the cranes by means of a Petri net. Hint: create for each crane a partial Petri net where the marking of one place represents the number of free crane positions to the left of the crane and the marking of another place serves similarly for the free positions to the right of the crane. For positions between the cranes the two places merge in one place. For readers interested in more details of this exercise we recommend the book of Abel (1990) and references for Chapters 7–9.

    b.   Outline how you could utilize the created Petri net for the control of the crane operation.

c.   Analyze the properties of the Petri net. How can the number of the crane movements for a given reloading scheme be optimized? Note that it is possible to consider genetic algorithms for that optimization

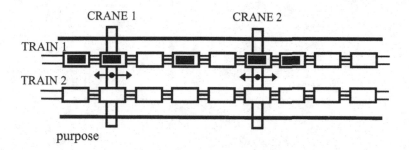

purpose

Figure 12.37. Train reloading station

13

Supervisory Control

13.1 Basic Notion

The basic notion of discrete event dynamic systems (DEDS) was introduced in Chapter 1. Finite automata and Petri nets have been studied as powerful models of DEDS for their behavior modeling and control. It has been shown that the states and transitions are key features of both models.

Consider an event-driven system being in its initial state. The complete behavior of the system is given by all possible event strings beginning in the initial state. The event strings are formally described by Equations (1.11) and (1.12). Such a system description is the core of a DEDS model proposed by Wonham and Ramadge (1987) and Ramadge and Wonham (1989) in the form

$$S = \left(\Sigma, \Sigma^*, L, K \right) \tag{13.1}$$

where Σ is a finite non-empty set of events, Σ^* is the set of all sequences (strings) that can be generated from the elements of Σ and the empty sequence $\tilde{\varepsilon}$, $L \subseteq \Sigma^*$ is a formal language comprising all event sequences realizable in the underlying real system, and $K \subset L$ is a sub-language of L, where K specifies the required behavior of the system.

The assumption for the model at Equation (13.1) has been adopted that only one event can occur at a discrete time point. The model at Equation (13.1) describes the behavior of a system in terms of the event strings; but considers no system states. It represents a different approach as usually the state machine models do.

In DEDS two types of the control can be distinguished, namely process control and supervisory control. The considerations developed in Chapters 1 and 4 will be used to characterize these two types. Equation (4.10) determines a control action, which is a function of states and the states depend on the event string starting in a system state and developing from the first event up to the $(v-1)$-st event at the $(v-1)$-st discrete time point. At the time point $\tau_{a_{v-1}}$ the system reaches the state when function **f** in Equation (4.10) determines the value of the control variable $\mathbf{w}\!\left(\tau_{a_{v-1}}+\Delta\tau\right)$. In the overall system *SYST* in Figure 4.1, setting **w** is considered as a new state. In system S (considering refined structure decomposition), $\mathbf{w}\!\left(\tau_{a_{v-1}}+\Delta\tau\right)$ represents an input activity effecting the behavior of system *S*. We can see that it is a system point of view.

In the case when for each $k=1,2,...,v\text{-}2$,

$$\mathbf{s}\!\left(\tau_{a_{k+1}}\right)=\mathbf{pcontr}\!\left(\mathbf{w}\!\left(\tau_{a_k}+\Delta\tau_{a_k}\right)\right)\!,\ \tau_{a_k}+\Delta\tau_{a_k}<\tau_{a_{k+1}} \tag{13.2}$$

where **pcontr** is a vector function, we speak about the process control. In the case when $\mathbf{w}\!\left(\tau_{a_k}+\Delta\tau_{a_k}\right)$ only delimits possible states and the particular next state depends on the next event in S, we speak about the supervisory control. The range of the possibilities is a matter of system autonomy. The function of the supervisory control is to keep the system within some subsets of the state set *Q*. It means that

$$\left\{set\ of\ admissible\ states\ in\ \tau_{a_{k+1}}\right\}=\mathbf{supvis}\!\left(\mathbf{w}\!\left(\tau_{a_k}+\Delta\tau_{a_k}\right)\right).$$

From the point of view of the model at Equation (13.1), S is to be ruled so that only strings of sub-language *K* occur in it.

There are two kinds of requirements to be laid on an event-driven system. The first is characterized by defining prohibited states. The second is characterized by requirements to preserve some event strings without violating the state prohibition.

A required system behavior cannot be achieved if a controllable agent does not exist in the system. A subset $\Sigma_c\subseteq\Sigma$ of controllable events serves for that. A controllable event can be allowed or prohibited by a control system. This is not possible with the uncontrollable events of set $\Sigma_u\subseteq\Sigma$. Together we have

$$\Sigma=\Sigma_c\cup\Sigma_u,\quad \Sigma_c\cap\Sigma_u=\varnothing \tag{13.3}$$

The supervisory control keeps the system within limits, enabling only event strings of language *K* to occur. It means that the range of the system autonomy or freedom is adjusted by the supervisory control *via* the controllable events. When $\Sigma_u\neq\varnothing$, the interventions into the controlled system have to anticipate possible occurrence of uncontrollable events that could violate the behavior limits given by the sub-language *K*.

13.2 System Controllability

Assume the system to be specified by Equations (13.1) and (13.3) and the required behavior by language $K\subset L$. Another possible way, in this context, to express the restriction on the event strings is the prohibition of certain states and preservation of certain event strings as always possible. This will be explained by using examples.

The behavior restriction determined by states and event strings may not always be easy to transform into language K or vice versa. An important question concerning K is whether the language can be secured by the supervisory control. It is known as the question of the DEDS controllability.

A common feature of real systems is the existence of an initial state. Any activity of a real system starts in this state. Suppose that an event string $\tilde{\sigma} = e_{i_1} e_{i_2} \ldots e_{i_k}$ has been realized starting from the initial state. Obviously, all partial event strings are also realized in the system, namely

$$\tilde{\sigma}_1 = e_{i_1}, \ \tilde{\sigma}_2 = e_{i_1} e_{i_2}, \ \tilde{\sigma}_3 = e_{i_1} e_{i_2} e_{i_3}, \ \ldots, \tilde{\sigma}_{k-1} = e_{i_1} e_{i_2} \ldots e_{i_{k-1}} \qquad (13.4)$$

Such strings are related to the prefix languages. A string $\tilde{u}$ is a prefix of a string $\tilde{v} \in \Sigma^*$ if $\tilde{v} = \tilde{u}\tilde{w}$ and $\tilde{w} \in \Sigma^*$. It follows that $\tilde{w}$ can be the empty string denoted $\tilde{\varepsilon}$, i.e., $\tilde{w} = \tilde{\varepsilon}$; thus a word $\tilde{v}$ is a prefix to itself. If $\tilde{v}$ is an event string in a real system, then all its prefixes have to occur in the system.

The prefix language $\overline{L}$ of language $L \in \Sigma^*$ is

$$\overline{L} = \left\{ \tilde{u} \mid \tilde{u}\tilde{w} \in L, \tilde{w} \in \Sigma^* \right\} \qquad (13.5)$$

From Equation (13.5) we have

$$L \subseteq \overline{L} \qquad (13.6)$$

i.e., all strings of L pass over in language $\overline{L}$ because $\tilde{w}$ can be $\tilde{\varepsilon}$.

If language L describes a real system, then L should contain all its prefixes, i.e.,

$$\overline{L} \subseteq L \qquad (13.7)$$

From Equations (13.6) and (13.7), a language L describing the real system must meet the condition

$$L = \overline{L} \qquad (13.8)$$

A language with the property at Equation (13.8) is called a prefix closed language.

Example 13.1. Consider a language L over the event set $\Sigma = \{\alpha, \beta\}$ containing the empty string $\tilde{\varepsilon}$ and strings beginning with α or β whereby in each string α and β alternate. The language has the following strings:

$$L = \{\tilde{\varepsilon}, \alpha, \beta, \alpha\beta, \beta\alpha, \alpha\beta\alpha, \beta\alpha\beta, ...\}$$

Obviously, the prefix language $\overline{L}$ of L is the same as L:

$$\overline{L} = \{\tilde{\varepsilon}, \alpha, \beta, \alpha\beta, \beta\alpha, \alpha\beta\alpha, \beta\alpha\beta, ...\}$$

such that

$$L = \overline{L}$$

The conclusion is that L is a prefix closed language.

Example 13.2. Let a language L describe the events of a palette buffer with a capacity equal to three palettes. The empty buffer represents the initial state. The event set is $\Sigma = \{\omega, \psi\}$, where ω means one palette insertion and ψ one palette pick-up from the buffer. The events occur irregularly and asynchronously. The number of insertions $|ins|$ has to be always greater or at least equal to the number of the pick-ups $|pck|$

$$|ins| \geq |pck|$$

The set of strings of L is

$$L = \{\tilde{\varepsilon}, \omega, \omega\omega, \omega\omega\omega, \omega\psi, \omega\omega\psi, \omega\omega\psi\psi, \omega\omega\psi\omega, \omega\psi\omega, \omega\omega\omega\psi, ...\}$$

Thus it is a prefix closed language.

Properties of prefix formal languages can be used to check the supervisory controllability of systems. A language K is said to be controllable with respect to a language L iff

$$\overline{K}\Sigma_u \cap L \subseteq \overline{K} \text{, by assumption } \Sigma_u \neq \varnothing \tag{13.9}$$

where $\overline{K}\Sigma_u$ is the set of all strings $\tilde{v}\tilde{w}$ given by the concatenation of all strings $\tilde{v} \in \overline{K}$ and all strings consisting of one uncontrollable event, *i.e.,* $\tilde{w} \in \Sigma_u$.

The controllability property at Equation (13.9) is induced by the fact that for an event string $\tilde{\sigma}_k = e_{i_1} e_{i_2} ... e_{i_k}$ starting in the initial state, the system goes through all prefixes of $\tilde{\sigma}_k$. Formally stated it means that in the system the following event strings have been realized:

$$\tilde{\sigma}_{k-1} = e_{i_1} e_{i_2} ... e_{i_{k-1}}, \quad \tilde{\sigma}_{k-2} = e_{i_1} e_{i_2} ... e_{i_{k-2}}, \quad ... \quad \tilde{\sigma}_2 = e_{i_1} e_{i_2}, \quad \tilde{\sigma}_1 = e_{i_1} \tag{13.10}$$

First, Equation (13.9) means that the considered strings are limited to be from the realizable language L and therefore there is the set intersection operation $\cap L$ in Equation (13.9). Second, the strings realizable in a system S and belonging to language K as well as all prefixes of these strings (which altogether are strings of language $\overline{K}$) with a supplemented uncontrollable event should still be the strings of $\overline{K}$. The reason for that is that no control can prevent an uncontrollable event to occur. Note that there is nothing said about the closure property of K. Generally speaking, K can be controllable with respect to L without property $K = \overline{K}$. In other words, there can be strings of $\overline{K}$ not belonging to K (it is not the case in languages based on real systems). For those strings the condition at Equation 13.9 has to hold, too. Such considerations are rather theoretical and we concentrate in Section 13.3 later on the case of the prefix-closed languages K.

For controllability by definition the following proposition holds: K is controllable if and only if $\overline{K}$ is controllable. It is understandable because $\overline{\overline{K}} = \overline{K}$.

Controllable events are used to keep the system within K. However, uncontrollable events represent a frequent and serious problem in supervisory control.

A natural question is how to resolve the situation when K is not controllable. First consider a more general case without condition $K = \overline{K}$. In such a situation we seek a language closest to K, which is controllable and prefix closed just following the property of DEDS. Such a language is called the supremal controllable sublanguage to K and is defined as

$$\sup C(K) = \cup\{J : J \subseteq K \text{ and } J \text{ is controllable with respect to } L\} \qquad (13.11)$$

Here the symbol "$\cup$" denotes the set union of all controllable languages J or more strictly the set union of all their strings. Then obviously J satisfies

$$\overline{J}\Sigma_u \cap L \subseteq \overline{J} \qquad (13.12)$$

The fixpoint characterization of $\sup C(K)$ underlies the theoretical reasoning that makes for the supervisory control resolution. We explain it below.

Let Λ be the set of all languages over the event set $\Sigma = \Sigma_c \cup \Sigma_u$ and let Ω be an operator

$$\Omega : \Lambda \to \Lambda \qquad (13.13)$$

such that for some given languages, $M, L \in \Lambda$ and $K \subset L$

$$\Omega(M) = K \cap \text{supremum}\left\{T : T \subseteq \Sigma^*, T = \overline{T}, T\Sigma_u \cap L \subseteq \overline{M}\right\} \qquad (13.14)$$

Over the event set Σ, operator $\Omega(M)$ transforms a language M to another one with respect to the given languages L and K. The obtained language is derived from the supremum of the languages (in the sense of the number of strings of languages T) that are prefix-closed and all its strings (including prefixes of T delimited by K) with a supplemented event of Σ_u are realizable (the strings are included in L) and contained in $\overline{M}$, as well as in K.

Wonham and Ramadge proved (1987) the following proposition.

Proposition 13.1. Given languages L, K and operator $\Omega(M)$ over the event set $\Sigma = \Sigma_c \cup \Sigma_u$ by Equation (13.14). Denote $S = \sup C(K)$. Then

$$S = \Omega(S) \tag{13.15}$$

and for every M such that $M = \Omega(M)$:

$$M \subseteq S \tag{13.16}$$

Obviously, the operator Ω is purposely chosen. It includes an important property for the controllability, namely, $T = \overline{T}$. In looking for a supremal controllable language for a given language K with respect to L, it considers just prefix-closed languages.

According to Proposition 13.1 $\sup C(K)$ is the fixpoint of operator Ω and any fixpoint of Ω, i.e., $M = \Omega(M)$ is a sublanguage of S from Equation (13.16). V (13.16) valid for $M = \Omega(M)$ supports the idea of using a limit process to find $\sup C(K)$. The authors of the above-mentioned paper have proved that an iteration process as follows:

$$M_0 = K \tag{13.17}$$
$$M_1 = \Omega(M_0) \tag{13.18}$$
$$M_2 = \Omega(M_1) \tag{13.19}$$
$$\vdots$$
$$M_j = \Omega(M_{j-1}) \tag{13.20}$$

converges to S in the sense that

$$\lim_{j \to \infty} M_j = M_{LIM} \tag{13.21}$$

exists and

$$S \subseteq M_{LIM} \tag{13.22}$$

If languages L and K in Equation (13.14) are regular then, in addition,

$$S \supseteq M_{LIM} \Rightarrow S = M_{LIM} \tag{13.23}$$

In what follows we confine ourselves to the regular languages and their controllability as an important case with respect to the real systems encountered in practice.

A task of the following example is used to elucidate the above mentioned concepts.

Example 13.3. A production layout is depicted in Figure 13.1. It consists of the cells A1 and A2 located on the plant's first floor, the connecting area A3, and the cells B0 up to B4 on the second floor. Two mobile robots are moving in the layout performing various transport and production tasks. We are going to study the behavior of the robots in the described production system using a formal language approach. Our considerations are restricted on robot R1. The case for R2 is similar. We need to define the event set: it is the set of controllable and uncontrollable transfers between cells including transfers between cells and the platform A3, as well. They are listed in Table 13.1 where, for example, $A1A3_c$ means a controllable event that a robot at room A1 enters A3; and $A1A2_u$ means an uncontrollable event that a robot at room A1 enters A2.

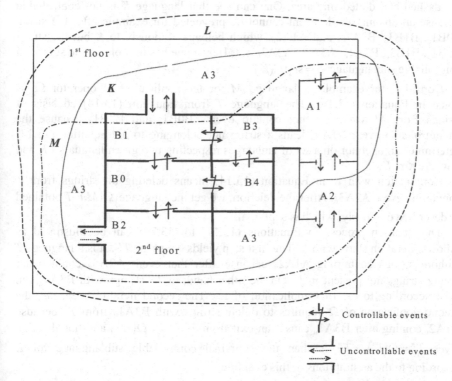

Figure 13.1. Illustration of the relation between languages *L*, *K*, and *M*

Table 13.1. Controllable and uncontrollable events for the layout in Figure 13.1

Controllable events
A1A3$_c$, A3A1$_c$, B1B3$_c$, B3B1$_c$, B1B0$_c$, B0B1$_c$, B3B4$_c$, B4B3$_c$, B0B2$_c$, B2B0$_c$, B0B4$_c$, B4B0$_c$
Uncontrollable events
A1A2$_u$, A2A1$_u$, A2A3$_u$, A3A2$_u$, A3B1$_u$, B1A3$_u$, A3B2$_u$, B2A3$_u$, A1B3$_u$, B3A1$_u$

All formal languages considered in this example for the system description are given by the strings generated from the position of robot R1 in cell B0. If the robot returns to B0, the string is assumed to be finished. The next movements represent new strings. Clearly, all considered languages are prefix closed ones.

The set Λ of languages over the event set $\Sigma = \Sigma_c \cup \Sigma_u$ specified in Table 13.1 is defined according to Equation (13.14). Let language L correspond to all possible movements of robot R1 in the layout of Figure 13.1 from the initial position as stated before. The movements and transfer strings occur within the dashed line area in Figure 13.1. Let a language $K \subset L$ be given corresponding to the solid line area encompassing the movements of this language and a language M corresponding to the dashed and dotted line area. One can see that language K is not controllable because an uncontrollable event cannot be prevented for example after the string B0B1$_c$, B1B3$_c$, B3A1$_u$, and A1A2$_u$, which belongs obviously to K but the string B0B1$_c$, B1B3$_c$, B3A1$_u$, A1A2$_u$, A2A3$_u$ not. Hence one has to look for the supremal controllable sublanguage $S = \sup C(K)$.

Consider, for example, a language M for the application of operator Ω as given in Equation (13.14). The language T from Equation (13.14) consists of strings from M, which do not contain events B0B1$_c$ and B3B1$_c$ because the uncontrollable event B1A3$_u$ leads a string not belonging to M. Equation (13.14) determines an operator on a set of languages respecting two given languages K and L where $K \subset L$.

Conjunction with K in Equation (13.14) means deleting all strings from T containing event A2A3$_u$. After the deletion, we get the language $\Omega(M)$. T obtained as described obviously, which is supremal.

The iteration process in Equations (13.17) to (13.20) can be illustrated as follows. Iteration starts with K. The first step yields language T'. It is a language T without strings containing A1A2$_u$ because after that event A2A3$_u$ could occur giving strings not present in K. On the other hand, in T' strings with B3A1$_u$ can exist according to the first application of Ω. The second iteration step, *i.e.*, the second application of Ω requires to delete strings with B3A1$_u$ from T' because A1A2$_u$ coming after B3A1$_u$ causes an exit from T' ($T' = \Omega(K)$) and this deletion gives $T'' = \Omega(T')$. T'' is then the supremal controllable sublanguage for K according to the assumptions of this example.

13.3 Supervisory Control Solution Based on Finite Automata

Regular languages play a decisive role in practical applications of the supervisory control. As treated in Chapter 5 every regular language corresponds to a finite automaton, called the generator in this context. In other words, every finite automaton $A = (\Sigma, Q, q_0, \delta, F)$ generates a regular language L and a regular marked language L_m (see Definition 5.2).

Proposition 13.2. A language generator generates languages L for which $L = \overline{L}$.

Proof. The proof follows directly from the definitions of the language generator and the prefix closure property.

A language L_m may not be the prefix-closed language.

As mentioned in Section 5.1, a simple labeled directed mathematical graph corresponds one-to-one to a language generator $A = (\Sigma, Q, q_0, \delta, F)$.

For a finite automaton A, the accessible set Q_{ac} is

$$Q_{ac} = \left\{ q : q \in Q \text{ and } \hat{\delta}(q_0, \tilde{\eta}) = q \text{ for some } \tilde{\eta} \in \Sigma^* \right\} \tag{13.24}$$

and the co-accessible set is

$$Q_{co} = \left\{ q : q \in Q \text{ and } \hat{\delta}(q, \tilde{\eta}) \in F \text{ for some } \tilde{\eta} \in \Sigma^* \right\} \tag{13.25}$$

A finite automaton A is said to be accessible if $Q_{ac} = Q$, and co-accessible if $Q_{co} = Q$. It is said to be the trim iff

$$Q_{ac} = Q_{co} = Q \tag{13.26}$$

A marked language has the following property for a trim.

Proposition 13.3. Let a generator of a formal language L and a marked formal language L_m be a trim. Then

$$L = \overline{L_m} \tag{13.27}$$

Proof. First, $L_m \subseteq L$ and each string of $\overline{L_m}$ belongs to L according to the generation rules for L and L_m, that is $\overline{L_m} \subseteq L$. But second also $L \subseteq \overline{L_m}$ because each string of L belongs to $\overline{L_m}$ as each string of L can be prolonged to reach a state of F (the trim property). $L_m \subseteq L$ and $L \subseteq \overline{L_m}$ results in Equation (13.27).

The trim property means that some required system states (set F) can be reached from each state. It is useful to use the trim property under the supervisory control solution when it is required to preserve the system state reachability. This is contained in the following proposition.

Now consider a question: what does it mean when a generator of a formal language is accessible but not co-accessible? It means that there are states, or at least one state, from which it is not possible to reach the set F. We say that such a generator or finite automaton is blocking. Obviously, it is not a trim. It yields that such strings, from which F cannot be reached, are not contained in $\overline{L_m}$ and, therefore, Equation (13.27) is not valid, i.e., $L \neq \overline{L_m}$. As far as all strings of $\overline{L_m}$ belongs to L, it yields $\overline{L_m} \subset L$. The last expression is a necessary and sufficient condition a finite automaton to be blocking. The states from which the set F cannot be reached are:

a. States, from which no transition to another state exists, the so-called deadlocks
b. States, from which transitions to another states exist (but no path to F), forming the so-called livelocks

Figure 13.2 illustrates the blocking property. The generator (finite automaton) depicted in Figure 13.2 is accessible and not co-accessible. The states $q_3, q_4 \in F$ are not reachable from q_5 and q_5, which are forming a livelock, and q_7 is a deadlock.

By the supervisory control solution it is often a task to find $\sup C(K)$ and its corresponding generator to be not blocking. It is meant to find such a supremal sublanguage for K, which is a trim.

Prior to formulating the proposition, consider A that is not a trim. Then the trim component $Tr(A)$ of the automaton $A = (\Sigma, Q, q_0, \delta, F)$ is a finite automaton given by

$$Tr(A) = (\Sigma, Q_{tr}, q_0, \delta_{tr}, F \cap Q_{tr}) \qquad (13.28)$$

where $Q_{tr} \neq \varnothing$, $Q_{tr} = Q_{ac} \cap Q_{co}$, $\delta_{tr} = \delta$ for set $\Sigma \times Q_{tr}$; and δ_{tr} is a partial function with respect to set $\Sigma \times Q$.

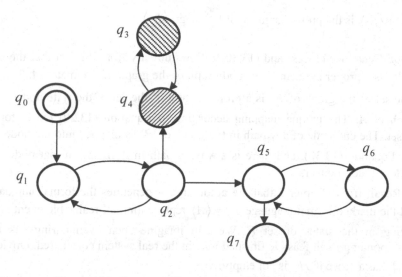

Figure 13.2. An accessible and not co-accessible finite automaton

Proposition 13.4. Let $A = (\Sigma, Q, q_0, \delta, F)$ and $B_j = (\Sigma, P_j, p_{0j}, \xi_j, H_j)$ be trim finite automata. The automaton A generates the formal language L and the marked formal language L_m that is denoted as $\langle A \rangle$ indicating the relation with automaton A. Let automaton B_j generate the marked language $J_j = \langle B_j \rangle \subset \langle A \rangle$; further let the mathematical graph corresponding to B_j be a proper subdigraph of the mathematical graph corresponding to A such that p_{0j} corresponds to q_0. It means that the subdigraph includes the initial state, which corresponds to q_0. The event sets in both automata are the same and consist of the controllable and uncontrollable events: $\Sigma = \Sigma_c \cup \Sigma_u, \Sigma_c \cap \Sigma_u = \varnothing$. Then there exists the unique function

$$h_j : P_j \to Q \tag{13.29}$$

satisfying the following equality:

$$h_j \left[\hat{\xi}_j (p_{0j}, \tilde{\rho}) \right] = \hat{\delta}(q_0, \tilde{\rho}) \tag{13.30}$$

for each string

$$\tilde{\rho} \in \overline{\langle B_j \rangle} \tag{13.31}$$

where $\overline{\langle B_j \rangle}$ is the prefix language of language $\langle B_j \rangle$.

Proof. Equations (13.29) and (13.30) follows directly from the fact that the graph of B_j is a proper mathematical subdigraph of the graph of automaton A. The state node set of the graph of B_j is a proper subset of the set of the state nodes of the graph of A. The unique mapping according to Equation (13.29) exists for this subset. The end node of any path in the graph of B_j is mapped into the node of A, *i.e.*, Equation (13.30); and there is always a path in A ending in that node. As a matter of fact A covers B_j.

Recall from Chapter 5 that the automaton A generates the formal language L and the marked formal language $L_m = \langle A \rangle$ representing all realizable event strings ending in the states of set F. We can imagine such event strings as those corresponding to all feasible finished jobs in the real system considered. Obviously, $\langle B_j \rangle$ has a sense if H_j is not empty.

Proposition 13.5. Let $A = (\Sigma, Q, q_0, \delta, F)$ and $B_j = (\Sigma, P_j, p_{0j}, \xi_j, H_j)$ be trim automata as assumed in Proposition 13.4. The operator Ω for given languages L_m and $J_0 \subset L_m$ (L_m and J_0 stands for L and K, respectively, in Equation (13.14)) is determined by the transformation of automaton B_j to B_{j+1} described in details below. Ω is such that B_{j+1} is a trim automaton. The transformation gives the searched supremum of controllable language T. The supremum is generated by the trim automaton resulting from the transformation. Define the transformation such that it goes through an intermediate automaton B'_j, *i.e.*, B_j is transformed to B'_j and then B'_j to B_{j+1}. The automata generate the following languages $J_j = \langle B_j \rangle, J'_j = \langle B'_j \rangle$, $J_{j+1} = \langle B_{j+1} \rangle$.   Thus, $J_{j+1} = \Omega(J_j)$ where $\Omega(J_j) = J_0 \cap \sup\{T : T \subseteq \Sigma^*, T = \overline{T}, T\Sigma_u \cap L_m \subseteq \overline{J_j}\}$. An iteration process for i=0, 1, 2, ... has been anticipated as shown below. For all subsequent iterations of the iteration process the following setting $K = J_0 = \langle B_0 \rangle$ is taken as language K in Equation (13.14). The automata transformation realizing Ω is a reduction, which is carried out in two steps:

1. The states of automaton B_j are mapped by a partial injection mapping (one-to-one mapping) to the states of automaton B'_j. Such states are mapped, which fulfil the conditions described below. The partial mapping is a subset of the one-to-one correspondences $p_{0,j} \leftrightarrow p'_{0,j}$, $p_{1,j} \leftrightarrow p'_{1,j}$, ..., *etc.*, where $p_{0,j}, p_{1,j}, ... \in P_j$, $p'_{0,j}, p'_{1,j}, ... \in P'_j$. A string $\tilde{\eta}$ satisfies

$$\tilde{\eta} \in J_{j+1} = \Omega(J_j) \text{ and } p'_{k,j} \in P'_j \tag{13.32}$$

if and only if $\tilde{\eta} \in J_j$ and for each prefix $\tilde{\pi}$ of $\tilde{\eta}$ holds

$$p'_{k,j} = \hat{\xi}'_j(p'_{0,j}, \tilde{\pi}) \Rightarrow \Sigma(h'_j(p'_{k,j})) \cap \Sigma_u \subseteq \Sigma(p'_{k,j}) \tag{13.33}$$

where $\Sigma(p'_{k,j})$ denotes a subset of Σ for which the partial function ξ'_j is defined in state $p'_{k,j}$. Similarly it is for $\Sigma(h'_j(p'_{k,j}))$ and function δ. Proposition 13.4 holds for function h'_j. The states not satisfying Equation (13.33) are cancelled from set P'_j and function ξ'_j is adjusted correspondingly to the state cancellation. In the graphical representation the arcs connected to the cancelled state nodes are removed, as well. If $p'_{0,j}$ is not cancelled and $H'_j \neq \varnothing$ then the automaton

$$B'_j = \left(\Sigma, P'_j, p'_{0,j}, \xi'_j, H'_j \right) \tag{13.34}$$

is obtained; and otherwise it is not defined.

2. If B'_j is defined, then the trim component of B'_j is constructed as follows:

$$B_{j+1} = \left(\Sigma, P_{j+1}, p_{0,j+1}, \xi_{j+1}, H_{j+1} \right) = Tr(B'_j) =$$
$$Tr\left(\left(\Sigma, P'_j, p'_{0,j}, \xi'_j, H'_j \right) \right) \tag{13.35}$$

where the partial injection mapping of the states from set P'_j is given as before. The mapping yields a subset of correspondences $p'_{0,j} \leftrightarrow p_{0,j+1}$, $p'_{1,j} \leftrightarrow p_{1,j+1}$, ... where $p'_{0,j}, p'_{1,j}, ... \in P'_j$, $p_{0,j+1}, p_{1,j+1}, ... \in P_{j+1}$. Here, some of the correspondences do not exist because of the reduction of the states of B'_j in order for B_{j+1} to be a trim automaton. B_{j+1} is defined if the correspondence for the initial states $p'_{0,j} \leftrightarrow p_{0,j+1}$ is preserved on the way to obtain the trim; otherwise B_{j+1} is not defined. B_{j+1} generates language $J_{j+1} = \Omega(J_j)$ and the graph of B_{j+1} is a proper mathematical subgraph of B_j.

Proof. This proposition concerns the controllability. The operator $\Omega(J_j)$ provides supremum of the prefix closed languages, which fulfils the condition $(T\Sigma_u \cap L_m) \subseteq \overline{J_j}$. Parameters of the operator are languages L_m and $J_0 \subseteq L_m$ (in comparison with Equation (13.14)). The concatenation of any uncontrollable event to the strings of T does not break the membership of the strings in J_j. The conjunction $\cap L_m$ in the condition ensures the restriction of strings only to the strings of language $\langle A \rangle$. The supremal language T is obtained by meeting the condition at Equation (13.33). It cancels those states in which uncontrollable events may occur and realize prohibited strings. It is the first step towards creating operator Ω and possible canceling of some states; the next step in completing operator Ω is to create the trim automaton. The creation of the trim automaton is substantiated because the set of strings is limited to the strings of J_0. Namely, this limitation can spoil the trim property. Strings, which do not end in the states of H'_j, are to be removed. Possibly some additional states are to be cancelled next along with their related arcs, as well. The iterative application of Ω gives the supremal controllable sublanguage. It is based on Proposition 13.1 and the fact that all languages generated by finite automata are regular and Equation (13.23) holds.

By Proposition 13.4 we have established a way for determining operator Ω and are prepared to formulate an algorithm for finding the supremal sublanguage given a language. The algorithm is given below:

Algorithm 13.1.
1. Let $A = (\Sigma, Q, q_0, \delta, F)$ be a trim automaton generating the language L_m.
2. The variable j is set to $j = 0$.
3. Let $B_0 = (\Sigma, P, p_0, \xi, H)$ be a trim automaton generating the language of the required behavior $J_0 = K$.
4. For automaton B_j, automaton B_{j+1} is constructed satisfying the onditions at Equations (13.33) and (13.35). B_{j+1} generates language J_{j+1}.
5. If B_{j+1} cannot be constructed, go to Step 9.
6. If $B_{j+1} = B_j$ go to Step 8.
7. Increase j by one, i.e., $j := j + 1$ and go to Step 4.
8. End. B_j generates the supremal sublanguage for the required language K.
9. End. B_{j+1} cannot be constructed.

If the algorithm ends with step 8, a solution has been found. In each state the resulting automaton B_j determines which controllable events are allowed and which are prohibited. A control automaton can be constructed for that purpose. It is illustrated in the following example.

Example 13.4. For the purpose of surveillance, a sea region is divided into sections according to the shape of the seabed as it is schematically shown in Figure 13.3. Two submarines A and B cruise and guard the sea region. A has its own separate repair port – section No. 5, its home harbor is section No. 2, while B has home and repair harbor in section No. 4. Due to the seabed form and the sea underwater streams, as well as different size of the submarines, possible and controllable transits of the submarines between sections are those shown in Figure 13.3.

Meeting of the submarines in any section of the guarded sea region is prohibited for safety reasons. At the beginning of the inspection, A is in section No. 5 and B in section No. 4. As for event strings it is required that in any situation each submarine can start from its home harbor and return there. Such a requirement is meant in the sense that a submarine has possibly to wait until the other submarine undertakes some admissible transits. Thus the start or return is not immediately feasible but in no case does a deadlock occur when no transit is possible because of the violation of the "no-meeting" condition. A stronger condition could be put on the guard system, *i.e.,* the transits of a submarine to start or return are always and immediately realizable without waiting for the other submarine.

In this example the supervisory control of the submarine guard system can be solved using finite automaton modeling. The task of the supervisory control is to prevent the violation of the requirements described above.

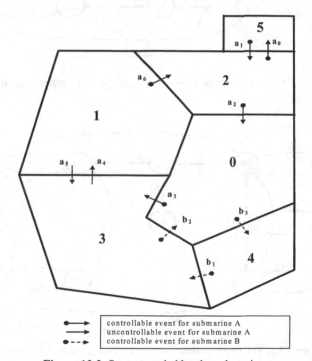

Figure 13.3. Space guarded by the submarines

The finite automaton specifying all possible and feasible transits in the studied system is shown in Figure 13.4. Considering the automaton as a language generator we can see that it generates L_m when the final state set is $F = \{\underline{24}\}$ and $q_0 = \{\underline{54}\}$. A state notation $\underline{ij}$ means that submarine A is in the i-th and B in the j-th section, respectively. In other words an automaton state is a combination of submarine positions. Such an automaton is called the shuffle automaton and is designed by combining two sub-automata. The choice of state $\underline{24}$ for F represents the requirement that transit event strings always exist enabling each submarine to return to its home harbor. Automaton A in Figure 13.4 is accessible and co-accessible.

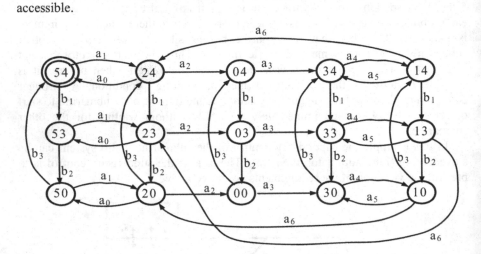

Figure 13.4. The finite automaton representing all possible states and transits in the submarine system

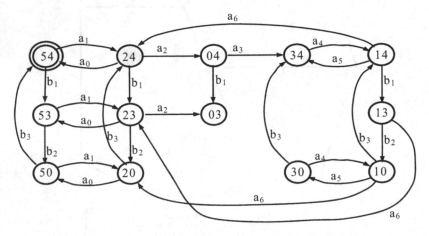

Figure 13.5. The finite automaton after removing the prohibited states

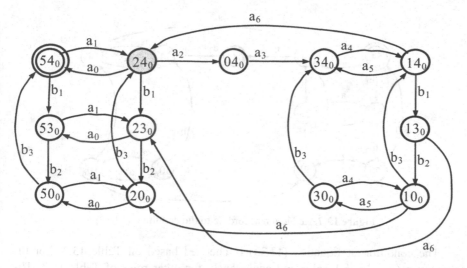

Figure 13.6. The trim automaton

Now let finite automaton B_0 specifying the required behavior be constructed. It is done by a reduction of automaton A. Obviously the forbidden states are $\underline{00}$ and $\underline{33}$. They are removed from A as shown in Figure 13.5. The automaton is accessible but not co-accessible. The state $\underline{03}$ must be removed in order to keep the co-accessibility of the automaton (Figure 13.6). The automaton in Figure 13.6 is a trim and it is a finite automaton denoted B_0 in Algorithm 13.1. Recall that by Proposition 13.5 both automata A and B_j should be trims in order to be able to form the operator Ω by the finite automata reduction.

Now, the operator Ω is given first by the changeover from automaton B_0 to B_0' and by changeover from automaton B_0' to B_1 in the sense of Proposition 13.5. By comparing with Equation (13.29), the function h_0 is obviously

$$h_0\left(\underline{54}_0\right)=\underline{54}, \quad h_0\left(\underline{24}_0\right)=\underline{24}, \quad h_0\left(\underline{04}_0\right)=\underline{04}, \quad h_0\left(\underline{34}_0\right)=\underline{34},$$
$$h_0\left(\underline{14}_0\right)=\underline{14}, \quad h_0\left(\underline{53}_0\right)=\underline{53}, \quad h_0\left(\underline{23}_0\right)=\underline{23}, \quad h_0\left(\underline{13}_0\right)=\underline{13},$$
$$h_0\left(\underline{50}_0\right)=\underline{50}, \quad h_0\left(\underline{20}_0\right)=\underline{20}, \quad h_0\left(\underline{30}_0\right)=\underline{30}, \quad h_0\left(\underline{10}_0\right)=\underline{10}.$$

Function h_0' is

$$h_0'\left(\underline{54}_0'\right)=\underline{54}, \quad h_0'\left(\underline{24}_0'\right)=\underline{24} \quad etc.$$

as Figure 13.7 specifies.

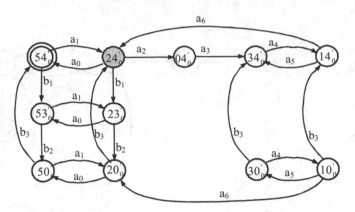

Figure 13.7. Reduction according to the condition (13.35)

The condition at Equation (13.33) is checked based on Table 13.2. For the empty set $\varnothing \subseteq \{a_3, b_1\}$ holds and analogously for other rows of Table 13.2. The condition is not fulfilled in one row of the table, namely for state $\underline{13}_0'$. There $a_5 \not\subseteq \{a_6, b_2\}$ and state $\underline{13}_0'$ should be cancelled (Figure 13.7) in order to approach the finite automaton, which results from the application of the operator Ω.

Table 13.2. The transition table for checking the condition at Equation (13.33)

	$04_0'$	$10_0'$	$13_0'$	$14_0'$	$20_0'$	$23_0'$	$24_0'$	$30_0'$	$34_0'$	$50_0'$	$53_0'$	$54_0'$	$\Sigma(h_0(p_j')) \cap \Sigma_u$	$\subseteq \Sigma(p_j')$
$04_0'$			b_1						a_3				$\varnothing$	a_3, b_1
$10_0'$				b_3	a_6			a_5					a_5	a_5, a_6, b_3
$13_0'$		b_2				a_6							a_5	a_6, b_2
$14_0'$			b_1				a_6		a_5				a_5	a_5, a_6, b_1
$20_0'$							b_3			a_0			$\varnothing$	a_0, b_3
$23_0'$					b_2						a_0		$\varnothing$	a_0, b_2
$24_0'$	a_2					b_1						a_0	$\varnothing$	a_0, a_2, b_1
$30_0'$		a_4							b_3				a_4	a_4, b_3
$34_0'$				a_4									a_4	a_4
$50_0'$					a_1							b_3	$\varnothing$	a_1, b_3
$53_0'$						a_1				b_2			$\varnothing$	a_1, b_2
$54_0'$							a_1				b_1		$\varnothing$	a_1, b_1

The next step is to preserve the trim property. The states $30_0'$ and $10_0'$ are not accessible and are to be cancelled. According to Figure 13.8, h_1 is defined $h_1\left(54_1\right)=\underline{54}, \quad h_1\left(24_1\right)=\underline{24}$, etc.

The result of the last step is depicted in Figure 13.8 and Tables 13.3–13.5.

Then, the above steps are repeated. A table corresponding to Table 13.2 is Table 13.3. As there is no more reduction of the automaton, the algorithm stops. The automaton depicted in Figure 13.8 generates the supremal controllable sublanguage of language B_0.

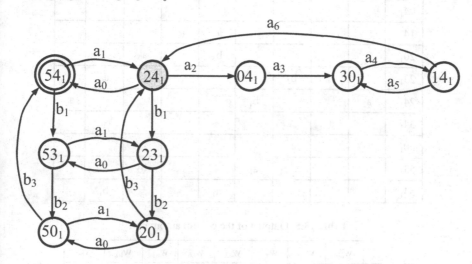

Figure 13.8. Finite automaton B_1 generating language J_1

Table 13.3. The transition table for the next step of Algorithm 13.1.

	$04_1'$	$14_1'$	$20_1'$	$23_1'$	$24_1'$	$34_1'$	$50_1'$	$53_1'$	$54_1'$	$\Sigma\!\left(h_0\!\left(p_j'\right)\right)\cap\Sigma_u \subseteq \Sigma\!\left(p_j'\right)$
$04_1'$					a_3					a_3, b_1
$14_1'$					a_6	a_5			a_5	a_5, a_6, b_1
$20_1'$					b_3		a_0			a_0, b_3
$23_1'$			b_2				a_0			a_0, b_2
$24_1'$	a_2		b_1					a_0		a_0, a_2, b_1
$34_1'$		a_4							a_4	a_4
$50_1'$		a_1						b_3		a_1, b_3
$53_1'$			a_1				b_2			a_1, b_2
$54_1'$					a_1		b_1			a_1, b_1

The supervisory control is performed through the controllable events. In each possible system state the allowed or prohibited events are set up. For example let the system be in state $\underline{04}_1$. The study of state $\underline{04}$ in Figures 13.8 and 13.4 yields that event b_1 is to be prohibited and a_3 is to be allowed.

Table 13.4. States of the control automaton

	04_1	14_1	20_1	23_1	24_1	34_1	50_1	53_1	54_1
04_1						a_3			
14_1					a_6	a_5			
20_1					b_3		a_0		
23_1			b_2					a_0	
24_1	a_2			b_1					a_0
34_1		a_4							
50_1			a_1						b_3
53_1				a_1			b_2		
54_1					a_1			b_1	

Table 13.5. Outputs of the control automaton

	W_{a0}	W_{a1}	W_{a2}	W_{a3}	W_{a6}	W_{b1}	W_{b2}	W_{b3}
04_1				1		0		
14_1					1	0		
20_1	1		0					1
23_1	1		0				1	
24_1	1		1			1		
34_1						0		
50_1		1						1
53_1		1					1	
54_1		1				1		

A deterministic finite automaton with outputs can be constructed by setting up the controllable events. Such an automaton would have the same topology as B_1 with outputs for the event control. It can be minimized by finding the state classes equivalent from the output point of view. Denote the automaton outputs w_{ai} and w_{bi}. If $w_{ai} = 1, w_{bi} = 1$, events a_i and b_i are allowed, while in the inverse case they are prohibited. The free outputs or the so-called "don't care" are purposely completed with suitable values.

The equivalent classes are

$$C1 = \{\underline{24}_1, \underline{54}_1\}, \quad C2 = \{\underline{04}_1, \underline{14}_1, \underline{20}_1, \underline{23}_1, \underline{34}_1, \underline{50}_1, \underline{53}_1\}$$

(Table 13.6).

A suitable completion of the output function is given in Table 13.5. The minimum control automaton is depicted in Figure 13.9. The outputs $w_{a0}, w_{a1}, w_{a3}, w_{a6}, w_{b2}, w_{b3}$ are permanently set to logical one.

Table 13.6. Completion of the "don't care"

	$\underline{04}_1$	$\underline{14}_1$	$\underline{20}_1$	$\underline{23}_1$	$\underline{24}_1$	$\underline{34}_1$	$\underline{50}_1$	$\underline{53}_1$	$\underline{54}_1$	w_{a0}	w_{a1}	w_{a2}	w_{a3}	w_{a6}	w_{b1}	w_{b2}	w_{b3}
$\underline{04}_1$						a_3				1	1	0	1	1	0	1	1
$\underline{14}_1$				a_6	a_5					1	1	0	1	1	0	1	1
$\underline{20}_1$				b_3		a_0				1	1	0	1	1	0	1	1
$\underline{23}_1$			b_2				a_0			1	1	0	1	1	0	1	1
$\underline{24}_1$	a_2			b_1					a_0	1	1	1	1	1	1	1	1
$\underline{34}_1$		a_4								1	1	0	1	1	0	1	1
$\underline{50}_1$			a_1						b_3	1	1	0	1	1	0	1	1
$\underline{53}_1$				a_1			b_2			1	1	0	1	1	0	1	1
$\underline{54}_1$					a_1			b_1		1	1	1	1	1	1	1	1

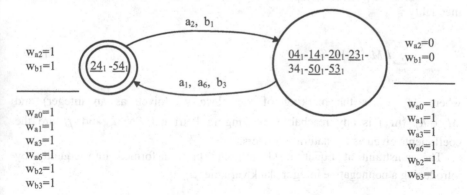

Figure 13.9. The supervisory control finite automaton with outputs

13.4 Supervisory Control Solution with *P*-invariants

A method based on Petri nets and place invariants (Moody and Antsaklis 1998; Yamalidou *et al.* 1996) appeared to be able to cope with the state space dimensionality. The dimensionality problem can appear in the method which uses the finite automaton described in the previous section or the reachability graph of the Petri net to be discussed later. The *P*-invariant method does not require the state space construction, is computationally very effective and can be used for large and complex systems. On the other hand the method neither resolves the problem of uncontrollable transitions nor the problem of the required accessibility of the home places directly. A partial elimination of these drawbacks is possible by transforming the problem into the proper system constraints. There is also the possibility to transform the constraints given in other forms into the form required by the method.

The method using the place invariants, *P*-invariants for short, of a Petri net enables design of the feedback supervisory controller. The idea is to enhance the given Petri net describing the system so that the enhanced Petri net contains the required *P*-invariants. The *P*-invariants are derived from the linear constraints imposed on the Petri net markings, which represent the supervisory requirements and restrictions imposed on the system under supervisory control. It was shown in Section 8.10 that a place invariant determines the set of places for which the weighted sum of tokens remains constant for all markings in the reachability set $R_{PN}(\mathbf{m}_0)$ of a Petri net *PN*.

Consider a Petri net *PN* with n places and m transitions representing the controlled system. Let $\mathbf{N}_\Delta$ be the Δ-incidence matrix of this net. Control requirements for *PN* are realized by a Petri net supplement representing a Petri net controller. The Δ-incidence matrix of the supplemented control part of the Petri net is denoted by $\mathbf{N}_{\Delta c}$.

A control requirement imposed on the system is formulated in the form of inequality

$$\sum_{i=1}^{n} l_i M_r(p_i) \le \beta, \tag{13.36}$$

where $M_r(p_i)$ is the marking of the place p_i (given as an integer) and $M_r \in R_{PN}(\mathbf{m}_0)$ is any reachable marking in Petri net *PN*, l_i and β are the coefficients given as the natural numbers.

The constraint at Equation (13.36) can be transformed into equality by introducing a nonnegative integer slack variable μ_{rc}:

$$\sum_{i=1}^{n} l_i M_r(p_i) + \mu_{rc} = \beta \tag{13.37}$$

where μ_{rc} represents a marking of the supplementary control place p_c, which holds the extra tokens to meet the equality. Values of μ_{rc} are chosen to preserve the equality for all reachable markings for the extended Petri net. The described procedure can be applied to more constraints given in the form of Equation (13.36) leading to more supplemented places. Then the structure of the supplemented control part of the Petri net is to be determined. As a matter of fact it is necessary to determine the arcs connecting the supplemented places with the transitions of the Petri net *PN* and the initial marking of the control places. This is based on the determination of the place invariants. A constraint in the form of equality introduces a place invariant into the Petri net, *i.e.*, the weighted sum of tokens in all places of *PN* and in the supplementary place p_c are constant and equal β. The number of supplementary places of the controller part of the Petri net is equal to the number of constraints to be enforced. Every place of the controller part adds one row to the incidence matrix $\mathbf{N}_{As}$ of the whole controlled system. Thus, $\mathbf{N}_{As}$ is composed of two matrices, namely $\mathbf{N}_A$ of the system to be controlled and $\mathbf{N}_{Ac}$ of the controller net.

The arcs connecting the controller places to the original Petri net of the system can be computed by Equation (8.29) defining place-invariants

$$\mathbf{N}_{As}^T \mathbf{i}_P = \mathbf{0} \Rightarrow \mathbf{i}_P^T \mathbf{N}_{As} = \mathbf{0}^T, \ \mathbf{0}^T = \underbrace{(0,0,...,0)}_{m-times} \tag{13.38}$$

where the unknowns are the elements of the new rows of the matrix $\mathbf{N}_{As}$, *i.e.*, the elements of matrix $\mathbf{N}_{Ac}$ and vector $\mathbf{i}_P$ is the desired *P*-invariant according to Equations (8.41) and (13.37) and it is obviously given by

$$\mathbf{i}_P^T = (l_1 \, ... \, l_n \, 1) \tag{13.39}$$

The variable μ_{rc} has a coefficient one.

All constraints in the form of Equation (13.36) can be aggregated in a matrix where the number of constraints is n_c:

$$\mathbf{L}\,\mathbf{m}_r \le \mathbf{b} \tag{13.40}$$

The matrix equality is

$$\mathbf{L}\,\mathbf{m}_r + \mathbf{m}_r^c = \mathbf{b}, \tag{13.41}$$

where $\mathbf{L}$ is an $n_c \times n$ matrix of non-negative integers containing the constraint coefficients, $\mathbf{b}$ is an $n_c \times 1$ vector of non-negative integers, $\mathbf{m}_r$ is a vector variable representing reachable markings of the original Petri net with n components (for places of the original Petri net), $\mathbf{m}_r^c$ is an $n_c \times 1$ vector variable of non-negative integers that represents markings of the supplementary controller places, and n_c is the number of constraints. For all constraints we have

$$(\mathbf{L} \quad \mathbf{I}) \mathbf{N}_{\Delta s} = (\mathbf{0}) = (\mathbf{L} \quad \mathbf{I}) \begin{pmatrix} \mathbf{N}_\Delta \\ \mathbf{N}_{\Delta c} \end{pmatrix}$$

$$\Rightarrow \quad \mathbf{L}.\mathbf{N}_\Delta + \mathbf{N}_{\Delta c} = (\mathbf{0})$$

(13.42)

where $\mathbf{L}$ is a matrix whose rows are the transposed P-invariant vectors, $\mathbf{I}$ is an $n_c \times n_c$ identity matrix and $(\mathbf{0})$ is an $n_c \times m$ zero matrix. The matrix $\mathbf{N}_{\Delta c}$ contains arcs that connect the controller net places with the transitions of the original Petri net PN and is given by

$$\mathbf{N}_{\Delta c} = -\mathbf{L} \mathbf{N}_\Delta$$

(13.43)

The initial marking of the supplementary controller part of the extended Petri net is calculated from the place invariant conditions that are initially met:

$$\mathbf{L} \mathbf{m}_0 + \mathbf{m}_{c0} = \mathbf{b}$$

(13.44)

where $\mathbf{m}_0$ is the initial marking of the Petri net PN and $\mathbf{m}_{c0}$ is the searched initial marking of the supplementary places, *i.e.*,

$$\mathbf{m}_{c0} = \mathbf{b} - \mathbf{L} \mathbf{m}_0$$

(13.45)

The described control is maximally permissive from the point of view of possible transition firings in the controlled Petri net. Consider Δ - incidence matrix $\mathbf{N}_\Delta$ and assume its rank $r < n$. Then the Petri net PN has $k = (n - r)$ P-invariants because the homogenous equation at Equation (8.29) has k basic linearly independent non-zero solutions (each solution is a vector). The invariants represent bindings or restrictions contained in the original PN. The Δ - incidence matrix of the extended Petri net is

$$\mathbf{N}_{\Delta s} = \begin{pmatrix} \mathbf{N}_\Delta \\ \mathbf{N}_{\Delta c} \end{pmatrix} = \begin{pmatrix} \mathbf{N}_\Delta \\ -\mathbf{L} \mathbf{N}_\Delta \end{pmatrix}$$

(13.46)

The rows of the lower matrix part are linear combinations of the rows of $\mathbf{N}_\Delta$. The rank of $\mathbf{N}_{\Delta s} = k$, *i.e.*, it is equal to the rank of $\mathbf{N}_\Delta$. Thus there are no new bindings due to the control extension of the Petri net PN.

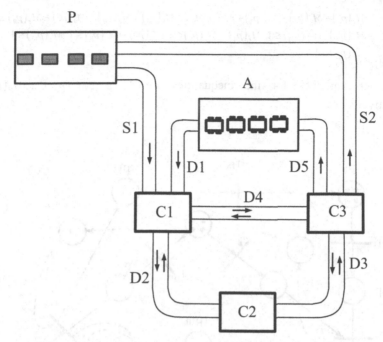

Figure 13.10. A manufacturing layout with the transport system

Example 13.5. A technological layout is schematically illustrated in Figure 13.10. Its transportation part works as follows. Four palettes are in room P and four automatic guided vehicles in room A at the beginning. The palettes are cleaned in P after some number of uses. The accumulators of the vehicles are loaded when necessary in room A. The palettes are transported to room C1 by belt conveyor S1, which has capacity of four palettes. The capacity of S2 is four palettes, too. The vehicles go one by one to room C1 where a palette is mounted on a vehicle. Only one vehicle can be in one-way corridor D1 at a time. The same holds true for D5. Each vehicle with a palette on it can move between the manufacturing cells as the arrows indicate. The manufactured parts are transported with the help of vehicles with the palettes. After some transportation a palette is taken from the vehicle and transported back in P *via* the belt conveyor S2 while the vehicle goes to room A for the accumulator reloading.

Only one vehicle with a palette can pass through bidirectional corridors D2 up to D4. The number of the vehicles can be maximally two in rooms C1 and C2 together including the corridor between the rooms. Moreover in the corridor D2 can be only one vehicle. Analogously maximally three vehicles can be in rooms C2 and C3 together with corridor D3 and one vehicle in corridor D3. Maximally one vehicle can be in corridor D4.

The Petri net describing the transportation part of the system is represented by solid lines in Figure 13.11.

The inequalities expressing the condition of the system behavior given above are

$$M_r(p_3)+M_r(p_4)+M_r(p_5)+M_r(p_6)\le 2,\ M_r(p_6)+M_r(p_7)+M_r(p_8)+M_r(p_9)\le 3$$
$$M_r(p_{10})+M_r(p_{11})\le 1,\ M_r(p_4)+M_r(p_5)\le 1,\ M_r(p_7)+M_r(p_8)\le 1,\ M_r(p_{15})\le 1,$$
$$M_r(p_{12})\le 1$$

The slack variables for the inequalities are $M_r(p_{c1}), M_r(p_{c2}), ..., M_r(p_{c7})$, respectively.

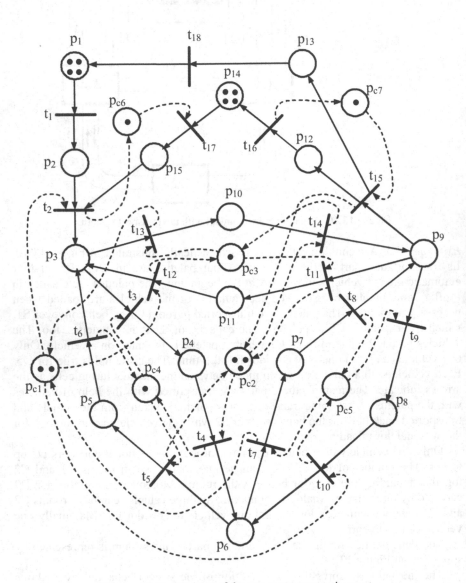

Figure 13.11. Petri net of the system transportation part

The controller Petri net structure is given by the Δ-incidence matrix $\mathbf{N}_{\Delta c}$. It specifies a supplementary structure, which binds the system behavior. It is drawn with dashed lines in the Petri net in Figure 13.11. The Petri net is a base for the system supervisory control. Suppose that the passages of vehicles and palettes are controllable. The Petri net transitions correspond to the passages. The transition firings are guarded by places $p_{c1}, \ldots, p_{c7}$. If a transition in the Petri net is not fireable the corresponding passage is not allowed. Using the Petri net of Figure 13.11, a control program for the control of the passages can be written.

The Δ-incidence matrix $\mathbf{N}_{\Delta c}$ is

$$\mathbf{N}_{\Delta c} =$$

$$-\mathbf{L}\,\mathbf{N}_\Delta = -\begin{pmatrix}
0 & 0 & 1 & 1 & 1 & 1 & 0 & 0 & 0 & 0 & 0 & 0 & 0 & 0 & 0 & 0 \\
0 & 0 & 0 & 0 & 0 & 1 & 1 & 1 & 1 & 1 & 0 & 0 & 0 & 0 & 0 & 0 \\
0 & 0 & 0 & 0 & 0 & 0 & 0 & 0 & 0 & 1 & 1 & 0 & 0 & 0 & 0 & 0 \\
0 & 0 & 0 & 1 & 1 & 0 & 0 & 0 & 0 & 0 & 0 & 0 & 0 & 0 & 0 & 0 \\
0 & 0 & 0 & 0 & 0 & 0 & 1 & 1 & 0 & 0 & 0 & 0 & 0 & 0 & 0 & 0 \\
0 & 0 & 0 & 0 & 0 & 0 & 0 & 0 & 0 & 0 & 0 & 0 & 0 & 0 & 0 & 1 \\
0 & 0 & 0 & 0 & 0 & 0 & 0 & 0 & 0 & 0 & 0 & 0 & 1 & 0 & 0 & 0
\end{pmatrix} \times$$

$$\begin{pmatrix}
-1 & 0 & 0 & 0 & 0 & 0 & 0 & 0 & 0 & 0 & 0 & 0 & 0 & 0 & 0 & 0 & 0 & 1 \\
1 & -1 & 0 & 0 & 0 & 0 & 0 & 0 & 0 & 0 & 0 & 0 & 0 & 0 & 0 & 0 & 0 & 0 \\
0 & 1 & -1 & 0 & 0 & 1 & 0 & 0 & 0 & 0 & 0 & 1 & -1 & 0 & 0 & 0 & 0 & 0 \\
0 & 0 & 1 & -1 & 0 & 0 & 0 & 0 & 0 & 0 & 0 & 0 & 0 & 0 & 0 & 0 & 0 & 0 \\
0 & 0 & 0 & 0 & 1 & -1 & 0 & 0 & 0 & 0 & 0 & 0 & 0 & 0 & 0 & 0 & 0 & 0 \\
0 & 0 & 0 & 1 & -1 & 0 & -1 & 0 & 0 & 1 & 0 & 0 & 0 & 0 & 0 & 0 & 0 & 0 \\
0 & 0 & 0 & 0 & 0 & 0 & 1 & -1 & 0 & 0 & 0 & 0 & 0 & 0 & 0 & 0 & 0 & 0 \\
0 & 0 & 0 & 0 & 0 & 0 & 0 & 0 & 1 & -1 & 0 & 0 & 0 & 0 & 0 & 0 & 0 & 0 \\
0 & 0 & 0 & 0 & 0 & 0 & 0 & 1 & -1 & 0 & -1 & 0 & 0 & 1 & -1 & 0 & 0 & 0 \\
0 & 0 & 0 & 0 & 0 & 0 & 0 & 0 & 0 & 0 & 0 & 0 & 1 & -1 & 0 & 0 & 0 & 0 \\
0 & 0 & 0 & 0 & 0 & 0 & 0 & 0 & 0 & 0 & 1 & -1 & 0 & 0 & 0 & 0 & 0 & 0 \\
0 & 0 & 0 & 0 & 0 & 0 & 0 & 0 & 0 & 0 & 0 & 0 & 0 & 0 & 1 & -1 & 0 & 0 \\
0 & 0 & 0 & 0 & 0 & 0 & 0 & 0 & 0 & 0 & 0 & 0 & 0 & 0 & 1 & 0 & 0 & -1 \\
0 & 0 & 0 & 0 & 0 & 0 & 0 & 0 & 0 & 0 & 0 & 0 & 0 & 0 & 0 & 1 & -1 & 0 \\
0 & -1 & 0 & 0 & 0 & 0 & 0 & 0 & 0 & 0 & 0 & 0 & 0 & 0 & 0 & 0 & 1 & 0
\end{pmatrix} =$$

$$\begin{array}{c}
p_{c1} \\ p_{c2} \\ p_{c3} \\ p_{c4} \\ p_{c5} \\ p_{c6} \\ p_{c7}
\end{array}
\begin{pmatrix}
0 & -1 & 0 & 0 & 0 & 0 & 1 & 0 & 0 & -1 & 0 & -1 & 1 & 0 & 0 & 0 & 0 & 0 \\
0 & 0 & 0 & -1 & 1 & 0 & 0 & 0 & 0 & 0 & 1 & 0 & 0 & -1 & 1 & 0 & 0 & 0 \\
0 & 0 & 0 & 0 & 0 & 0 & 0 & 0 & 0 & 0 & -1 & 1 & -1 & 1 & 0 & 0 & 0 & 0 \\
0 & 0 & -1 & 1 & -1 & 1 & 0 & 0 & 0 & 0 & 0 & 0 & 0 & 0 & 0 & 0 & 0 & 0 \\
0 & 0 & 0 & 0 & 0 & 0 & -1 & 1 & -1 & 1 & 1 & 0 & 0 & 0 & 0 & 0 & 0 & 0 \\
0 & 1 & 0 & 0 & 0 & 0 & 0 & 0 & 0 & 0 & 0 & 0 & 0 & 0 & 0 & 0 & -1 & 0 \\
0 & 0 & 0 & 0 & 0 & 0 & 0 & 0 & 0 & 0 & 0 & 0 & 0 & 0 & -1 & 1 & 0 & 0
\end{pmatrix}$$

A correct required behavior depends on the formulation of the inequalities at Equation (13.40). If for example we put the restrictions as follows

$$M(p_3) + M(p_6) \leq 2, \quad M(p_6) + M(p_9) \leq 3$$

While the rest are given as before, the P-invariant method gives a structure that includes dead-locks. Seemingly the inequalities specify the same behavior but the control structure is not good. We postpone the supervisory control for the erroneous inequalities to the reader.

The presented method does not directly solve the case of uncontrollable events. It can be resolved by a suitable set of inequalities at Equation (13.40). The problem of proper inequalities will be treated in the following example.

Example 13.6. Consider Example 13.5 with the same behavior requirements. It is required that in every situation each submarine can immediately start from its home position or to return there. As before, transits a_4 and a_5 are uncontrollable, and the other transitions are controllable (in comparison with Figure 13.2). The system Petri net is represented *via* solid arcs in Figure 13.12. Obviously, to prevent the submarines' meeting requires the following inequalities:

$$M_r(p_{A0}) + M_r(p_{B0}) \leq 1, \quad M_r(p_{A3}) + M_r(p_{B3}) \leq 1 \tag{13.47}$$

where the place p_{A0} corresponds to the presence of the submarine A in section 0, and similarly for other places. The prohibition of uncontrollable event a_5 when submarine B is accidentally in section 3 requires that sections 1 and 3 together contain just one submarine. Moreover, in the situation when submarine A is in section 0 and B in section 3 the system is in a deadlock because B cannot return immediately to its home section 4 unless it meets A. The last problem can be coped with by requirement that if A is in section 1, there cannot be B in section 3, and if A is in 0, B cannot be in section 3, and finally if B is in 3, A cannot be in 1 or 0. Expressing it by an inequality we have

$$M_r(p_{A1}) + M_r(p_{B3}) + M_r(p_{A0}) \leq 1 \tag{13.48}$$

Now Equations (13.47) and (13.48) completely cover requirements of the admitted states or events and the deadlock prevention. The solution with the P-invariants is

$$\mathbf{N}_{\Delta c} = -\mathbf{L}\,\mathbf{N}_{\Delta} =$$

$$-\begin{pmatrix} 0 & 0 & 1 & 0 & 0 & 0 & 0 & 1 \\ 0 & 0 & 0 & 1 & 0 & 0 & 1 & 0 \\ 0 & 0 & 1 & 0 & 1 & 0 & 1 & 0 \end{pmatrix} \begin{pmatrix} 1 & -1 & 0 & 0 & 0 & 0 & 0 & 0 & 0 & 0 \\ -1 & 1 & -1 & 0 & 0 & 0 & 1 & 0 & 0 & 0 \\ 0 & 0 & 1 & -1 & 0 & 0 & 0 & 0 & 0 & 0 \\ 0 & 0 & 0 & 1 & -1 & 1 & 0 & 0 & 0 & 0 \\ 0 & 0 & 0 & 0 & 1 & -1 & -1 & 0 & 0 & 0 \\ 0 & 0 & 0 & 0 & 0 & 0 & 0 & -1 & 0 & 1 \\ 0 & 0 & 0 & 0 & 0 & 0 & 0 & 1 & -1 & 0 \\ 0 & 0 & 0 & 0 & 0 & 0 & 0 & 0 & 1 & -1 \end{pmatrix} =$$

$$\begin{pmatrix} 0 & 0 & -1 & 1 & 0 & 0 & 0 & 0 & -1 & 1 \\ 0 & 0 & 0 & -1 & 1 & -1 & 0 & -1 & 1 & 0 \\ 0 & 0 & 1 & 1 & -1 & 1 & 1 & -1 & 1 & 0 \end{pmatrix}$$

where the columns of **L** and the rows of $\mathbf{N}_{\Delta}$ correspond to the places p_{A5}, p_{A2}, p_{A0}, p_{A3}, p_{A1}, p_{B4}, p_{B3}, and p_{B0}, , respectively, and the columns of $\mathbf{N}_{\Delta}$ correspond to the transitions $t_{a0}-t_{a2}$ and $t_{b1}-t_{b3}$ respectively. The supplementary control places are p_{c1}, p_{c2}, and p_{c3} corresponding respectively to the rows of matrix **L**. The control part of the Petri net is depicted with dashed lines in Figure 13.12.

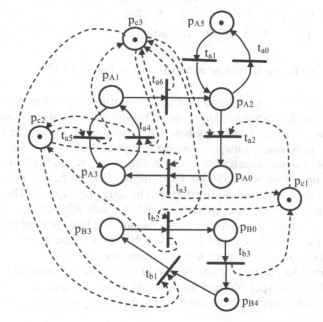

Figure 13.12. Petri net for the submarine system

Impact of the requirement inequalities on the solution is visible from the next reasoning. The inequalities at Equations (13.47) and (13.48) can be equivalently substituted by two inequalities as follows

$$M_r(p_{A0}) + M_r(p_{A3}) + M_r(p_{A1}) + M_r(p_{B3}) \leq 1$$
$$M_r(p_{A0}) + M_r(p_{B0}) \leq 1$$

In this case

$$\mathbf{N}_{\Delta c} = -\mathbf{L}\,\mathbf{N}_\Delta = -\begin{pmatrix} 0 & 0 & 1 & 1 & 1 & 0 & 1 & 0 \\ 0 & 0 & 1 & 0 & 0 & 0 & 0 & 1 \end{pmatrix} \times$$

$$\begin{pmatrix} 1 & -1 & 0 & 0 & 0 & 0 & 0 & 0 & 0 & 0 \\ -1 & 1 & -1 & 0 & 0 & 0 & 1 & 0 & 0 & 0 \\ 0 & 0 & 1 & -1 & 0 & 0 & 0 & 0 & 0 & 0 \\ 0 & 0 & 0 & 1 & -1 & 1 & 0 & 0 & 0 & 0 \\ 0 & 0 & 0 & 0 & 1 & -1 & -1 & 0 & 0 & 0 \\ 0 & 0 & 0 & 0 & 0 & 0 & 0 & -1 & 0 & 1 \\ 0 & 0 & 0 & 0 & 0 & 0 & 0 & 1 & -1 & 0 \\ 0 & 0 & 0 & 0 & 0 & 0 & 0 & 0 & 1 & -1 \end{pmatrix} =$$

$$\begin{pmatrix} 0 & 0 & -1 & 0 & 0 & 0 & 1 & -1 & 1 & 0 \\ 0 & 0 & -1 & 1 & 0 & 0 & 0 & 0 & -1 & 1 \end{pmatrix}$$

The Petri net for the described case is shown in Figure 13.13.

Example 13.7. Figure 13.14 shows landing and starting fields of an airport. Q1 and Q2 are the air parking sections. On commands from the control tower airplanes can transfer from Q1 or Q2 to parking airspace S consisting of sections S1 through S4. In S can be only one airplane. Transfers between sections S1 through S4 are uncontrollable events. Transfers between sections A1, K, and A2 of the landing field I are controllable events. So are those for the landing field II and the starting field. In sections A1, K, and A2 together can be only one airplane. The same holds for the landing field II. Each of sections R1, K, and R2 can host at most one airplane. In T1 and T2 together is allowed one airplane. The airport traffic is subdued to the control requirements given below. Of course, the goal of the supervisory control is to avoid the airplane collisions.

The Petri net for the P-invariant method of the supervisory control solution is in Figure 13.15. The respective matrices are given in a tabular form as shown in Tables 13.7–13.9. Control places are connected to the system Petri net model with dashed lines.

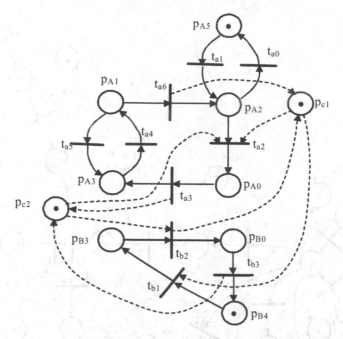

Figure 13.13. Modified Petri net for the submarine system

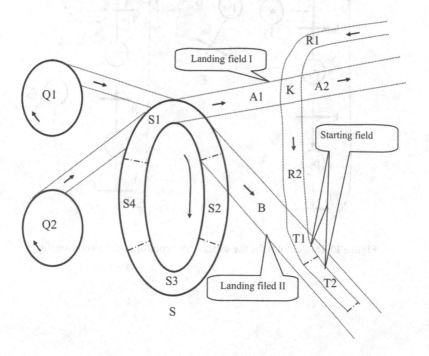

Figure 13.14. Organization of the airport traffic

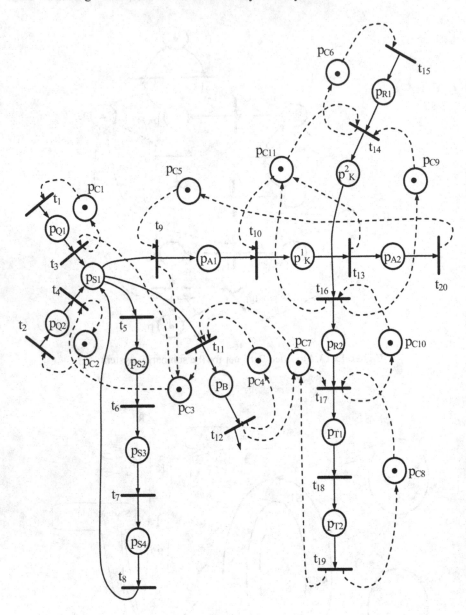

Figure 13.15. Petri net for the supervisory control of the airport traffic

Control requirements:	Corresponding control place:
$p_{Q1} \leq 1$	p_{C1}
p_{C1}	p_{C2}
$p_{S1} + p_{S2} + p_{S3} + p_{S4} \leq 1$	p_{C3}
$p_B \leq 1$	p_{C4}
$p_{A1} + p_K^1 + p_{A2} \leq 1$	p_{C5}
$p_{R1} \leq 1$	p_{C6}
$p_B + p_{T1} + p_{T2} \leq 1$	p_{C7}
$p_{T1} + p_{T2} \leq 1$	p_{C8}
$p_K^2 \leq 1$	p_{C9}
$p_{R2} \leq 1$	p_{C10}
$p_K^1 + p_K^2 \leq 1$	p_{C11}

Place markings are for simplicity denoted equally as corresponding places.

Table 13.7. Matrix N_Δ

	t_1	t_2	t_3	t_4	t_5	t_6	t_7	t_8	t_9	t_{10}	t_{11}	t_{12}	t_{13}	t_{14}	t_{15}	t_{16}	t_{17}	t_{18}	t_{19}	t_{20}
p_{Q1}	1	0	−1	0	0	0	0	0	0	0	0	0	0	0	0	0	0	0	0	0
p_{Q2}	0	1	0	−1	0	0	0	0	0	0	0	0	0	0	0	0	0	0	0	0
p_{S1}	0	0	1	1	−1	0	0	1	−1	0	−1	0	0	0	0	0	0	0	0	0
p_{S2}	0	0	0	0	1	−1	0	0	0	0	0	0	0	0	0	0	0	0	0	0
p_{S3}	0	0	0	0	0	1	−1	0	0	0	0	0	0	0	0	0	0	0	0	0
p_{S4}	0	0	0	0	0	0	1	−1	0	0	0	0	0	0	0	0	0	0	0	0
p_B	0	0	0	0	0	0	0	0	0	0	1	−1	0	0	0	0	0	0	0	0
p_{A1}	0	0	0	0	0	0	0	0	1	−1	0	0	0	0	0	0	0	0	0	0
p_K^1	0	0	0	0	0	0	0	0	0	1	0	0	−1	0	0	0	0	0	0	0
p_{A2}	0	0	0	0	0	0	0	0	0	0	0	0	1	0	0	0	0	0	0	−1
p_{R1}	0	0	0	0	0	0	0	0	0	0	0	0	0	−1	1	0	0	0	0	0
p_K^2	0	0	0	0	0	0	0	0	0	0	0	0	0	1	0	−1	0	0	0	0
p_{R2}	0	0	0	0	0	0	0	0	0	0	0	0	0	0	0	1	−1	0	0	0
p_{T1}	0	0	0	0	0	0	0	0	0	0	0	0	0	0	0	0	1	−1	0	0
p_{T2}	0	0	0	0	0	0	0	0	0	0	0	0	0	0	0	0	0	1	−1	0

Table 13.8. Matrix L

	p_{Q1}	p_{Q2}	p_{S1}	p_{S2}	p_{S3}	p_{S4}	p_B	p_{A1}	p^1_K	p_{A2}	p_{R1}	p^2_K	p_{R2}	p_{T1}	p_{T2}
p_{C1}	1	0	0	0	0	0	0	0	0	0	0	0	0	0	0
p_{C2}	0	1	0	0	0	0	0	0	0	0	0	0	0	0	0
p_{C3}	0	0	1	1	1	1	0	0	0	0	0	0	0	0	0
p_{C4}	0	0	0	0	0	0	1	0	0	0	0	0	0	0	0
p_{C5}	0	0	0	0	0	0	0	1	1	1	0	0	0	0	0
p_{C6}	0	0	0	0	0	0	0	0	0	0	1	0	0	0	0
p_{C7}	0	0	0	0	0	0	1	0	0	0	0	0	0	1	1
p_{C8}	0	0	0	0	0	0	0	0	0	0	0	0	0	1	1
p_{C9}	0	0	0	0	0	0	0	0	0	0	0	1	0	0	0
p_{C10}	0	0	0	0	0	0	0	0	0	0	0	0	1	0	0
p_{C11}	0	0	0	0	0	0	0	0	1	0	0	1	0	0	0

Table 13.9. Matrix $N_{\Delta C} = -L\, N_\Delta$

	t_1	t_2	t_3	t_4	t_5	t_6	t_7	t_8	t_9	t_{10}	t_{11}	t_{12}	t_{13}	t_{14}	t_{15}	t_{16}	t_{17}	t_{18}	t_{19}	t_{20}
p_{C1}	-1	0	1	0	0	0	0	0	0	0	0	0	0	0	0	0	0	0	0	0
p_{C2}	0	-1	0	1	0	0	0	0	0	0	0	0	0	0	0	0	0	0	0	0
p_{C3}	0	0	-1	-1	0	0	0	0	1	0	1	0	0	0	0	0	0	0	0	0
p_{C4}	0	0	0	0	0	0	0	0	0	0	-1	1	0	0	0	0	0	0	0	0
p_{C5}	0	0	0	0	0	0	0	0	-1	0	0	0	0	0	0	0	0	0	0	1
p_{C6}	0	0	0	0	0	0	0	0	0	0	0	0	0	1	-1	0	0	0	0	0
p_{C7}	0	0	0	0	0	0	0	0	0	0	-1	1	0	0	0	0	-1	0	1	0
p_{C8}	0	0	0	0	0	0	0	0	0	0	0	0	0	0	0	0	-1	0	1	0
p_{C9}	0	0	0	0	0	0	0	0	0	0	0	0	0	-1	0	1	0	0	0	0
p_{C10}	0	0	0	0	0	0	0	0	0	0	0	0	0	0	0	-1	1	0	0	0
p_{C11}	0	0	0	0	0	0	0	0	0	-1	0	0	1	-1	0	1	0	0	0	0

The initial marking of control places is

$$\mathbf{m}_{c0} =$$
$$\mathbf{b} - \mathbf{L}\,\mathbf{m}_0 =$$
$$(1\ 1\ 1\ 1\ 1\ 1\ 1\ 1\ 1\ 1\ 1)^{\mathrm{T}} - (0\ 0\ 0\ 0\ 0\ 0\ 0\ 0\ 0\ 0\ 0)^{\mathrm{T}} =$$
$$(1\ 1\ 1\ 1\ 1\ 1\ 1\ 1\ 1\ 1\ 1)^{\mathrm{T}}.$$

The system of conditions can be simplified. The conditions $p_B + p_{T1} + p_{T2} \le 1$, $p_B \le 1$ and $p_{T1} + p_{T2} \le 1$ can be equivalently substituted by condition $p_B + p_{T1} + p_{T2} \le 1$. Similarly $p_K^2 \le 1$ and $p_K^2 + p_K^2 \le 1$ by $p_K^2 + p_K^2 \le 1$. Eight conditions give the same result as the original eleven ones.

The discussed method is widely used. Its recent application is to synthesize the deadlock control policies in (Uzam and Zhou 2006).

13.5 Supervisory Control Solution with Reachability Graph

The Petri net reachability graphs offer a good tool for the study and design of the supervisory control. In what follows their use in supervisory control will be described. The Petri nets interpreted for DEDS control will be used.

Let a Petri net interpreted for the control PCN be given. Petri nets interpreted for the control were treated in Section 7.5. For the given PCN a function $\psi : T \to LOG$ is defined where LOG is a set of the logical expressions whose value can be true or false, and T is the set of the transitions of PCN. In this section we assume the logic expressions are Boolean propositions in the form $w_i = = 1$ where w_i is a Boolean variable. One such a proposition is associated with each transition so that $w_i = = 1$ is associated with transition t_i, $t_i \in T$. If $w_i = 1$ then the value of the proposition $w_i = = 1$ is true. Transition t_i in turn is associated with event e_i. The function ς is empty (see Section 7.5). Transition t_i is fireable if the actual marking fulfils the firing conditions valid for PNC, which includes a condition that the logical proposition associated with t_i is true.

The first step of the method for the solution of the supervisory control of a DEDS is design of PNC specifying the complete and possible system behavior. In terms of the model Equation (13.1) the behavior is specified by the language L. The strings or words of events of language L are given by the set of all possible transition firing sequences from the initial marking of the given Petri net PNC. All Boolean variables w_i have value $w_i = 1$ for the specification of language L by PNC. The event string $\tilde{\sigma} = e_{i_1} e_{i_2} ... e_{i_p}$ clearly corresponds to transition string $\tilde{\kappa} = t_{i_1} t_{i_2} ... t_{i_p}$ in the time points given by $\tilde{\tau} = \tau_{i_1} \tau_{i_2} ... \tau_{i_p}$.

For language K the values w_i determine whether the corresponding transitions of the Petri net are fireable in a particular situation represented by an actual marking. The words of K are given by the restriction through the supervisory control. A function $F : R_{PNC}(\mathbf{m}_0) \to 2^W$ can be defined where $W = \{w_1, w_2, ..., w_m\}$ is the set of the variables associated with the Petri net transitions that have value 1 for the actual marking while the other have value 0. As a matter of fact the variables of set W control the fireability of the transitions depending on the marking.

The Petri net reachability graph (see Section 8.2) is useful for the supervisory control solution. Each arc of the reachability graph is labeled with a transition. The transition is fireable and a change of marking arises if the associated variable w_i is true, *i.e.,* $w_i = 1$. It is assumed in the method that the Petri net, which models the given system, is bounded so that the reachability graph can be composed.

The second step of the method is assure by means of a sublanguage K the system behavior, which is represented by the model at Equation (13.1). Using the Petri net *PNC* as the system model the behavior is specified by means of inadmissible system states and/or a set of the graph nodes that can be reachable by oriented paths from any reachable and admissible graph node in the reachability graph. As explained above, inadmissible states can be represented by inadmissible markings of the system modeling Petri net. It follows naturally that no oriented path of the required connection can contain an inadmissible marking. The nodes to be always accessible are called the home states or home markings. A general case is a requirement that at least one element of the set of the home markings should be reachable from any reachable marking. These can be distinguished by mulitple sets of the home markings denoted as *H1, H2, ..., Hz.* The sense of the accessibility results from the system function and the supervisory requirements. It was discussed in the connection with the P-invariant methods studied in the previous section. More will be understandable from the example discussed below.

The third step of the method can be modified according to the behavior requirements. The first version is suitable for the case when only inadmissible states are defined and no achievability of home nodes is required. The second more complete version satisfies both requirements. The third step varies depending on the computational problem with respect to the state space dimension. The problem can be formulated in terms of Petri nets as a problem of the cardinality of its reachability set. If it is computationally acceptable the reachability graph for the given Petri net is constructed. The procedure using the reachability graph will be described below. The procedure covers both system requirements: avoidance of the inadmissible states and/or preserving of the paths to the specified home states.

The third step can be modified for the case when the reachability graph is too large and its computation is practically impossible. The predecessors of the inadmissible markings are computed and if they are connected to the inadmissible markings with uncontrollable arcs, the predecessors are put into the set of inadmissible markings, too. Possible transition activations are searched for in the actual state (marking). Those transitions are further analyzed which do not lead into an inadmissible marking. The required paths to the home markings should be checked from the admissible next markings in order to exclude the deadlocks. If *via* a reasonable computation one finds that the path to the home marking exists, the next marking is finally allowed. In the opposite case the next marking is denoted as inadmissible. The supervisor should decide if in that case the system can undergo the risk of the deadlock in some of the successor states. The supervisor decision can be based on heuristic approaches.

In the rest of this section we will study the case when it is possible to construct the reachability graph for a given bounded Petri net starting from the given initial marking, and the construction is computationally acceptable. As explained above, each arc of the graph is labeled with the transition whose activation causes the transfer from one marking to the next. The arc is also associated with the control variable that can enable or disable the transition and accordingly can enable or disable the corresponding event. The uncontrollable transitions are specified and they are associated with control variables having a constant value of 1 that means they are permanently enabled..

The method can be divided into the following three parts.

Part 1
We denote the set of inadmissible markings as *IA*. The set *IA* initially consists of the *a priori* given forbidden markings and of the dead markings if there are any. A dead marking is characterized by the property that no arc is going out of it. Then we add to *IA* all those markings from which an arc labeled with uncontrollable transitions goes to an inadmissible marking from *IA*.

Part 2
The set of not allowed arcs denoted as *NA* is further formed. To the set belong those arcs that are labeled with the controllable transitions and go in an inadmissible node belonging to the set *IA*. The so-called *NA*-dead-markings are searched for and put in the set *IA*. The *NA*-dead-marking is the marking from which only not allowed (belonging to the set *NA*) arcs go out.

Part 3
Each element of the reachability set is analyzed and its outgoing allowed arcs are checked. If there is an oriented path leading into an element of the given set or sets of the home markings, the arc remains allowed; otherwise it is not allowed and is placed into the set *NA*. After that the conditions on inadmissibility of the nodes are repeated continuing from Part 1 because it is possible that the new not allowed arcs cause some nodes to become inadmissible. The procedure is repeated until no new and not allowed arcs and no new inadmissible markings spring up.

An algorithm resolving the supervisory control is described concisely as follows.

> *Step 1*. Construct the reachability graph for the supervisory control assuming an bounded Petri net interpreted for the control is given. Label each arc with the corresponding activated transition.
>
> *Step 2*. Create empty sets *IA* and *NA*, respectively.

Step 3. Put in set *IA* the inadmissible markings according to the system requirements.

Step 4. Add into set *IA* dead markings (those with no arcs going out at all).

Step 5. Add into *IA* the markings from which an arc labeled with an uncontrollable transition goes in an inadmissible marking.

Step 6. Put into set *NA* the allowed arcs labeled with a controllable transition and going in an inadmissible marking.

Step 7. Add *NA*-dead-markings into *IA*. The *NA*-dead-marking is the marking from which no outgoing arcs are allowed, *i.e.,* the arcs belonging to *NA*.

Step 8. If in Steps 4 through 6 a new not-allowed arc arises and/or in Step 7 a new inadmissible marking arises continue with Step 4; and otherwise with Step 9.

Step 9. Check if from any admissible marking (not an element of *IA*), there exist, an oriented path leading to at least one element of the home markings *H1*, the same into *H2*, ..., up to *Hz*. Put the admissible marking not fulfilling the condition into *IA*.

Step 10. If in Step 9 a marking was added to *IA* continue with Step 4; and otherwise end.

Example 13.8. A manufacturing cell with mobile robots for the handling of the manufactured parts is depicted in Figure 13.16. Two robots R1 and R2 transfer parts of two kinds A and B within the cell. Robot R1 picks up a part A from the input conveyor C1 and transports it into rooms S2–S4 to machines M1–M5. The doorway to S5 is not passable for R1. Analogously it is for the movements of robot R2 as Figure 13.16 shows. Manufacturing runs according to an actual technological scheme including the processing at machines and movements of the robots with semi-products.

Whether the movements are managed by an operator or control unit, the task of the supervisory control is to prevent the forbidden situations which are the meetings of the robots in the same room. On the other hand the return of R1 to S1 and R2 to S5 in each position of the robots should be possible under the assumption that robots can realize movements, which loosen the movements of the other robot. For example, one robot moves back to a room and the second can continue in its route.

The control agent used in the system determines the back tracking maneuvers. It is important that the system leads to no deadlock. The initial position of the robots is indicated in Figure 13.16.

The Petri net serving for the supervisory control design is in Figure 13.17. The reachability graph is in Figure 13.18.

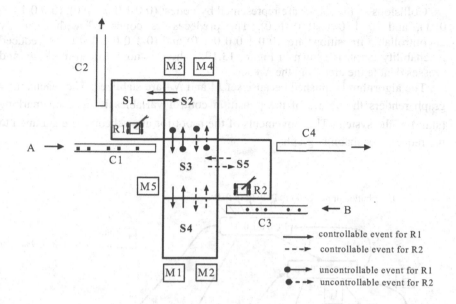

Figure 13.16. A manufacturing cell with mobile robots

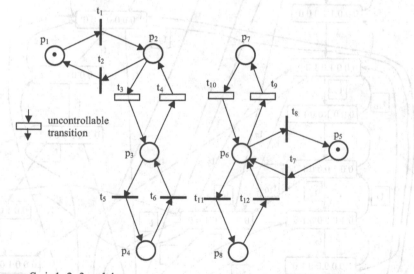

p_i: R1 at room S_i, i=1, 2, 3 and 4;
p_5: R2 at room S_5; p_6: R2 at room S_3; p_7: R2 at room S_2; and p_8: R2 at room S_4.
Transitions represent the movement of a robot from one room to another.
Hollow-bar transitions are uncontrollable while solid-bar ones are controllable.

Figure 13.17. Petri net for the manufacturing system (two separate Petri nets are in fact two automata)

Collisions of the robots are represented by nodes: (0 0 1 0 0 1 0 0), (0 0 0 1 0 0 0 1), and (0 1 0 0 0 0 1 0). The predecessors connected with them by uncontrollable transitions are: (0 0 1 0 0 0 1 0) and (0 1 0 0 0 1 0 0). A reduced reachability graph is shown in Figure 13.19. There is no NA-dead-marking, and crosses denote the arcs from the NA set.

The algorithm is finished because set IA and NA are stablized. The reachability graph renders the values of the transition control variables w_i in each marking (state) of the system. The movements of the robot for any schedule are subdued to the reduced reachability graph.

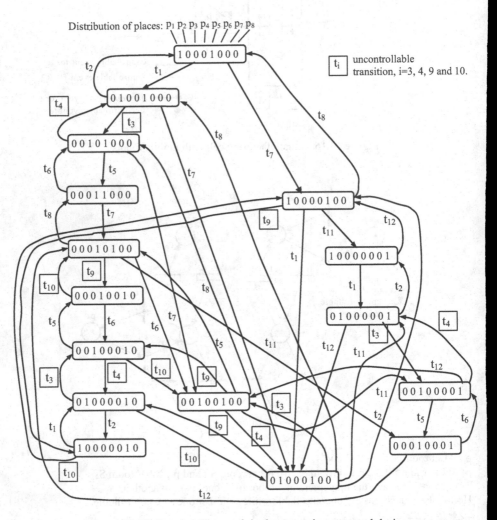

Figure 13.18. Reachability graph in the supervisory control design

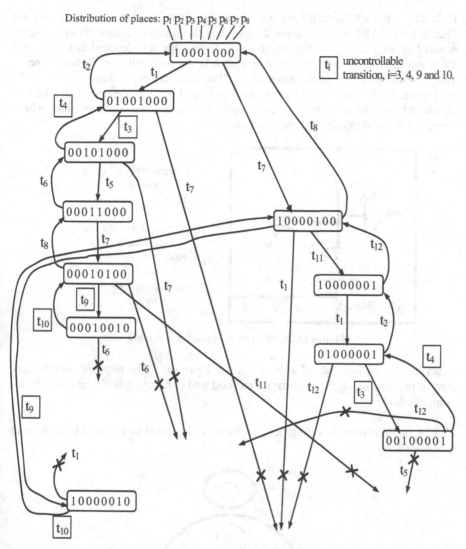

Distribution of places: p_1 p_2 p_3 p_4 p_5 p_6 p_7 p_8

Figure 13.19. The reduced reachability graph

13.6 Problems and Exercises

13.1. Propose supervisory control for the robotized manufacturing system depicted in Figure 7.18 (Exercise 7.2) using P-invariant method. For the robot movements assume the same requirements as in Exercise 7.2.

Give a basic idea of how the supervisory control could be realized from the technical and programming point of view.

13.2. Two robots R1 and R2 are moving in the space divided into rooms (Figure 13.20). Initially R1 is in the room 2 and R2 in the room 4. Doors between rooms denoted A are passable in the directions of arrows for R1, denoted B for R2. All doors except for the movements A13 and A31 are controllable. Robots execute transfers of objects within the given space. One such task is defined by a start and a final room. Tasks are given by a supervisory control level while a particular path is elaborated by a local control system. Supervisory control level determines which doors can be used in every actual situation.

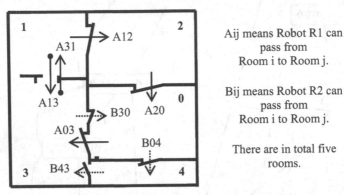

Aij means Robot R1 can
pass from
Room i to Room j.

Bij means Robot R2 can
pass from
Room i to Room j.

There are in total five
rooms.

Figure 13.20. Movement space for two robots

Describe the movements of robots with a Petri net. The required supervisory control propose using the P-invariant method and compare it with the reachability graph method.

13.3. A popular story of eating philosophers is illustrated in Figure 13.21. A meal is prepared.

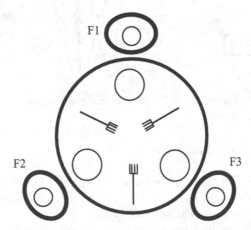

Figure 13.21. Three philosopher problem

Each philosopher has one fork right and one left. When a philosopher wants to eat he takes forks one by one and afterwards drops them on the table. Philosophers are stubborn - if they want to eat they do not give warning. A philosopher needs both forks left and right to eat; otherwise he waits for free ones. Solve the supervisory control problem saying what forks cannot be picked up in order to prevent deadlock when all philosophers want to eat.

Represent the behavior of philosophers using the Petri net and solve the supervisory control problem with the reachability graph of the Petri net.

13.4. Consider the transportation system in Figure 7.20 of Exercise 7.5. It is reasonable for transportation control to divide the vehicle tracks into sections. A crossing is always a separate section. Let only the entry to the first section of each track be a controllable event. Solve the supervisory control for such an arrangement of the system.

13.5. A transportation system using three automatic guided vehicles (AGV) in a manufacturing plant is depicted in Figure 13.22. The vehicle tracks are divided into sections. Only one AGV can be present in a section – to go through or to stop there. No AGV can stop in any crossing. Each section is separated by two control points using sensors.

A way to solve the AGV control is the supervisory control approach. As far as the supervisory control is preventing the AGVs from collisions and deadlocks, the personnel doesn't have to take care about contemporary state of the system during the programming of new transportation requests. Use a Petri subnet for each AGV as it is shown in Figure 13.23.

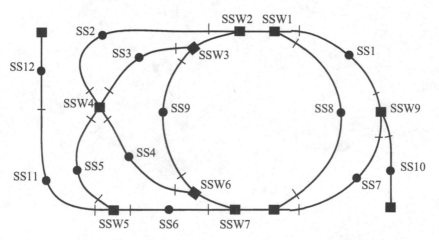

Figure 13.22. An AGV transportation system operating in a manufacturing plant

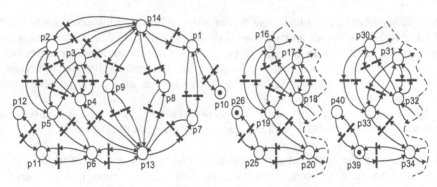

Figure 13.23. Petri net for the transportation system

Places p2, p16, p30 in the respective subnets correspond to the presence of an AGV in the same section, namely the section S2.

Formulate the solution of the supervisory control preventing collisions and deadlocks using the reachability graph of the Petri net consisting of the three subnets according to Figure 13.23. The nodes of the graph can be calculated comparatively easily by enumerating combinations of the AGV positions by a computer program. The forbidden nodes serve to avoid AGV collisions by performing the planned routes of the vehicles. What do you propose for planning at least near to optimal routing of the vehicles from given the initial section to the required goal.

14

Job Scheduling

14.1 Problem Formulation

Job scheduling or operation scheduling is a typical problem frequently appearing within DEDS. The core of the problem consists in how to achieve an optimal distribution of jobs or operations among the processing units or servers available in the system under various criteria. In other words, the problem is the optimal allocation of the system resources (Frankovič and Budinská 1998).

Typical environments in which a scheduling problem occurs are flexible manufacturing systems, distributed computer systems, database systems, and other. For example, flexible manufacturing systems (FMS) usually consist of product processing or machining units, measuring and testing equipments, transportation facilities, manipulators and robots, intermediate storages, input and output devices. Various methods have been developed for scheduling problems (Engell 1989; Li *et al.*, 1995; Zhou and Venkatesh, 1998).

It has been discussed earlier in this book that process control means control of the basic processes at the level responsible for direct control. It is the control level or layer closest to the system processes. A hierarchically higher level is the co-ordination level of the basic processes. Here, the co-ordination is considered as a selection of servers performing basic processing, if there are more options. For example, in flexible manufacturing systems it is a selection of production units if there are more options to realize a prescribed technological recipe. One of co-ordination aims is to accomplish the required jobs in the minimum time span. For this aim it is necessary to know the duration times of the scheduled operations. Other scheduling optimization criteria can be, *e.g.,* the maximum utilization of the resources and minimum tardiness of the required operations. A problem related to the job scheduling is routing of semi-products to servers according to a chosen scheduling.

The first step in the solution of a job scheduling problem is a system specification using a suitable tool. The specification has to bring about such an abstraction that enables to present and solve the problem. An efficient specification is based on Petri nets (Lee and F. DiCesare, 1994, 1995; Xiong and Zhou 1998; Zhou and Venkatesh, 1998) and max-plus algebra (Moßig and Rehkopf, 1996).

The first approach to be dealt with in this chapter is oriented on Petri net specification (Section 14.2). Obviously, from the nature of the scheduling problems, some extension with respect to time relations is necessary. Some researchers associate time with Petri net transitions. In Section 14.2 another approach is presented by utilizing a certain kind of place timing. Another approach is the max-plus algebra (Section 14.3).

For our purposes, consider a system specified as

$$SYST = (S, \Omega, op, T) \tag{14.1}$$

where

S: a set of servers $S = \{S_1, S_2, ..., S_{|S|}\}$;

Ω: a set of all different operations realizable in system $SYST$ by servers,

op: $S \rightarrow 2^{\Omega}$ is a mapping of the set S into the set of the operation subsets,

$tu : S \times \Omega \rightarrow N^+$ is a function mapping a particular operation performed at a particular server into a positive integer representing the number of time units consumed by the operation.

Let the i-th individual subset of the co-domain of the function op be denoted as

$$\Omega_i = \{{}^i\kappa_1, {}^i\kappa_2, ..., {}^i\kappa_{|\Omega_i|}\}, \quad {}^i\kappa_1, {}^i\kappa_2, ..., {}^i\kappa_{|\Omega_i|} \in \Omega \tag{14.2}$$

where Ω_i denotes the set of operations available at the server S_i, $i = 1, 2, ..., |S|$. The operations are from the set Ω; then the function op can be expressed as

$$op(S_i) = \Omega_i \tag{14.3}$$

From Equations (14.2) and (14.3) it is possible to write

$$tu(S_i, {}^i\kappa_k) = {}^i\tau_k, i = 1, 2, ..., |S|, k = 1, 2, ..., |\Omega_i| \tag{14.4}$$

The system flexibility is due to the fact that an operation can alternatively be performed on different system resources (servers). Thus

$$\Omega_{i_1} \cap \Omega_{i_2} \neq \varnothing \text{ for some } i_1 \neq i_2 \tag{14.5}$$

In such a case times of the same operation performed on different servers may not be the same.

Various processing procedures can be realized in the system $SYST$. In FMS terminology it means various technological work-plans. Let the p-th processing procedure be defined by a sequence of operations

$$\tilde{O}_p = o_{p1} o_{p2} ... o_{pr} \tag{14.6}$$

where $o_{p1}, o_{p2}, ..., o_{pr} \in \Omega$ are operations taken from the set Ω defined in Equation (14.1). Each procedure is determined by its own sequence, *e.g.*, the *p*-th procedure by $\tilde{O}_p$. There are several alternatives how to use the servers in a case when more servers can perform the same operation o_{pj}. It is the case when the required operation is included in several subsets Ω_i. It is assumed that each operation depends only on one preceding operation, and a next operation can start after the preceding operation has been accomplished..

A system *SYST* can be additionally completed with the input servers $X = \{X_1, X_2, ..., X_I\}$ and the output servers $Y = \{Y_1, Y_2, ..., Y_O\}$. Availability of an object to be processed can be specified by means of an input server. Analogously, output servers are used for outputs. A particular scheduling task can be defined as follows. A part available at an input server passes through a prescribed processing procedure realized on a chosen set of servers realizing a sequence of operations. Finally, the part appears at an output server ready for a next use.

The scheduling problem formulation as presented above will be illustrated on an example. Figure 14.1 shows a manufacturing system with three servers, two inputs and two outputs. Let a technological work-plan be realized on the described manufacturing system. The work-plan is specified in Table 14.1. There are two kinds of products to be produced. The notation S_1 / S_2 means an optional realization of the first technological step for the job J_1 either on the server S_1 or on the server S_2. Figure 14.1 depicts possible transportation routes of parts.

Through X_1, X_2 parts are fed in a random sequence. When the manufacturing is finished, the manufactured parts are placed at the output servers Y_1, Y_2. For each job and each step of a particular work-plan operations are assigned so that the set of servers is $S = \{S_1, S_2, S_3\}$ and the set of operations is $\Omega = \{\omega_1, \omega_2, \omega_3, \omega_4\}$. The same operation can be prescribed in different jobs. Keeping in mind the notation at Equation (14.2) we have

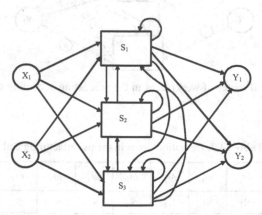

Figure 14.1. Technological layout of the manufacturing cell

$$\Omega_1 = \left\{{}^1\kappa_1, {}^1\kappa_2, {}^1\kappa_3\right\} = \left\{\omega_1, \omega_3, \omega_4\right\}$$
$$\Omega_2 = \left\{{}^2\kappa_1, {}^2\kappa_2, {}^2\kappa_3\right\} = \left\{\omega_1, \omega_2, \omega_4\right\} \tag{14.7}$$
$$\Omega_3 = \left\{{}^3\kappa_1, {}^3\kappa_2, {}^3\kappa_3\right\} = \left\{\omega_2, \omega_3, \omega_4\right\}$$

Order of elements in equal sets of Equation (14.7) determines the element correspondence. Processing procedures are given by the following sequences:

$$\tilde{O}_1 = o_{11}\, o_{12} = \omega_1\, \omega_2 = \left({}^1\kappa_1 \text{ or } {}^2\kappa_1\right)\left({}^1\kappa_2 \text{ or } {}^3\kappa_1\right)$$
$$\tilde{O}_2 = o_{21}\, o_{22} = \omega_3\, \omega_4 = \left({}^1\kappa_2 \text{ or } {}^3\kappa_2\right)\left({}^1\kappa_3 \text{ or } {}^2\kappa_3 \text{ or } {}^3\kappa_1\right) \tag{14.8}$$

Table 14.1. A particular work-plan to be realised in the manufacturing cell

Step	Job J_1		Step	Job J_2	
	Operation	Available at		Operation	Available at
1	ω_1	$S_1/\,S_2$	1	ω_3	$S_1/\,S_3$
2	ω_2	$S_2/\,S_3$	2	ω_4	$S_1/\,S_2/\,S_3$

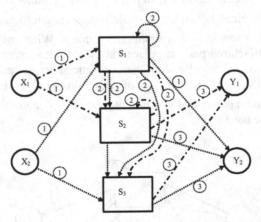

Figure 14.2. Optional routes of workpieces in the processing according to the work-plan in Table 14.1

Table 14.2. Operation times in the manufacturing cell

Operation	ω_1	ω_1	ω_2	ω_2	ω_3	ω_3	ω_4	ω_4	ω_4
Server	S_1	S_2	S_2	S_3	S_1	S_3	S_1	S_2	S_3
Duration	3	4	3	2	4	2	3	4	4

Figure 14.2 shows optional transfer of workpieces according to the work-plan given in Table 14.1. The operation times are given in Table 14.2.

14.2 Job Scheduling and Petri Nets

Job scheduling can be solved using Petri nets. There are many ways to use Petri nets for this purpose. Two approaches based on the scheduling analysis presented in the previous section will be presented. Timed Petri nets are always necessary.

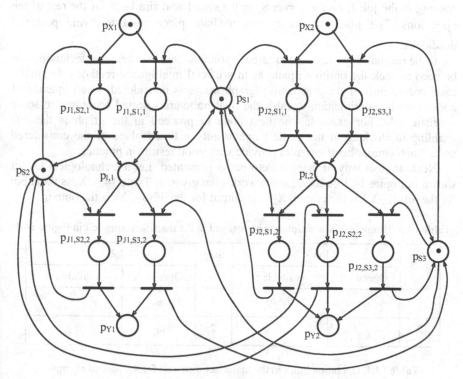

Figure 14.3. Petri net representation of the scheduling solution

In the first approach, the binary timed safe Petri nets are considered, in which the scheduled operations are associated with timed places (Lee and DiCesare 1995). The timed places are mapped to times equal to the respective operation durations. A token is blocked after arriving in the timed place during the corresponding operation time. By convention, the operation time is expressed by the number of time units. After expiration of the time the token is free for transition firing under standard firing rules. Figure 14.3 shows a Petri net for the manufacturing cell depicted in Figure 14.1 and for the work-plan given in Table 14.1. The work-plan is cyclically repeated. It is assumed that there is sufficient stock of parts at the input. The cycle is finished when both parts to be processed are at the output. Only then does a new manufacturing cycle start. An alternative for a production optimization is to minimize the time (work-span) from start to end of one cycle. The optimal solution of scheduling chosen from possible routings can be achieved by means of the reachability graph. A time account has to be accomplished for the timed places in the reachability graph. A change in the Petri net marking is possible when the blocking time for some place has expired and some transition is fireable.

The places p_{X1}, p_{X2} stand for the input servers. Presence of a token in one of them indicates that a part is available for processing. An analogous notation is used for output. A timed place is denoted by $p_{J1,S1,1}$; it corresponds to an operation running in the job J_1 on the server S_1 in the step 1 and similarly for the rest of the operations. The place $p_{I,1}$ is an intermediate place used to avoid potential deadlocks.

If the reachability graph is too large, a suitable heuristic decision technique can be used to seek the optimum path, as in artificial intelligence methods. In such a case, only a part of the reachability graph nodes is considered. The sequence of nodes and the continuation towards the final node are selected according to some heuristic rule. For example, the next node to proceed in the graph is the one enabling unblocking at the earliest opportunity a timed place of the considered node. Such similar heuristic rules provide very good results in practice.

Next, another way of using a Petri net is presented. Let a technological layout shown in Figure 14.1 be used for the work-plan given in Table 14.3. X_1 is the input for the job J_1, Y1 is the output, X_2 is the input for the job J_2, Y_2 is the output.

Table 14.3. Example of a work-plan to be realised in the manufacturing cell in Figure 14.1

Step	Job J_1		Step	Job J_2	
	Operation	Available at		Operation	Available at
1	ω_1	S_1/ S_2	1	ω_3	S_1
2	ω_2	S_3	2	ω_4	S_3

Table 14.4. Operation times in the manufacturing cell for the second example

Operation	ω_1		ω_2	ω_3	ω_4
Server	S_1	S_2	S_3	S_1	S_3
Duration	4	3	3	1	2

For this case operation times are given in Table 14.4. There are two kinds of parts to be processed simultaneously available at the input. A part of the first (second) kind is processed into the first (second) kind of product.

As in the previous example, the optimum manufacturing is considered in terms of the optimum of one production cycle. The servers perform the following different operations in the respective steps:

$$\Omega_1 = \left\{ {}^1\kappa_1, {}^1\kappa_2 \right\} = \left\{ \omega_1, \omega_3 \right\}$$
$$\Omega_2 = \left\{ {}^2\kappa_1 \right\} = \left\{ \omega_1 \right\} \tag{14.9}$$
$$\Omega_3 = \left\{ {}^3\kappa_1, {}^3\kappa_2 \right\} = \left\{ \omega_2, \omega_4 \right\}$$

and processing is executed by the following sequences:

$$\tilde{O}_1 = o_{11}\, o_{12} = \omega_1\, \omega_2 = \left({}^1\kappa_1 \text{ or } {}^2\kappa_1 \right) {}^3\kappa_1$$
$$\tilde{O}_2 = o_{21}\, o_{22} = \omega_3\, \omega_4 = {}^1\kappa_2\ {}^3\kappa_2 \tag{14.10}$$

A special class of Petri nets supporting operation scheduling analysis and solution is defined in the following where the timed Petri net for the scheduling problem is given by the 6-tuple

$$TPNS = \left(P, T, F, \hat{W}, M_0, c, INH \right) \tag{14.11}$$

where P, T, F, M_0 are defined as usual. The function $c : P \rightarrow N^+$ associates capacities with places, $INH \subseteq P \times T$ is a set of inhibitors disabling transition firings if the source node of an inhibitor has at least one token. Timing consists in that the marking is a function of places and of the discrete time $M : P \times \Theta \rightarrow N$, $\Theta = \left\{ \vartheta_1, \vartheta_2 .... \right\}$, $0 < \vartheta_1 < \vartheta_2 < ....$ are discrete time points. $\hat{W}$ is a specially defined function

$$\hat{W} : F \rightarrow \left\{ c(p_i) \right\} \quad \text{for } p_i \text{ and } \forall t_j \in T \text{ for which } \left(t_j, p_i \right) \in F,$$
$$F \rightarrow \left\{ 1 \right\} \quad \text{for } \forall \left(p_i, t_j \right) \in F$$

Firing of a transition t_j in the $TPNS$ Petri net is enabled iff

$$0 = \mathbf{m}_{\vartheta_k} + \mathbf{t}_j \le \mathbf{c}, \ \mathbf{m}_{\vartheta_k} \in R_{TPNS}(\mathbf{m}_0) \tag{14.12}$$

where the actual marking M in the discrete time point ϑ_k is aggregated into the vector $\mathbf{m}_{\vartheta_k}$; $\mathbf{t}_j$ is a vector associated with the transition t_j; $R_{TPNS}(\mathbf{m}_0)$ is a reachability set. If the transition t_j is firable (enabled), a new marking is obtained according to the vector equation

$$\mathbf{m}_{\vartheta_{k+1}} = \mathbf{m}_{\vartheta_k} + \mathbf{t}_j \tag{14.13}$$

Consider a system $SYST$ according to Equation (14.1), which can be represented by the $TPNS$ Petri net. The Petri net construction is as follows.

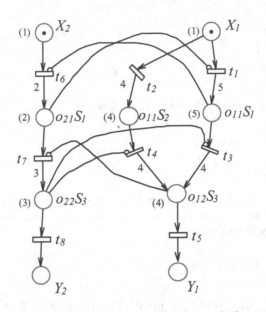

Figure 14.4. The Petri net representing the required processing

Elements of the operation sequences $\tilde{O}_p$ are associated with the TPNS places. For each $o_{pv} \in \Omega_i$, $v = 1,2,...,r$; $i = 1,2,...,|S|$, a place denoted as $o_{pv}S_i$ is put into the net. The place is given a capacity equal to $\tau_{ik} + 1$, where τ_{ik} is given by Equation (14.4). The arcs connecting the place $o_{pv}S_i$ with all places $o_{p(v+1)}S_i$ *via* post-transitions are added to the net. According to the TPNS definition weights of the arcs are 1 and $(\tau_{ik} + 1)$, respectively. Input and output servers are associated with corresponding places. The *TPNS* Petri net construction is depicted in Figure 14.4.

Inhibitors are used to prevent a transition firing when a token is in the respective place. Not to prevent the firing would be considered as a misuse of the already occupied server.

The first step of the scheduling problem solution procedure is creation of the system model *SYST* defined by Equation (14.1). Operation times τ_{ik} are expressed in terms of multiples of the basic sampling time period $\Delta\tau$, whereby $\Delta\tau$ is chosen as large as possible with respect to a sufficiently accurate representation of the system dynamics.

The model *SYST* at Equation (14.1) is transformed into the *TPNS* Petri net. The initial marking represents input availability of the objects to be processed. It is non-zero for the places associated with the input servers. Initial marking of places $o_{pv}S_i$ is set to zero as well. Fireable transitions in the created *TPNS* Petri net are fired in discrete time points, which are multiples of the basic time period $\Delta\tau$, i.e., $0, \Delta\tau, 2\Delta\tau, ..., k\Delta\tau, ...$ Prior to firing the transitions, the marking of each place

$o_{pv}S_i$ is decreased by one if $M(o_{pv}S_i) > 1$, otherwise it remains unchanged. After the marking is decreased, the fireable transitions are fired and a new marking is obtained.

The obtained *TPNS* may contain conflicts reflecting existence of various possibilities of how to perform the required operation sequences. Consider the minimum time span criterion applied to the operation sequences. Hence the goal is to find out preferences among the possible ways of performing the operation sequence with respect to the specified cost function.

The optimization problem can be solved using the reachability graph of the *TPNS*. The nodes of the graph correspond to the reachable markings $\mathbf{m}_{\vartheta_k}$. Possible paths in the reachability graph corresponding to the individual required operation sequences can be analyzed in order to select the best solution with respect to the chosen criterion.

The described approach will be illustrated on an FMS layout with a required work-plan defined in Table 14.3. As mentioned earlier, processing of the next pair of input parts can start when both products are ready at the output. The goal is to find the processing procedure requiring minimum overall time for processing one pair of the parts. The reachability graph of the *TPNS* Petri net is in Figure 14.5. There are three possible paths specifying the operation scheduling. The first, which starts with the activation of transition t_1, is the most time consuming. The other two are better with respect to the minimum work-span criterion. The path starting with t_6 is the best.

In this section an approach to the scheduling problem solution using a special class of Petri nets is presented. It constitutes a framework to cope with the problem. The drawback of the solution is a tremendous increase of the nodes number in the reachability graph to be analyzed. One possibility to avoid this difficulty is to apply heuristic searches as mentioned in the description of the first approach in this section. The case of the dynamic processing of inputs during processing of the previous ones can be solved following the presented framework, too.

14.3 Job Scheduling Based on the Max-plus Algebra

The job scheduling problem in Sections 14.1 and 14.2 was studied under the restriction that each operation depends on one preceding and finished operation. But frequently there can be the dependence on more operations. Such dependence can be presented in a graphical form. Figure 14.6 shows an example. Circles represent operations and arc weights given as real numbers represent the necessary durations of the preceding operations, after which the next operation can start. For example, the second operation can start when time equal to 3 from the start of the first operation elapsed and time equal to 2 from the start of the third operation elapsed. Note that the example is adopted from Moßig and Rehkop (1996).

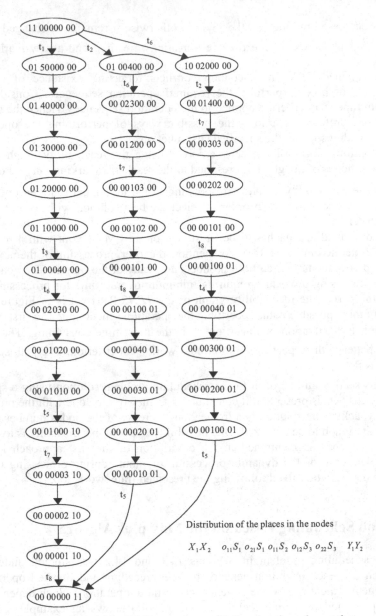

Distribution of the places in the nodes

$X_1 X_2 \quad o_{11} S_1 \; o_{21} S_1 \; o_{11} S_2 \; o_{12} S_3 \; o_{22} S_3 \quad Y_1 Y_2$

Figure 14.5. A reachability graph considering duration of operations

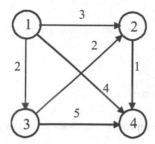

Figure 14.6. Graphical representation of operation dependence

Note that there are no cycles in the graph in Figure 14.6. Such a situation will be supposed in the sequel until another assumption is made.

The Petri net approach can be less effective if there is the operation dependence described above. The so-called max-plus algebra may be more productive. The max-plus algebra is used as a tool for modeling the time development operations and the scheduling problem can be solved on the simulation basis trying different operation setup.

The max-plus algebra $[R_{max}, \oplus, \otimes]$ is defined by:

a. The fundamental set R_{max}

$$R_{max} = R \cup \{-\infty, +\infty\} \tag{14.14}$$

where R is the set of real numbers; $-\infty, +\infty$ are additional elements, for which

$$-\infty < x < +\infty, \quad \forall x \in R \tag{14.15}$$

b. The binary operation $\oplus$, defined as the maximum of two real numbers – elements of R_{max}, whereby the inequality at Equation (14.15) provides the result of the operation $\oplus$ for elements $-\infty, +\infty$.
    The operation $\oplus$ is commutative, associative, with the neutral element $-\infty$, and is idempotent ($\forall a \in R_{max}, a \oplus -\infty = -\infty \oplus a = a$).

c. The binary operation $\otimes$, defined as

$$a \otimes b = a + b, \quad \forall a, b \in R \tag{14.16}$$

where $+$ is the usual operation over the field of the real numbers $[R, +, .]$. For the additional elements $-\infty, +\infty$ holds:

$$a \otimes -\infty = -\infty \otimes a = -\infty$$
$$a \otimes +\infty = +\infty \otimes a = +\infty$$
$$+\infty \otimes +\infty = +\infty \tag{14.17}$$
$$-\infty \otimes -\infty = -\infty$$
$$+\infty \otimes -\infty = -\infty \otimes +\infty = -\infty$$

d. The operation $\otimes$ is commutative, associative, and distributive over $\oplus$, *i.e.*,

$$\forall a, b, c \in R_{max}, \quad (a \oplus b) \otimes c = (a \otimes c) \oplus (b \otimes c),$$
$$c \otimes (a \oplus b) = (c \otimes a) \oplus (c \otimes b)$$

The following does not hold:
$$(a \otimes b) \oplus c = (a \oplus c) \otimes (b \oplus c)$$
for example, if

$$a = 2, b = 3, c = 6, (2 \otimes 3) \oplus 6 = \max[(2+3),6] = 6 \neq (2 \oplus 6) \otimes (3 \oplus 6) =$$
$$\max[2,6] + \max[3,6] = 6 + 6 = 12$$

The neutral element in the operation $\otimes$ is 0, as it is in the field of real numbers. The following holds $-\infty + 0 = -\infty, +\infty + 0 = +\infty$.

From the viewpoint of algebraic structure theory the max-plus algebra $[R_{max}, \oplus, \otimes]$ as defined above is a special commutative field (sometimes simply called the field). It is because $[R_{max}, \oplus, \otimes]$ is a ring where $[R_{max}, \otimes]$ is a monoid and $[R_{max} - \{-\infty, +\infty\}, \otimes]$ is a commutative group. In other words, $[R, \otimes]$ is a commutative group.

The max-plus algebra can be extended to the fundamental set of matrices M. Let the matrix entries be elements of R_{max}. The operation $\oplus$ then

$$(\mathbf{A} \oplus \mathbf{B})_{ij} = a_{ij} \oplus b_{ij} \tag{14.18}$$

where a_{ij} is the entry in the i-th row and the j-th column of the matrix $\mathbf{A}$, $\mathbf{A} \in (R_{max})^{m \times r}, \mathbf{B} \in (R_{max})^{m \times r}$.

The operation $\otimes$ is given by

$$(\mathbf{A} \otimes \mathbf{B})_{ij} = \bigoplus_{k=1}^{r} (a_{ik} \otimes b_{kj}) = \max_{k}(a_{ik} \otimes b_{kj}), \quad \mathbf{A} \in (R_{max})^{m \times r}, \mathbf{B} \in (R_{max})^{r \times n}$$
$$\tag{14.19}$$

The max-plus algebra $[M,\oplus,\otimes]$ has the same properties as $[R_{\max},\oplus,\otimes]$ except for the neutral element in $\oplus$ being the matrix

$$\mathbf{N} = (n_{ij}), \quad n_{ij} = -\infty \tag{14.20}$$

The neutral element in $\otimes$ is the matrix

$$\mathbf{I} = (i_{ij}), \quad i_{ij} = \begin{cases} 0 \text{ for } i = j \\ -\infty \text{ for } i \neq j \end{cases} \tag{14.21}$$

The dependence of an operation on the others can be expressed by the equation

$$x_i = \max_j\left(a_{ij} + x_j\right) \tag{14.22}$$

where x_j is the start time of the j-th operation preceding the i-th operation, a_{ij} is the time necessary to run the j-th operation before the i-th operation starts. Because the i-th operation depends on several operations, the start of x_i is given by Equation (14.22) as the maximum time of the preceding operation starts and the operation durations.

The diagrams in Figure 14.7 illustrate the meaning of Equation (14.22). The 3rd operation can start only when the times a_{31} and a_{32} of the 1st and 2nd operations respectively have elapsed. The starting point of the 3rd operation is then

$$x_3 = \max(x_1', x_2') = x_3 \tag{14.23}$$

Using the max-plus algebra

$$x_i = \left(a_{ij_1} \otimes x_{j_1}\right) \oplus \left(a_{ij_2} \otimes x_{j_2}\right) \oplus .... \oplus \left(a_{ij_k} \otimes x_{j_k}\right) \tag{14.24}$$

Moreover, it is possible to extend Equation (14.22) by an additional time condition, namely that the i-th operation cannot start before a given time point u_i. It can be either $u_i > x_i$ where x_i is calculated by Equation (14.22) and then the i-th operation starts in u_i, or $u_i < x_i$ and then the i-th operation starts in x_i. Extension of Equation (14.22) is as follows:

$$x_i = \max_j(a_{ij} + x_j, u_i) \tag{14.25}$$

In terms of the max-plus algebra

$$x_i = \left(a_{ij_1} \otimes x_{j_1}\right) \oplus \left(a_{ij_2} \otimes x_{j_2}\right) \oplus .... \oplus \left(a_{ij_k} \otimes x_{j_k}\right) \oplus u_i \tag{14.26}$$

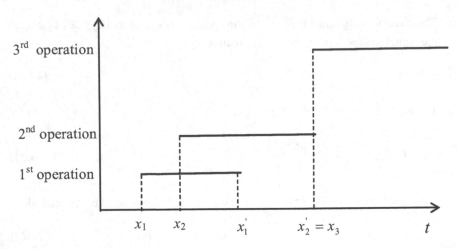

Figure 14.7. Dependence of operations

Using the matrix the dependence of n operations can be expressed as follows:

$$
\begin{pmatrix} {}^{r}x_1 \\ {}^{r}x_2 \\ \cdot \\ \cdot \\ \cdot \\ {}^{r}x_n \end{pmatrix} = \left(\begin{pmatrix} a_{11} & a_{12} & \cdots & a_{1n} \\ a_{21} & a_{22} & \cdots & a_{2n} \\ \cdot & & & \cdot \\ \cdot & & & \cdot \\ \cdot & & & \cdot \\ a_{n1} & a_{n2} & \cdots & a_{nn} \end{pmatrix} \otimes \begin{pmatrix} {}^{r-1}x_1 \\ {}^{r-1}x_2 \\ \cdot \\ \cdot \\ \cdot \\ {}^{r-1}x_n \end{pmatrix} \right) \oplus \left(\begin{pmatrix} b_{11} & b_{12} & \cdots & b_{1n} \\ b_{21} & b_{22} & \cdots & b_{2n} \\ \cdot & & & \cdot \\ \cdot & & & \cdot \\ \cdot & & & \cdot \\ b_{n1} & b_{n2} & \cdots & b_{nn} \end{pmatrix} \otimes \begin{pmatrix} u_1 \\ u_2 \\ \cdot \\ \cdot \\ \cdot \\ u_n \end{pmatrix} \right)
$$

$$(14.27)$$

or shortly

$$
{}^{r}x = \left(A \otimes {}^{r-1}x \right) \oplus \left(B \otimes u \right)
\tag{14.28}
$$

where r denotes the r-th step of the iterative operation dependence. Recall that the dependence is without feedback loops, *i.e.,* without cycles in the graph representation. One starts with the step for $r = 2$ and $r - 1 = 1$ and continues with $r := r + 1$. If there is no dependence between the j-th and i-th operation, the entry a_{ij} in the matrix A is equal $-\infty$. According to Equation (14.27) dependent operations are associated with the rows of the matrix (index i). Entries of matrix B are 0 or $-\infty$. In the former case an additional time condition is put into force, in the latter there is no additional condition on the start. If there is no forced starting condition u_i is set to zero as illustrated in the accompanying example. Equation (14.28) represents the state equation of the system with operation dependence.

In the example the state equation without additional starting conditions (Figure 14.6) is considered:

$$\begin{pmatrix} ^r x_1 \\ ^r x_2 \\ ^r x_3 \\ ^r x_4 \end{pmatrix} = \begin{pmatrix} -\infty & -\infty & -\infty & -\infty \\ 3 & -\infty & 2 & -\infty \\ 2 & -\infty & -\infty & -\infty \\ 4 & 1 & 5 & -\infty \end{pmatrix} \otimes \begin{pmatrix} ^{r-1} x_1 \\ ^{r-1} x_2 \\ ^{r-1} x_3 \\ ^{r-1} x_4 \end{pmatrix} \tag{14.29}$$

The neutral element $-\infty$ for the operation $\oplus$ is used when there is no dependence between two operations. The starting time point for the first operation is set to $^1 x_1 = 0$; e.g.,

$$^r x_2 = \left(3 \otimes^{r-1} x_1\right) \oplus \left(-\infty \otimes^{r-1} x_2\right) \oplus \left(2 \otimes^{r-1} x_3\right) \oplus \left(-\infty \otimes^{r-1} x_4\right) = \left(3 \otimes^{r-1} x_1\right) \oplus \left(2 \otimes^{r-1} x_3\right)$$

is a dependence, which can be verified by the graph in Figure 14.6.

The development of the state equation is possible considering n operations in the system

$$\begin{aligned}
^n x &= \left(\mathbf{A} \otimes {}^{n-1} x\right) \oplus \left(\mathbf{B} \otimes \mathbf{u}\right) \\
&= \left(\mathbf{A} \otimes \left(\left(\mathbf{A} \otimes {}^{n-2} x\right) \oplus \left(\mathbf{B} \otimes \mathbf{u}\right)\right)\right) \oplus \left(\mathbf{B} \otimes \mathbf{u}\right) \\
&= \left(\left(\mathbf{A}^2 \otimes {}^{n-2} x\right) \oplus \left(\mathbf{A} \otimes \mathbf{B} \otimes \mathbf{u}\right)\right) \oplus \left(\mathbf{B} \otimes \mathbf{u}\right) \\
&= \left(\left(\mathbf{A}^2 \otimes \left(\left(\mathbf{A} \otimes {}^{n-3} x\right) \oplus \left(\mathbf{B} \otimes \mathbf{u}\right)\right)\right) \oplus \left(\mathbf{A} \otimes \mathbf{B} \otimes \mathbf{u}\right)\right) \oplus \left(\mathbf{B} \otimes \mathbf{u}\right) \\
&= \left(\left(\mathbf{A}^3 \otimes {}^{n-3} x\right) \oplus \left(\mathbf{A}^2 \otimes \left(\mathbf{B} \otimes \mathbf{u}\right)\right) \oplus \left(\mathbf{A} \otimes \mathbf{B} \otimes \mathbf{u}\right) \oplus \left(\mathbf{B} \otimes \mathbf{u}\right)\right) \\
&\quad \vdots \\
&= \left(\left(\mathbf{A}^n \otimes {}^0 x\right) \oplus \left(\mathbf{A}^{n-1} \otimes \left(\mathbf{B} \otimes \mathbf{u}\right)\right) \oplus \left(\mathbf{A}^{n-2} \otimes \left(\mathbf{B} \otimes \mathbf{u}\right)\right) \oplus \ldots \oplus \left(\mathbf{B} \otimes \mathbf{u}\right)\right) \\
&= \left(\mathbf{A}^n \otimes {}^0 x\right) \oplus \left(\mathbf{A}^{n-1} \oplus \mathbf{A}^{n-2} \oplus \ldots \oplus \mathbf{I}\right) \otimes \mathbf{B} \otimes \mathbf{u}
\end{aligned} \tag{14.30}$$

Entries of the matrix $\mathbf{A}^n$ are the maximum weight sums of the paths of the length n between the column and row operation pairs of the graph nodes. As there are no cycles in the graph

$$\mathbf{A}^n = \mathbf{N} = \begin{pmatrix} -\infty & -\infty & -\infty & -\infty \\ \cdot & \cdot & \cdot & \cdot \\ \cdot & \cdot & \cdot & \cdot \\ -\infty & \cdot & \cdot & -\infty \end{pmatrix} \tag{14.31}$$

so that

$$^n x = \left(\mathbf{A}^{n-1} \oplus \mathbf{A}^{n-2} \oplus \ldots \oplus \mathbf{I}\right) \otimes \mathbf{B} \otimes \mathbf{u} \tag{14.32}$$

In our example

$$^{4}\mathbf{x} = \left(\mathbf{A}^{3} \oplus \mathbf{A}^{2} \oplus \mathbf{A} \oplus \mathbf{I}\right) \otimes \mathbf{B} \otimes \mathbf{u} \tag{14.33}$$

$$^{4}\mathbf{x} = \begin{pmatrix} 0 & -\infty & -\infty & -\infty \\ 4 & 0 & 2 & -\infty \\ 2 & -\infty & 0 & -\infty \\ 7 & 1 & 5 & 0 \end{pmatrix} \otimes \begin{pmatrix} 0 & -\infty & -\infty & -\infty \\ -\infty & 0 & -\infty & -\infty \\ -\infty & -\infty & 0 & -\infty \\ -\infty & -\infty & -\infty & 0 \end{pmatrix} \otimes \begin{pmatrix} 0 \\ 0 \\ 0 \\ 0 \end{pmatrix} \tag{14.34}$$

where the first operation starts in time 0 and setting the entries of $\mathbf{u}$ to zero means that there is no forced starting condition applied. According to Equation (14.34)

$$^{4}\mathbf{x} = \begin{pmatrix} ^{4}x_{1} \\ ^{4}x_{2} \\ ^{4}x_{3} \\ ^{4}x_{4} \end{pmatrix} = \begin{pmatrix} 0 \\ 4 \\ 2 \\ 7 \end{pmatrix} \tag{14.35}$$

Time developments of operations are depicted in Figure 14.8.

Let the first and the second operation has the forced starting points u_{1} and u_{2}, $u_{1} = 6$, $u_{2} = 11$. Then Equation (14.30) becomes

$$\begin{pmatrix} ^{4}x_{1} \\ ^{4}x_{2} \\ ^{4}x_{3} \\ ^{4}x_{4} \end{pmatrix} = \begin{pmatrix} 0 & -\infty & -\infty & -\infty \\ 4 & 0 & 2 & -\infty \\ 2 & -\infty & 0 & -\infty \\ 7 & 1 & 5 & 0 \end{pmatrix} \otimes \begin{pmatrix} 0 & -\infty & -\infty & -\infty \\ -\infty & 0 & -\infty & -\infty \\ -\infty & -\infty & -\infty & -\infty \\ -\infty & -\infty & -\infty & -\infty \end{pmatrix} \otimes \begin{pmatrix} u_{1} \\ u_{2} \\ 0 \\ 0 \end{pmatrix} \tag{14.36}$$

$$\begin{pmatrix} ^{4}x_{1} \\ ^{4}x_{2} \\ ^{4}x_{3} \\ ^{4}x_{4} \end{pmatrix} = \begin{pmatrix} 0 \otimes u \\ (4 \otimes u_{1}) \oplus u_{2} \\ 2 \otimes u_{1} \\ (7 \otimes u_{1}) \oplus (1 \otimes u_{2}) \end{pmatrix} \tag{14.37}$$

and the result is

$$\begin{pmatrix} ^{4}x_{1} \\ ^{4}x_{2} \\ ^{4}x_{3} \\ ^{4}x_{4} \end{pmatrix} = \begin{pmatrix} 6 \\ 11 \\ 8 \\ 13 \end{pmatrix} \tag{14.38}$$

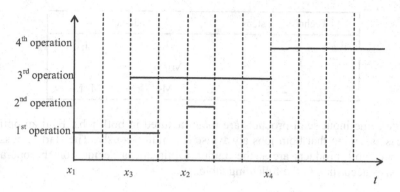

Figure 14.8. The development of operations

Figure 14.9 shows the operation diagrams in case of additional starting conditions.

Additional starting conditions can be utilized in the solution when the graph of the operation dependences illustrated in Figure 14.6 contains cycles. The values of the vector **u are** changing during the steps of the iterative process analogous to Equation (14.30) and depend on the operation starting points according to the function $^r\mathbf{u} = \mathbf{R}\ ^{r-1}\mathbf{x}$ where **R** is a transfer matrix. The reader can learn more in (Dorn and Moßig 1997), and from the references listed therein. Optimal scheduling solution can be found by simulation approaches based on models obtained with the described method and by changing the system structure.

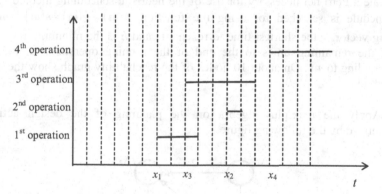

Figure 14.9. Development of operations under additional starting conditions

14.4 Problems and Exercises

14.1. An actual technological process in a manufacturing system with two processing machines is given in the following table.

Technologic steps	Jobs	
	J_1	J_2
1	M_1/M_2	M_2
2	M_2	M_1/M_2

The same input semi-products are manufactured in both jobs. Find an optimal process assuming that both jobs always start simultaneously. The start is possible whenever both products are at the output. Optimum is decided on the operation time basis according to the following table.

Operation	Duration
O_{111}	2
O_{112}	1
O_{122}	3
O_{212}	2
O_{221}	1
O_{222}	2

Create a Petri net necessary for use of the heuristic scheduling method where the schedule is searched for using the function $f(\mathbf{m}) = g(\mathbf{m}) + h(\mathbf{m})$. $\mathbf{m}$ is a marking vector, g the shortest time to reach $\mathbf{m}$, and h is the minimum time to be left of the remaining times to the end of the running operations in the state corresponding to $\mathbf{m}$. On about 10 nodes of the reachability graph show the use of the method.

14.2. Apply the max-plus algebra on the planning of the design activities characterized by the following figure.

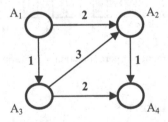

Figure 14.9. A graph of the design activities and their dependences

Durations in the graph is in time units, *e.g.*, months. Find when the whole design process could be on earliest finished using the max-plus algebra approach.

14.3. Duration of the processes in the distributed computer network and their dependences are depicted in Figure 14.10. Write the state equation of the process starts in term of the max-plus algebra. Calculate times when earliest can start separate processes.

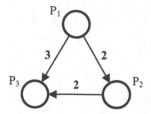

Figure 14.10. Dependences of the processes in the distributed computer network

References

References for Chapter 1
Bogdan S., Lewis F.L., Kovacic Z., Mireles J. Jr. (2006) Manufacturing Systems Control Design: A Matrix-based Approach. Springer, London.
Cao X. R. (1989) A comparison of the dynamics of continuous and discrete event systems. *Proc. IEEE*, Vol. 77, No 1, pp.7–12.
Cassandras C.G. (1993). Discrete event systems: Modeling and performance analysis. Irwin, Homewood, Boston.
Cassandras C.G., Lafortune S. (1999) Introduction to Discrete Event Systems, Springer, NY.
Frištacký N., *et al.* (1981) Programmable Logic Processors. SNTL Prague (in Czech),
Frištacký N., Kolesár M., Kolenička J., Hlavatý J. (1990) Logic circuits. Alfa Bratislava (in Slovak).
Ho Y.C. (1989) Dynamics of a discrete event system. Proc. IEEE, Vol. 77, No 1, pp. 3–6.
Ho Y.C. (1991) Discrete event dynamic systems: Analyzing complexity and performance in the modern world. IEEE Press, New York.
Ho Y.C., Cassandras C.G. (1983) A new approach to the analysis of discrete event dynamic systems. Automatica, Vol. 19, pp. 149–167.
Jafari M. (1995) Supervisory control specification and synthesis. In: Zhou M.C. (ed.) Petri nets in flexible and agile automation. Kluwer Academic Publishers, Boston, MA, 1995.
Jörgl H.P. (1993) Repetitorium regelungstechnik. R. Oldenbourg Verlag, Wien, München.
Kozák Š. (2002) Development of control engineering methods and their applications in industry. In: Proc. of the 5th International Scientific-Technical Conference "Process Control 2002". Kouty nad Desnou, Czech Republic, June 9–12, 2002, str. R218.
Manna Z., Pnueli A. (1991) The temporal logic of reactive and concurrent systems. Springer-Verlag, New York.
Marko H. (1986) Methoden der Systemtheorie. Springer Verlag, New York.
Wonham W.M., Ramadge P.J. (1987) On the supremal controllable sublangauge of a given language. SIAM Journal on Control and Optimization, Vol. 25, pp. 637–659.
Schnieder E. (1991) Braucht die Automatisierungstechnik eine Theorie? Automatisierungstechnik, Vol. 39, pp. 391–401.
Stanat, D., McAllister D. (1977) Discrete mathematics in computer science. Prentice-Hall, Englewood Cliffs, NJ.
Voigt, G., Cramer S. (1986) Diskontinuierliche technologische Prozesse. Akademie Verlag, Berlin.
Zhou, M.C. DiCesare F. (1993) Petri Net Synthesis for Discrete Event Control of Manufacturing Systems. Kluwer Academic Publishers, Boston, MA, 1993.

326   References

Zhou M.C. (ed.) (1995) Petri nets in flexible and agile automation. Kluwer Academic Publishers, Boston.
Zhou M.C., Venkatesh K. (1998) Modeling, simulation and control of flexible manufacturing Systems: A Petri net approach. World Scientific, Singapore.

References for Chapter 2
Deo N. (1974) Graph theory with applications to engineering and computer science. Prentice-Hall, Englewood Cliffs, NJ.
Harary F. (1969) Graph theory. Addison-Wesley, Reading, MA.
Liu C.L. (1977) Elements of discrete mathematics. McGraw-Hill, New York.

References for Chapter 3
Carroll J., Long D. (1989) Theory of finite automata with an introduction to formal languages. Prentice-Hall, Englewood Cliffs, NJ.
Hopcroft J.E., Ullman J.D. (1979) Introduction to automata theory, languages, and computation. Addison-Wesley, Reading, MA.

References for Chapter 4
Martins de Carvalho J.L. (1993) Dynamical systems and automatic control. Prentice-Hall, New York.
Voigt G., Cramer S. (1986) Diskontinuierliche technologische prozesse. Akademie Verlag, Berlin.

References for Chapter 5
Tornambé A. (1995) Discrete-event system theory. World Scientific, Singapore.
Krapp M. (1988) Digitale automaten. VEB Verlag Technik, Berlin.

References for Chapter 6
Hrúz B. (1994) Discrete event systems modeling and real-time control. Journal of Electrical Engineering, Vol. 45, No 10, pp. 363–370.
Zöbel D. (1987) Programmierung von echtzeitsystemen. R. Oldenbourg Verlag, München.

References for Chapters 7, 8 and 9
Abel D. (1990) Petri neze für inmgenieure. Springer Verlag, Berlin.
Češka M. (1994) Petriho sítě. Akademické nakladatelství CERM, Brno.
David R., Alla H. (1992) Petri nets and Grafcet. Prentice Hall, New York.
David R., Alla H. (1994) Petri nets for modeling of dynamic systems – a survey. Automatica, Vol. 30, No 2, pp. 175–202.
Ezpeleta J., Colom J. and Martinez J. (1995). A Petri net based deadlock prevention policy for flexible manufacturing system. IEEE Trans. Robot. Automat., vol. 11, no.5, pp. 173-184.
Gao M., Zhou M.C., Huang X., Wu Z. (2003) Fuzzy reasoning Petri nets. IEEE Trans. on Systems, Man, and Cybernetics: Part A, 33(3), pp. 314–324.
Gao M., Zhou M.C., Tang Y. (2004) Intelligent decision making in disassembly process based on fuzzy reasoning Petri nets," IEEE Trans. on Systems, Man, and Cybernetics: Part B, 34(5), 2029–2034.
Giua A. DiCesare F. (1993) Grafcet and Petri nets in manufacturing. In: Gruver W.A., Boudreaux J.C. (eds.) Intelligent manufacturing: Programming environments for CIM. pp. 153–76, Springer-Verlag.

Hanzálek Z. (1998a) A parallel algorithm for gradient training of feedforward neural network. Parallel Computing, No 24, pp. 823–839.

Hanzálek Z. (1998b) Algorithm modeling with Petri nets. IEEE Conference on System, Man and Cybernetics Piscataway, pp. 214–220.

Hirel C., Tuffin B., Trivedi K.S. (2000) SPNP: Stochastic Petri nets Version 6.0, 11th International Conference of TOOLS 2000, Schaumburg, US. Also, http://www.ee.duke.edu/~chirel/IRISA/spnpInstructions.html

Hrúz B. (1994) Discrete event systems modeling and real-time control. Journal of Electrical Engineeirng. Vol. 45, No 10, pp. 363–370.

Hudák Š. (1999). Reachability analysis of systems based on Petri nets. Academic Press Elfa, Ltd., Košice.

Jensen K. (1997) Coloured Petri nets.Vol. I., II., III., second edition, Springer, Berlin.

John K.-H., Tiegelkamp M. (2001) IEC 61131-3: Programming industrial automation systems. Springer, Berlin.

König R. and Quäck L. (1988) Petri-netze in der steuerungs- und digitaltechnik. Oldenbourg Verlag, München.

Jeng M.D. and Peng M.Y. (1999) Augmented Reachability Trees for 1-Place-Unbounded Generalized Petri Nets. IEEE Trans. on Systems, Man, and Cybernetics, 29(2), pp. 173-183.

Li Z. and Zhou M.C. (2004) Elementary Siphons of Petri Nets and Their Applications to Deadlock Prevention in Flexible Manufacturing Systems. IEEE Trans. on Systems, Man, and Cybernetics, 34(1), pp. 38-51.

Li Z. and Zhou M.C. (2006) Two-stage method to design liveness-enforcing Petri net supervisor for FMS. IEEE Trans. on Industrial Informatics, Vol. 2, No. 4, pp. 313-325.

Murata T. (1989) Petri nets: properties, analysis and applications. Proc. IEEE, Vol. 77, No 4, pp. 541–580.

Peterson J. (1981) Petri net theory and the modeling of systems. Prentice Hall, Englewood Cliffs, NJ.

Petri C.A. (1962) Kommunikation mit automaten. Thesis, Schriften des Instituts für Instrumentelle Mathematik, No 3, Bonn University.

Schnieder E. (ed.) (1992) Petrinetze in der automatisierungstechnik. R. Oldenbourg Verlag, München.

Starke P.H. (1990) Analyse von Petri-netz-modellen. B.G. Teubner, Stuttgart.

Svádová M., Hanzálek Z. (2001) Tool demonstration in ICATPN 2001, Newcastle upon Tyne, University of Newcastle, pp. 35–39.

Wang F.-Y., Gao Y., Zhou M.C. (2004) A modified reachability tree approach to analysis of unbounded Petri nets. IEEE Trans. on Systems, Man, and Cybernetics: Part B, Vol 34, No. 1, 303–308.

Wu N.Q. Necessary and sufficient conditions for deadlock-free operation in flexible manufacturing systems using a colored Petri net model, IEEE Trans. on Systems, Man, and Cybernetics, Part C, vol. 29, no. 2, pp. 192–204, 1999.

Wu N.Q., Zhou M.C. (2001) Avoiding deadlock and reducing starvation and blocking in automated manufacturing systems. IEEE Transactions on Robotics and Automation, vol. 17, no.5, pp. 657–668.

Wu N., Zhou M.C. (2004) Deadlock modeling and control of automated guided vehicle systems. IEEE/ASME Trans. on Mechatronics, 9(1), pp. 50–57.

Wu N., Zhou M.C. (2005) Modeling and deadlock avoidance of automated manufacturing systems with multiple automated guided vehicles. IEEE Trans. on Systems, Man, and Cybernetics: Part B, 35(6), pp. 1193–1202.

Zhou M.C., Leu M.C. (1991) Modeling and performance analysis of a flexible PCB assembly station using Petri nets. Transactions of the ASME. Journal of Electronic Packaging, Vol. 113, No 4, pp. 410–416.

Zhou M.C. (ed.) (1995) Petri nets in flexible and agile automation. Kluwer Academic Publishers, Boston.

Zhou M.C. Twiss E. (1996) Discrete event control design methods: A review. Preprints of 13th IFAC World Congress, San Francisco, CA, Vol. 9, pp. 401–409.

Zhou M.C., Venkatesh K. (1998) Modeling, simulation and control of flexible manufacturing Systems: A Petri net approach. World Scientific, Singapore

References for Chapter 10

Ajmone Marsan M., Balbo G., Conte G., Donatelli S., Franceschinis G. (1995) Modeling with generalized stochastic Petri nets. John Wiley & Sons, New York.

Asar A.U., Zhou M.C., Caudill R.J. (2005) Making Petri nets adaptive: a critical review. Proc. of IEEE Int. Conf. on Networking, Sensors and Control, Tucson, AZ, March 19–21, 2005, pp. 644–649.

Bause F., Kritzinger P.S. (1996) Stochastic Petri nets. Verlag Vieweg,, Wiesbaden.

Campos J., Chiola G., Colom J.M. Silva M. (1992). Properties and performance bounds for timed marked graphs. IEEE Trans. On Circuits and Systems –I, 39(5), pp. 386–401.

Cardoso J., Camargo H. (1999) Fuzziness in Petri nets. Physica-Verlag, Springer Verlag Company, Heidelberg.

Čapek J., Hanzálek Z. (2000) STPN model of physical and MAC layer of lonworks. Preprints of the IFAC Conference on Control Systems Design, Bratislava, June 18-20, 2000, pp.335–340.

Čapkovič F. (1993) Modelling and justifying discrete production processes by Petri nets. Computer Integrated Manufacturing Systems, Vol. 6, No 1, pp. 27–35.

Čapkovič F. (1994) Petri net-based approach to the maze problem solving. In: Balemi S., Kozák P., Smedinga R. (eds.) Discrete event systems: Modelling and control, Birkhäuser Verlag, Basel, pp. 173–179.

Čapkovič F. (1998) Knowledge-based control synthesis of discret event dynamic systems. In: Tzafestas S.G. (ed.) Advances in manufacturing, decision, control and information technology, Springer, London.

Chen S., Ke J., Chang J. (1990) Knowledge representation using fuzzy Petri nets. IEEE Trans. Knowledge and Data Engineering, Vol.2, No. 3, pp.311–319.

Desel J. (2000) Simulation of Petri net processes. Preprints of the IFAC Conference on Control Systems Design, Bratislava, June 18-20, 2000, pp.14-25.

Gao M., Zhou M.C., Huang X., Wu Z. (2003) Fuzzy reasoning Petri nets. IEEE Trans. on Systems, Man, and Cybernetics: Part A, Vol. 33(3), pp. 314–324.

Genrich H.J., Lautenbach K. (1981) System modeling with high-level Petri nets. Theoretical Computer Science, Vol. 13, 1981, North-Holland, pp.109–136.

Hanisch H.M. (1993) Analysis of place/transition nets with timed arcs and its application to batch process control. In: Marsan M.A. (ed.) Applications and theory of Petri nets. Lecture Notes in Computer Science 691, Springer, pp. 282–299.

Hirel C., Tuffin B., Trivedi K.S. (2000) SPNP: Stochastic Petri Nets. Version 6.0. In: Haverkort B., Bohnenkamp H., Smith C. (eds.) 11th international conference computer performance evaluation: modelling tools and techniques Schaumburg Il., USA. Lecture Notes in Computer Science 1786, Springer.

Hilion H.P., Prpth J.M. (1989) Performance evaluation of job-shop systems using timed-event-graphs. IEEE Trans. On Automatic Control, Vol. 34, pp.3–9.

Hrúz B., Mrafko L., Bielko V. (2002) A comparison of the AGV control solution approaches using Petri nets. Proc. of the 2002 IEEE International Conference on Systems, Man and Cybernetics, Yasmine Hammamet, Tunisia, October 6–9, 2002.

Jensen K. (1981) Colored Petri nets and the invariant method. Theoretical Computer Science, Vol. 14, pp. 317-336

Jensen K. (1997) Coloured Petri nets.Vol. I., II., III., second edition, Springer, Berlin.

Juhás G. (2000) A unified approach to modelling and control of a class of discrete event and hybrid systems via algebraically generalized Petri nets. Preprints of the IFAC Conference on Control Systems Design, Bratislava, June 18–20, 2000, pp. 349–354.

Li X., Yu W., Lara-Rosano F. (2000) Dynamic knowledge inference and learning under adaptive fuzzy Petri net framework. IEEE Trans. System, Man, and Cybernetics, Part C, Vol. 30(4), pp. 442–450.

Li Z., Zhou M.Z. (2006) Two-stage method to design liveness-enforcing Petri net supervisor for FMS. IEEE Trans. on Industrial Informatics, Vol. 2, No. 4, pp. 313–325.

Morioka S., Yamada T. (1991). Performance evaluation of marked graphs by linear programming. Int. J. of Systems Science, 22(9), 1541–1552.

Ribarič S., Bašič B.D. (1998) Fuzzy time Petri net primitives for processing fuzzy temporal knowledge. Proc. of the 9th Mediterranean Electrotechnical Conference, May 18-20, 1998, Tel-Aviv, Vol. 1, pp.549–553.

Struhar M. (2000) A fuzzy object Petri net model applied for control. PhD Dissertation, Faculty of Electrical Eng. and Information Technology, Slovak University of Technology, Bratislava.

Yeung D.S., Tsang E.C.C. (1998) A multilevel weighted fuzzy reasoning algorithm for expert systems. IEEE Trans. System, Man, and Cybernetics, Part A: Systems and Humans, Vol. 28, No.2, pp. 149-158.

Uzam M., Zhou M.C. (2006) An improved iterative synthesis method for liveness Enforcing supervisors of flexible manufacturing systems," Int. J. of Production Research, Vol. 44, No. 10, 1987–2030.

Wang J. (1998) Timed Petri nets: Theory and application, Kluwer Academic Publishers, Boston, MA.

Wu N.Q. (1999) Necessary and sufficient conditions for deadlock-free operation in flexible manufacturing systems using a colored Petri net model. IEEE Trans. on Systems, Man, and Cybernetics, Part C, vol. 29, no. 2, pp. 192–204.

Wu N.Q., Zhou M.C. (2001) Avoiding deadlock and reducing starvation and blocking in automated manufacturing systems. IEEE Transactions on Robotics and Automation, 17(5), pp. 657–668.

Wu N.Q., Zhou M.C. (2004) Modeling and deadlock control of automated guided vehicle systems. IEEE/ASME Transactions on Mechatronics, vol. 9, no. 1, pp. 50–57.

Wu N., Zhou M.C. (2005) Modeling and deadlock avoidance of automated manufacturing systems with multiple automated guided vehicles. IEEE Trans. on Systems, Man, and Cybernetics: Part B, 35(6), pp. 1193–1202.

Wu, N. and Zhou M.C. (2007) Deadlock and Blocking-free Shortest Routing of Bi-directional Automated Guided Vehicles. IEEE Trans. on Mechatronics, 12(1), pp. 63-72.

Wu, N. and Zhou M.C. (2007) Deadlock-free scheduling for semiconductor track systems based on resource-oriented Petri nets. OR Spectrum, 29(3), pp. 421–443.

Wu, N. and Zhou M.C. (2007) Deadlock modeling and control of semiconductor track systems using resource-oriented Petri nets," Int. J. of Production Research, 45(15), pp. 3439–3456.

Zhou M.C. and Venkatesh K. (1998) Modeling, simulation and control of flexible manufacturing Systems: A Petri net approach. World Scientific, Singapore.

Zhou M.C., Zurawski R. (1995) Introduction to Petri nets in flexible and agile automation. In: Zhou M.C. (ed.) Petri nets in flexible and agile automation, Kluwer Academic Publishers, Boston, MA, 1–42.

References for Chapter 11
Eshuis R. (2006) Statecharting Petri nets. Technical report (BETA working paper 153), Eindhoven University of Technology, available at http://is.tm.tue.nl/staff/heshuis/pn2sc-beta.pdf.
Fogel J. (1997) A statecharts approach to the modeling of discrete manufacturing systems. Proc. of the 7th Symposium on Computer Aided Control Systems Design, Gent, April 28–30, 1997.
Fogel J. (1998) A statecharts approach to the modelling and simulation of discrete manufacturing systems. In: Frankovič B. (Ed.) Control Theory and Applications, Veda, Bratislava, str. 33–44.
Harel D. (1987) Statecharts: A visual formalism for complex systems. Science of Computer Programming, North Holland, Vol. 8, pp.231–274.
Harel D., Pnueli A., Schmidt J.P., Sherman R. (1987) On the formal semantics of statecharts. Proc. of the 2nd IEEE Symp.on Logic in Computer Science, New York, pp.54–64.
Harel D., Politi M. (1998) Modeling reactive systems with statecharts: the STATEMATE Approach. McGraw-Hill.
Lee, J. S., Zhou M.C., and Hsu P. L. (1995) Statechart modeling and Web-based simulation of hybrid dynamic systems for e-Automation. *Journal of Chinese Institute of Industrial Engineers*, 22(1), pp. 19-27.
Schnabel M.K., Nenninger G.M., Krebs V.G. (1999) Konvertierung sicherer Petri-netze in statecharts. Automatisierungstechnik, Vol. 47, No 12, pp. 571–580.

References for Chapter 12
Abel D. (1990) Petri neze für inmgenieure. Springer Verlag, Berlin.
Brand K.P., Kopainsky J. (1988) Principles and engineering of process control with Petri nets. IEEE Trans. On Automatic Control, Vol. 33, No 2, pp. 138–149.
Ferrarini L. (1992) An incremental approach to logic controller design with Petri nets. IEEE Trans. on Systems, Man, and Cybernetics, Vol. 22, pp. 461–473.
Ferrarini, L. Narduzzi, M. and Tassan-Solet, M. (1994) A new approach to modular liveness analysis conceived for large logic controllers' design. IEEE Transactions on Robotics and Automation, 10(2), pp. 169 - 184.
Ferrarini L. (1995). Computer aided design of logic controllers with Petri nets. In: Zhou, M.C. (ed.) (1995) Petri nets in flexible and agile automation. Kluwer Academic Publishers, Boston, pp. 71–92.
Hrúz B. (1997) A design method of the conflict-free Petri net models for the manufacturing systems control. Proc. of the 2nd IFAC Workshop on New Trends in Design of Control Systems, Bratislava, Sept. 7-10, 1997, pp. 259–264.
Hrúz B., Jörgl H.P., Kopčok I.K., Kozák Š. (2000) Control of the bouncing ball laboratory experiment with an Allen–Bradley PLC. Proc. of IFAC/IEEE Symposium on Advances in Control Education ACE 2000. Sea World Nara Resort, Gold Coast, Australia, December17–19, 2000.
Hrúz B., Niemi A., Virtanen T. (1996) Composition of conflict-free Petri net models for control of flexible manufacturing systems. Proc. of the 13th IFAC World Congress, San Francisco, Vol. B, pp. 37–42.
Hrúz B., Ondráš J., Flochová J. (1997) Discrete event systems – an approach to education. Proc. of the 4th IFAC Symposium on Advances in Control Education, 14-16 July, 1997, Istanbul, pp. 283–288.

Mudrončík D., Zolotová I. (2000) Industrial programmable logic controllers. Publishing House Elfa, Ltd., Košice.

Murata T. (1989) Petri nets: properties, analysis and applications. Proc. IEEE, Vol. 77, No 4, pp. 541–580.

Niemi A. J., Ylinen R., Heikkilä A., Niemi E., Virtanen T. (1992) Automatic FMC with vision as test bed for control methods. Int. Journal of Advanced Manufacturing Technology, Vol. 7, pp. 353–659.

Quäck L. (1991) Aspekte der Modellierung und realisierung der steuerung technologischer prozesse mit Petri-netzen. Automatisierungstechnik, Vol. 39, No 4, pp. 116–120, No 5, pp. 158-164.

Reißenweber B. (1988) Programmieren mit PEARL. Oldenbourg Verlag, München.

Venkatesh K., Zhou M.C., Caudill R.J. (1995) Discrete event control design for manufacturing systems via ladder logic diagrams and Petri nets: A comparative study. In: Zhou, M.C. (ed.) (1995). Petri nets in flexible and agile automation, Kluwer Academic Publishers, Boston, pp. 265–304.

Werum W., Windauer H. (1989) Introduction to PEARL. Process and Experiment Automation Realtime Language. Vieweg, Braunschweig.

Zhou M.C., DiCesare F. (1991) Parallel and sequential mutual exclusions for Petri net modeling of manufacturing systems with shared resources. IEEE Trans. on Robotics and Automation, Vol. 7, pp. 515–527.

Zhou M.C. DiCesare F. (1993) Petri net synthesis for discrete event control of manufacturing systems. Kluwer Academic Publishers, Boston, MA, 1993.

Zhou M.C., DiCesare F., Desrochers A.A. (1992) A hybrid methodology for synthesis of Petri net models for manufacturing systems. IEEE Trans. on Robotics and Automation, Vol. 8, No 3, pp.350–361.

Zhou M.C., DiCesare F., Rudolph D.L. (1992) Design and implementation of a Petri net based supervisor for a flexible manufacturing system. Automatica, Vol. 28, pp. 1199–1208.

Zhou M.C., McDermott K., Patel P.A. (1993) Petri net synthesis and analysis of a flexible manufacturing system cell. IEEE Trans. on Systems, Man, and Cybernetics, Vol. 23, No 2, pp. 523-531.

Zhou M.C., Twiss E. (1998) Design of industrial automated systems via relay ladder logic programming and Petri nets. IEEE Trans. on Systems, Man, and Cyberentics, Part C: Applications and Reviews, Vol. 28, pp. 137-150.

Zhou M.C., Venkatesh K. (1998) Modeling, simulation and control of flexible manufacturing Systems: A Petri net approach. World Scientific, Singapore.

Zöbel D. (1987) Programmierung von echtzeitsystemen. R. Oldenbourg Verlag, München.

References for Chapter 13

Balemi S., Hoffman G.J., Gyugyi P., Wong-Toi H., Franklin G.F. (1993) Supervisory control of a rapid thermal multiprocessor. IEEE Trans. on Automatic Control, Vol. 38, pp. 1040–1059.

Flochová J., Hrúz B., Jirsák P. (1997) Program solution of supervisory control based on Petri nets. Proc. of the 1st IFAC Workshop on New Trends in Design of Control Systems, Bratislava, Sept. 7-10, 1997, pp.278-282.

Flochová J., Lipták R., Boel R.K. (2001) A MATLAB-based Petri net supervisory controller discrete event systems. Proc. of the IFAC Workshop on Programmable Devices and Systems, Gliwice, Poland, November 22-23, 2001, pp. 119-126.

Giua A., DiCesare F., Silva M. (1992) Generalized mutual exclusion constraints on nets with uncontrollable transitions. Proc. of the 1992 IEEE International Conference on Systems, Man, and Cybernetics, Chicago, pp. 974-979.

Harušťák M., Hrúz B. (2000) Supervisory control of discrete event systems and its solution with the Petri net P-invariants. Preprints of the IFAC Conference on Control Systems Design, Bratislava, June 18-20, 2000, pp.390–394.

Holloway L.E., Krogh B.H. (1990) Synthesis of feedback control logic for a class of controlled Petri nets. IEEE Trans. on Automatic Control, Vol. 35, No 5, pp. 514-523.

Hrúz B. (1994) The supervisory control problem solved via Petri nets. Proc. of the 1st IFAC Workshop on New Trends in Design of Control Systems, Bratislava, Sept. 7–10, 1997, pp.386–391.

Jafari M.A. (1995) Supervisory control specification and synthesis. In: Zhou, M.C. (ed.) (1995) Petri nets in flexible and agile automation, Kluwer Academic Publishers, Boston, pp. 337–368.

Moody J.O., Antsaklis P.J. (1998) Supervisory control of discrete event systems using Petri nets. Kluwer Academic Publishers, Boston.

Ramadge P.J., Wonham W.M. (1989) The control of discrete event systems. Proc. of the IEEE, Vol. 77, pp. 81–98.

Uzam M., Zhou M.C. (2006) An improved iterative synthesis method for liveness Enforcing supervisors of flexible manufacturing systems. Int. J. of Production Research, Vol. 44, No. 10, pp. 1987–2030.

Wonham W.M., Ramadge P.J. (1987) On the supremal controllable sublangauge of a given language. SIAM Journal on Control and Optimization, Vol. 25, pp. 637–659.

Yamalidou K., Moody J., Lemmon M., Antsaklis P. (1996) Feedback control of Petri nets based on place invariants. Automatica, Vol. 32, pp. 15–28.

References for Chapter 14

Dorn C., Moßig K. (1997) Erweiterter Steuerungsentwurf für ereignisdiskrete Systeme mit Hilfe der Max-Plus-Algebra. Automatisierungstechnik, pp. 407-413.

Engell S. (1989) Modeling and on-line scheduling of flexible manufacturing systems. Proc. of the IFAC/IFORS/IMACS Symposium on Large Scale Systems 89, August 29–31, 1989, Berlin, pp. 379–383.

Frankovič B, Budinská I. (1998) Single and multi machine scheduling of jobs in production system. In Tzafestas, S.G. (ed.) Advances in manufacturing, decision, control and information technology, Springer, London.

Hrúz B. (2000) A class of the timed Petri nets used for the solution of scheduling. Preprints of the IFAC Conference on Control Systems Design, Bratislava, June 18–20, 2000, pp.361–365.

Lee D.Y., DiCesare F. (1994) Scheduling flexible manufacturing systems using Petri nets and heuristic search. IEEE Trans. on Robotics and Automation, Vol. 10, pp. 123–132.

Lee D.Y., DiCesare F. (1995) Petri net-based heuristic scheduling for flexible manufacturing. In: Zhou M.C. (ed.) (1995) Petri nets in flexible and agile automation, Kluwer Academic Publishers, Boston, pp. 149–187.

Li S., Takamori T., Tadokoro S. (1995) Scheduling and re-scheduling of AGVs for flexible and agile manufacturing. In: Zhou, M.C. (ed.) (1995) Petri nets in flexible and agile automation, Kluwer Academic Publishers, Boston, pp. 189–205.

Moßig K., Rehkopf A. (1996) Einführung in die Max-Plus–algebra zur beschreibung ereignisdiskreter dynamischer prozesse. Automatisierungstechnik, Vol. 44, No 1, pp. 3–9.

Xiong, H.H., Zhou M.C. (1998) Scheduling of Semiconductor Test Facility via Petri Nets and Hybrid Heuristic Search. *IEEE Transactions on Semiconductor Manufacturing*, 11(3), pp. 384-393.

Zhou M.C., Venkatesh K. (1998) Modeling, simulation and control of flexible manufacturing Systems: A Petri net approach. World Scientific, Singapore.

Index